FRIENDLY BIOLOGY

Joey Hajda DVM MEd

Lisa B. Hajda MEd

First Edition

Published by Hideaway Ventures, 79372 Road 443, Broken Bow, NE 68822

For information regarding this publication please contact Joey or Lisa Hajda at the above address or visit our website: www.friendlybiology.com.

Copyright 2017 Hideaway Ventures, Joey or Lisa Hajda

All rights reserved. No portion of this book may be reproduced in any form, or by any electronic, mechanical or other means, without the prior written permission of the publisher.

Front cover photo credits: *Tramea onusta.* (Red Saddlebags dragonfly) Tim Hajda; *Papaver spp.* (poppy) Joey Hajda; *Thamnophis spp.* (gartersnake) Tim Hajda; and our granddaughter, Phoebe Jean, by her mother, Clara Williams, Clara Williams Photography.

To our dads, Frankie Hajda and
Loren E. Betz, with immense gratitude,
for first introducing us to living things in
our world.

Friendly Biology

Table of Contents

Lesson 1: Characteristics of Living Things..1

Lesson 2: A Little Chemistry...11

Lesson 3: Carbohydrates...29

Lesson 4: Lipids..47

Lesson 5: Proteins..63

Lesson 6: pH...83

Lesson 7: Cytology (Part 1)..99

Lesson 8: Cytology (Part 2)..127

Lesson 9: Cell Division (Mitosis)..141

Lesson 10: Chromosome Duplication..153

Lesson 11: Protein Synthesis...165

Lesson 12: Methods of Reproduction..185

Lesson 13: Genetics...211

Lesson 14: Taxonomy...235

Lesson 15: Kingdom Animalia..257

Lesson 16: Kingdom Animalia (Phylum Chordata)...281

Lesson 17: Kingdom Plantae...307

Lesson 18: Kingdom Monera and Viruses..323

Lesson 19: Kingdom Protista...331

Lesson 20: Kingdom Fungi..339

Lesson 21: Body Systems of Movement..345

Lesson 22: Body Systems of Nutrient Delivery...365

Lesson 23: Acquisition of Energy Sources..385

Lesson 24: Body Systems of Waste Management...395

Lesson 25: Body Systems of Growth and Development......................................405

Lesson 26: Body Systems of Reproduction...417

Lesson 27: Sensory Systems of the Body..425

Lesson 28: Ecology..439

Index..455

Lesson 1: Characteristics of Living Things

Let's begin by taking a look at the word "biology." The word "biology" is built from two parts: bio– and –ology. Bio– comes from the Latin term **bios** which means **life**. The **–ology** portion of biology refers to **the study of**. If we put the two parts together we get that **biology is the study of life**. There are many other terms which end in –ology. You might know the word geology which is the study of geos or the earth. Another familiar "-ology term" is entomology which is the study of insects. Archeology is the study of human artifacts. As we progress through this biology course, we will introduce you to many fields of studies related to biology. Knowing the scientific names given to these studies becomes quite handy, especially when dealing with medical terms such as hematology (the study of blood) or oncology (the study of tumors or cancer). For now, we'll concentrate on biology, the study of life.

> bios- = life
> -ology = study of
> *biology = study of life*

If our goal is to study living things, first we need to establish exactly what makes something a living thing. How do you know when something is alive or not alive? If you think about things in your home for example, how do you know that your kitchen table is not alive, but your pet goldfish is alive? Or how do you know that the sidewalk outside your house is not alive, but the flowers growing along side it are alive?

Most likely the first idea that comes to mind when determining whether something is alive or not is the fact that living things can move. Cats pounce, birds fly, people walk and run, fish swim and tiny worms wiggle. Yes, being able to move does give you an indication that the object in question is indeed alive. But, how does something like a plant—which you would say is alive—move? You might say, well, it moves when the wind blows. But a flag also moves when the wind blows and a flag is not a living thing. So, while plants are alive and, yes, plants move, their movement is not as visible to the observer as the movement of some other living things.

Have you ever noticed how a plant can adjust its position to catch more sunlight? If you live in a farming area or have a vegetable garden, you may be familiar with sunflowers. The flower of a sunflower moves throughout the day in an effort to gain the most exposure to the sun. Plants also move water from the soil and food they make in their leaves to be stored in their seeds or down in their roots. So, yes, plants do meet the requirement of movement to be considered alive. Correspondingly, it is highly unlikely that once you place your non-living kitchen table in your kitchen you will find it sitting in a different location when you return later. **Living things move.** Non-living things (for the most part) do not move on their own.

> Living things move.

Can you think of another thing that living things can do, but non-living things cannot? How about the idea that living things can have babies? This idea, like the idea of movement, is a readily acceptable idea for differentiating between things that are alive and those that are not alive. Things that are living make more living things: cows have calves, pigs have piglets and people have infants. Daisies and zinnias make seeds. The mold that grows on old food in your refrigerator makes spores which can produce more mold. Even the tiniest bacteria and viruses reproduce and, to our dismay, can do so at amazing rates. **Living things reproduce.** And, once again, your kitchen table, which is not alive, cannot make more kitchen tables!

> Living things reproduce.

So far, when determining whether something is living or not, we've said that living things move (on their own) and reproduce. Let's continue. For the third characteristic of a living thing, consider this question: what happens to you about 7 o'clock each morning, again at about noon and then sometime around 6 o'clock each evening? Got it? Sure, you get hungry, right? You find yourself wandering into the kitchen or thinking about what you can get yourself to satisfy that feeling of hunger. Likewise, your pets get hungry and need to be fed. On a less appetizing note, the little parasitic worms living *inside* your pets get hungry, too! Whether large or small, living things need to eat or, more precisely, **living things need a source of food or energy**.

This is one characteristic of living things in which plants have a big advantage over animals and humans. Plants have the ability to make their own food. When was the last time you saw some tulips or maple trees shopping down at the local supermarket? While plants may not require a food source outside themselves, they do require a source of energy (the sun in most cases) to make their food. Without sunlight, most plants simply cannot survive. We'll discuss this process, known as photosynthesis, in greater detail in a later lesson.

> Living things need a source of food or energy.

So far, we've said that living things move, living things reproduce and living things require a source of energy. A fourth idea to help us differentiate between living and non-living things is the fact that living things grow. A baby chick, while cute and fuzzy when a few days old, soon becomes a straggly "teenager" chicken just a few weeks later. A puppy grows into a dog and a kitten into a cat. You outgrow your clothes. The zinnia seeds you plant sprout and grow into a flowering bush. The mold on the old food in your refrigerator spreads all across the food and sometimes even up the sides of the container. **Living things grow and develop.** Things that are not alive show no signs of growing nor developing.

> **Living things grow and develop.**

Let's pause and review. We've described four ideas that helps us differentiate living things from non-living things: living things move, living things reproduce, living things require a food or energy source and living things grow and develop. There is one other thing that living things do that non-living things do not.

To demonstrate this fifth characteristic of living things, try the following activity. Perhaps you are sitting in a chair or at a desk as you are reading this book. Carefully and quietly, stand up and walk around behind the chair. Sneak up to your chair very quietly and when really close, shout, "BOO!" Watch the results. What? Nothing happens? Okay, let's try this again, but this time, choose your Mom, Dad, brother, sister or classmate and ask him or her to sit in the chair. Repeat the procedure you did with your empty chair. Go ahead—you have permission from your science teacher to conduct this "important" first lab activity. Write or draw a picture of the results of your experiment here:

Now, unless the person on which you chose to conduct your lab was very asleep or unconscious, it should be apparent that living things respond to things in their environment. The chair, being non-living, did absolutely nothing when you attempted to surprise it. However, the person you chose to surprise probably not only showed you that **living things respond to their environment**, but may have shown you that living things move, too! Good review, right? So, yes, living things respond to their environment. In this case, environment means everything found surrounding that living thing.

Again, you might say, "It's easy to see something large like an animal, for example, responding to its environment. What about plants?" Think about a houseplant that may have been forgotten about for several weeks. No one has watered or cared for it. What happens to it? Do its leaves begin to droop and become a different color? Does it begin to drop some of its leaves? How might this plant respond once it gets watered and cared for again? Sometimes the plant returns to its prior condition, sometimes, as you may be aware, it does not. The plant has responded to a change in its environment. In this case, the lack of water caused many changes in the plant (drooping and changing of color of its leaves). Then, when the water returned, again (hopefully!) the plant responded.

Likewise, you may have noticed that some houseplants will grow toward the sunlight coming in from a window. Recall our discussion of the sunflower as it moves its flower and leaves to face the sun. These plants are responding to their environment.

> Living things respond to their environment.

This brings us to the final topic for this first lesson and that is the idea of death. While not generally thought of as a pleasant topic, the very fact that living things are indeed alive requires that we must accept the fact that life does eventually stop for all living things. By some means or another, the essence that we call life *does* leave the once-living organism. We snuck in a new term there: organism. Organism just means one complete living thing. Death eventually comes to all organisms.

If we look back at the five characteristics we presented in this lesson that we use to differentiate between living and non-living things (movement, reproduction, requirement of a food or energy source and response to one's environment), we realize that these features are the very characteristics we use to determine if, indeed, life has left that living thing. We look for movement and response to our voice or actions. We watch for breathing, which, as we will learn later on, is a major factor in how living things produce energy. And then, with a living thing that is more simple like a plant or bacteria, for example, we note that it is no longer creating new growth or dividing into larger colonies. It is when these forms of evidence are no longer present we deem the once-living organism dead.

In the lessons that follow, we will examine in greater detail how life "happens" in living things. We will begin on the very, very small level and then progress to understand how complex organisms such as you and I carry on our lives. Life itself is truly an amazing thing to ponder.

Let's stop now and review the ideas we discussed in this first lesson. We outlined five characteristics of living things. We said that living things:

- Move
- Reproduce
- Require a source of food or energy
- Grow and develop
- Respond to their environment.

These features are common to all living things and it is the absence of these five characteristics that tells us that life has ceased for that organism. Turn the page now to view the Lesson 1 Lab Activity.

> Living things move, reproduce, require a source of energy, grow and develop and respond to their environment

Lesson 1 Lab Activity: Growing a Pet Potato Plant

IMPORTANT: READ ALL OF THESE INSTRUCTIONS FIRST BEFORE BEGINNING!

The best way to begin to appreciate the five features common to living things is to observe a living thing first-hand. A great way to do this is to grow a pet potato plant. Items you'll need include:

- Flower pot or bucket or tub. It should be able to hold approximately one cubic foot (or more) of potting soil.
- Potting soil
- Small potato
- Magnifying glass

Procedure:

1. Prepare the pot for your pet potato plant

   Begin by preparing your flower pot, bucket or tub for your potato plant. Whichever container you choose to use, make sure that it has holes in the bottom so that excess water can readily drain from the pot. It is also a good idea to place a container beneath your pot to catch this water so that it doesn't harm the surface on which it is placed.

2. Add potting soil to your pot.

   Next, take some of the potting soil and place it into the pot. Continue to fill the pot about 3/4ths full. We strongly suggest that you **do not** use soil or dirt directly from your yard or garden. Potting soil, purchased from your local garden center or discount store works much better for this activity.

3. Observe the potato

   Take your potato and make some observations. Look closely at the surface for some indented places where you might see a small white sprout. This little spout can be very tiny on some potatoes and large and robust on others depending upon where you potato has been stored. Be careful not to disturb this little sprout or eye as is it called by potato farmers or gardeners. If you find any soft or discolored spots on your potato consider using another one. These soft spots are locations where the potato has begun to be invaded by other tiny living creatures, most likely fungi and bacteria that are working to decompose (or rot!) the potato. If you have access to a digital camera, take a photo of your potato. A close-up shot of the eye would be ideal. As a reference, place a coin next to the eye of the potato as you take the photo.

The small white bump on this potato is the eye of the potato. There may be more than one present. Each can develop into a "new" potato plant.

This seed potato has been allowed to set out longer before being planted. Note how small roots have begun to form. It is ready to be planted!

4. Once you've located an eye on your potato, gently place the potato down into the soil. It does not matter how you place the potato into the soil. The roots will grow downward and the stems will grow upward. Fill the remainder of the pot with potting soil and gently press the soil down. Do not pack the soil tightly.

4. Water the potato

Using tap water, gently pour water down through the top of your pot until you see water escaping from the holes underneath. Once all of the soil has been moistened, stop pouring in water. Set your potato and pot in sunny window. You can grow your potato outdoors as long as the weather remains well above freezing. Note that a frost can be harmful to your potato plant.

6. Begin your potato journal

Create a journal to record the daily observations you make of your potato plant. Begin your first entry by telling how you prepared the potato pot. List all of the steps you followed. Then list the observations you made of the potato. Tell about the eyes and how many of them you found. If you were able to make photographs of your potato, include those in this first entry. Tell how you planted the potato. Finally, tell how you watered the potato. Visit your potato everyday and continue to make observations. Record your observations in your journal. You may not see much happening right away, but continue to observe the pot everyday.

Notes about growing potatoes: If you've never grown potatoes before, you may have been somewhat surprised that most potatoes are grown from potatoes themselves and not seeds. Potato farmers and gardeners purchase what are called "seed potatoes" which are small potatoes chosen specifically for growing another crop of potatoes. Some farmers plant these small potatoes whole while other may cut them into pieces with each piece having at least one eye present. This process of creating a new plant from part of the old "mother" plant is called vegetative propagation which is a form of reproduction where only one parent is involved. We'll discuss this means of reproduction in greater detail in Lesson 12.

6. Care for your pet potato plant.

There are two important things you much provide for your potato in order for it to grow happily: constant supply of sunshine and constant supply of water. A sunny window sill or outside porch where your potato can receive full sunlight is an ideal location for your potato plant. As we discussed earlier, potato plants are not hardy in cold weather and can be killed if left outdoors at temperatures near or below freezing.

An optional thing you can do for your pet potato plant is to provide it with some nutrients in the form of plant food or fertilizer. We recommend you use a plant food labeled for house plants such as Miracle Gro plant food. Be sure to follow the plant food label directions carefully and store the fertilizer away from small children or pets.

Continue making observations of your potato plant on a daily basis. Note when things appear or disappear. Use a ruler to make measurements. Continue taking photographs of all parts of your plant. Remember to use a coin as a size reference in your photos. Take good care of it. It may reward you later.

Here are some things you might expect to see with your potato:

- Flower buds and blooms may appear a few weeks after the plant has begun to grow. Examine them closely and take close-up photographs if possible.

- If your plant grows tall and spindly, it may be seeking more light. Try moving it to a sunnier location.

- Eventually, the leaves of your plant will begin to lose their deep green color as the plant moves more and more of its nutrients to the newly forming potatoes beneath the soil level. The plant will eventually begin to wither and appear to die. However, it is only following its normal life cycle and is relocating itself beneath the soil surface. Continue to water the plant as you have been throughout its growth period.

- When the plant appears to be almost "dead," you can think about digging for your potatoes. If indoors, take the pot outside. Pull the plant from the soil. Depending upon the success of your growing attempts, you should find quite an assortment of potatoes clinging to the roots of your plant. Some will be very small yet, while some may be quite large. See if you can find your original "seed" potato. What does it look like? Take a photo of your potato crop. Break them free from the roots and wash them well. Count how many your plant produced. Weigh them if you have a scale. At this point, your potatoes are referred to as "new" potatoes and they can be cooked without peeling. Boiling them and then serving with butter and a sprinkle of salt is very delicious. Share with your family.

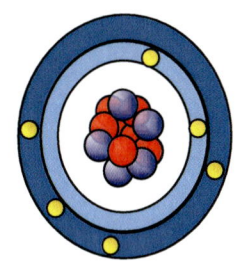

Lesson 2: A Little Bit of Chemistry

In Lesson 1, we discussed five characteristics that living things possess in order for them to be considered alive. We said that living things move, reproduce, require a source of energy, grow and develop and respond to their environment. In order for all of these things to happen, certain events must take place at the very small, microscopic and even sub-microscopic, or, atomic, level. To understand this, in this lesson, we'll take time to review some basic chemistry concepts regarding atoms, elements and the compounds they form.

The first idea to consider is that all things, whether living or not, are made up of tiny bits of matter known as atoms. Atoms are so tiny that, alone, they cannot be seen with the naked eye or even powerful microscopes. Much of what we know about atoms is based upon theories which have been thought and studied about for many, many years. These theories say that everything in this world as we know it is made up of atoms.

> Everything is made of small bits of matter known as atoms.

The theory says that atoms are designed similarly to the way our solar system is designed. There is a central portion, much like the sun in our solar system, known as the nucleus of the atom. The nucleus of the atom is made up of even smaller bits of matter known as subatomic particles. There are two kinds of subatomic particles found in the nucleus of an atom: neutrons and protons. Circling around the nucleus of an atom is a third type of subatomic particle known as electrons.

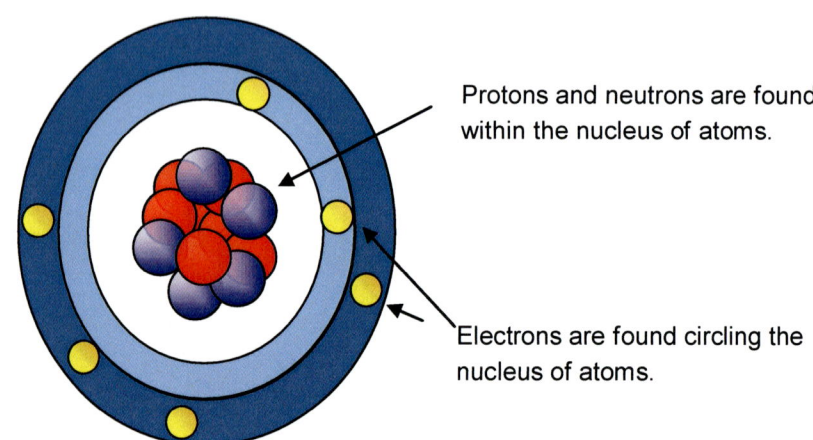

Protons and neutrons are found within the nucleus of atoms.

Electrons are found circling the nucleus of atoms.

Compared to the neutrons and protons within the nucleus, electrons are very light. Most of the mass of an atom is in its neutrons and protons. However, the electrons tell us a lot about the **behavior** of particular atoms. In other words, it is the way the electrons are arranged in atoms which predicts how the atom behaves around various other atoms.

Before we explore the electrons more deeply, we need to discuss the number of each subatomic particle (neutrons, protons and electrons) found in particular atoms. Chemists say that each element on the periodic table has its own number of protons, neutrons and electrons. The number found for each element is called the atomic number. Look at the periodic table on the next page.

Note that the atomic number is the whole number found in the upper left-hand corner of each square. Hydrogen (upper left side) with the symbol H, has an atomic number of 1. This means that atoms of hydrogen have one proton, one neutron and one electron (for the most part, as hydrogen is a sort of renegade when it comes to the number of electrons and may have up to three!).

Periodic Table of the Elements

Legend:
- Atomic Number → 1
- Element Symbol → H
- Element → Hydrogen
- Atomic Mass → 1.0080

1 H Hydrogen 1.0080																	2 He Helium 4.0026
3 Li Lithium 6.94	4 Be Beryllium 9.012											5 B Boron 10.811	6 C Carbon 12.0115	7 N Nitrogen 14.0067	8 O Oxygen 15.994	9 F Fluorine 18.994	10 Ne Neon 20.18
11 Na Sodium 22.9898	12 Mg Magnesium 24.31											13 Al Aluminum 26.9815	14 Si Silicon 28.086	15 P Phosphorus 30.974	16 S Sulfur 32.06	17 Cl Chlorine 35.453	18 Ar Argon 39.948
19 K Potassium 39.102	20 Ca Calcium 40.08	21 Sc Scandium 44.96	22 Ti Titanium 47.9	23 V Vanadium 50.94	24 Cr Chromium 51.996	25 Mn Manganese 54.938	26 Fe Iron 55.847	27 Co Cobalt 58.933	28 Ni Nickel 58.71	29 Cu Copper 63.546	30 Zn Zinc 65.37	31 Ga Gallium 69.72	32 Ge Germanium 72.59	33 As Arsenic 74.9216	34 Se Selenium 78.96	35 Br Bromine 79.909	36 Kr Krypton 83.80
37 Rb Rubidium 85.47	38 Sr Strontium 87.62	39 Y Yttrium 88.91	40 Zr Zirconium 91.22	41 Nb Niobium 92.91	42 Mo Molybdenum 95.94	43 Tc Technetium (99)	44 Ru Ruthenium 101.07	45 Rh Rhodium 102.91	46 Pd Palladium 106.4	47 Ag Silver 107.868	48 Cd Cadmium 112.40	49 In Indium 114.82	50 Sn Tin 118.69	51 Sb Antimony 121.75	52 Te Tellurium 127.60	53 I Iodine 126.904	54 Xe Xenon 131.30
55 Cs Cesium 132.91	56 Ba Barium 137.34	71 Lu Lutetium 174.97	72 Hf Hafnium 178.49	73 Ta Tantalum 180.95	74 W Tungsten 183.85	75 Re Rhenium 186.2	76 Os Osmium 190.2	77 Ir Iridium 192.22	78 Pt Platinum 195.09	79 Au Gold 196.97	80 Hg Mercury 200.59	81 Tl Thallium 204.37	82 Pb Lead 207.2	83 Bi Bismuth 208.98	84 Po Polonium (210)	85 At Astatine (210)	86 Rn Radon (222)
87 Fr Francium (223)	88 Ra Radium (226)	103 Lr Lawrencium (256)	104 Unq	105 Unp	106 Unh	107 Uns	108 Uno	109 Une									

However, the remaining elements each have the number of protons, neutrons and electrons equal to their atomic numbers. Helium has two of each particle, lithium has three, beryllium four, etc. To complete the "story," we need to mention that in some cases, the number of neutrons found in atoms may fluctuate up or down slightly. So, for the most part, the atomic number of an element tells us the number of protons, neutrons and electrons atoms of that element will have.

Let's return to our discussion of how the arrangement of electrons determines the behavior of atoms. Theories say that the electrons of atoms are arranged in layers or shells around the nucleus of the atom. On the first layer, nearest the nucleus of the atom, there are a maximum two electrons. After the first layer, there are up to eight electrons on each layer. The theories say that eight electrons is the maximum number of electrons that you'll ever find on a layer. The electrons fill up the layers nearest the nucleus first and then fill the next layer in an outward direction.

Let's look at an example of the element carbon. Look for carbon on the periodic table. The element symbol for carbon is C and can be found towards the right side of the table. Note that the atomic number for carbon is 6. From our discussion in the previous paragraph, this means an atom of carbon has six electrons circling about the nucleus. Two of these electrons will fill the first layer and then the remaining four will take their place in the second layer out.

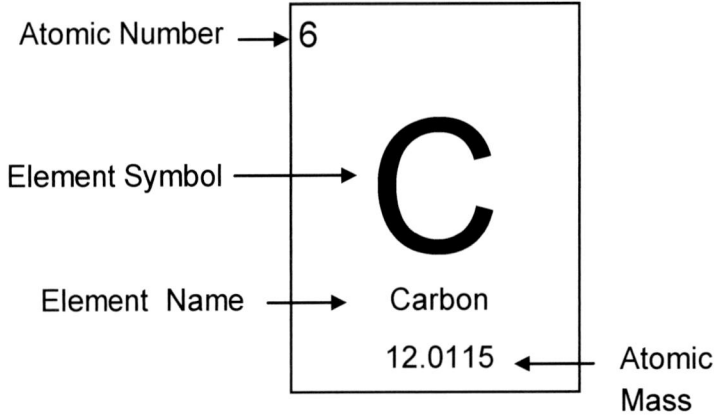

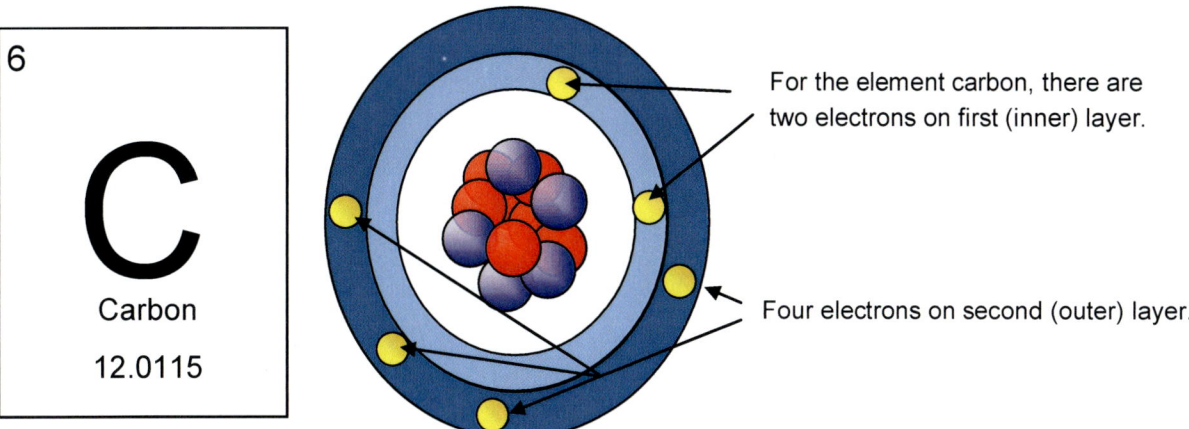

Let's look at another example. Find the element neon on the periodic table. The symbol for neon is Ne and can be found on the far right side of the table. Note that the atomic number for neon is 10, which means atoms of neon each have ten electrons. Once again, there will be two atoms on the first layer and the remaining eight will take their positions on the second layer. As we stated earlier, eight electrons is the maximum number of electrons that can be found on one layer, so the element neon has its outermost layer completely filled.

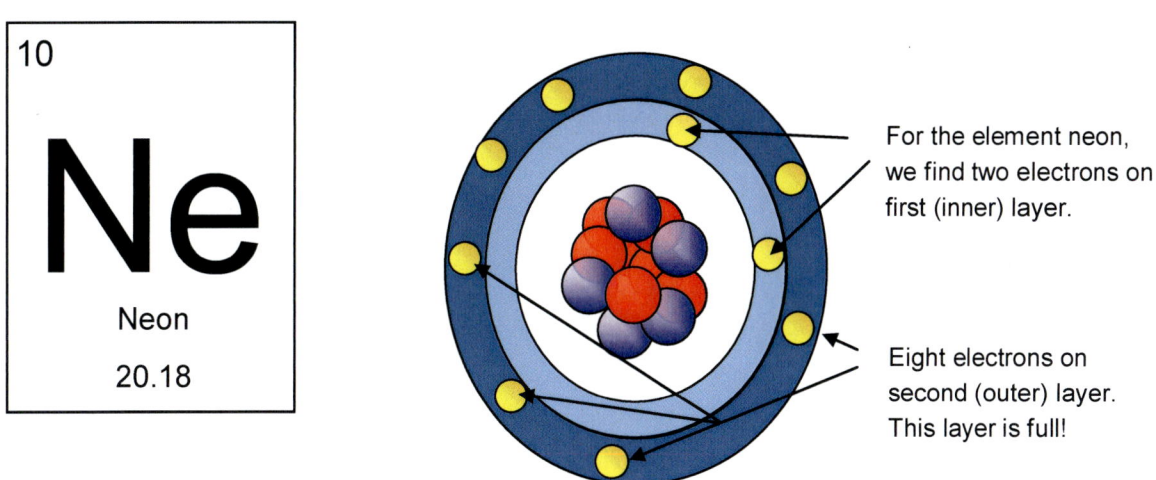

If we look at the next element on the table, sodium (Na, number 11), we find that it has two electrons on the first layer, eight on the second (total of 10, so far) and then we move to the third layer where there would be the one final electron.

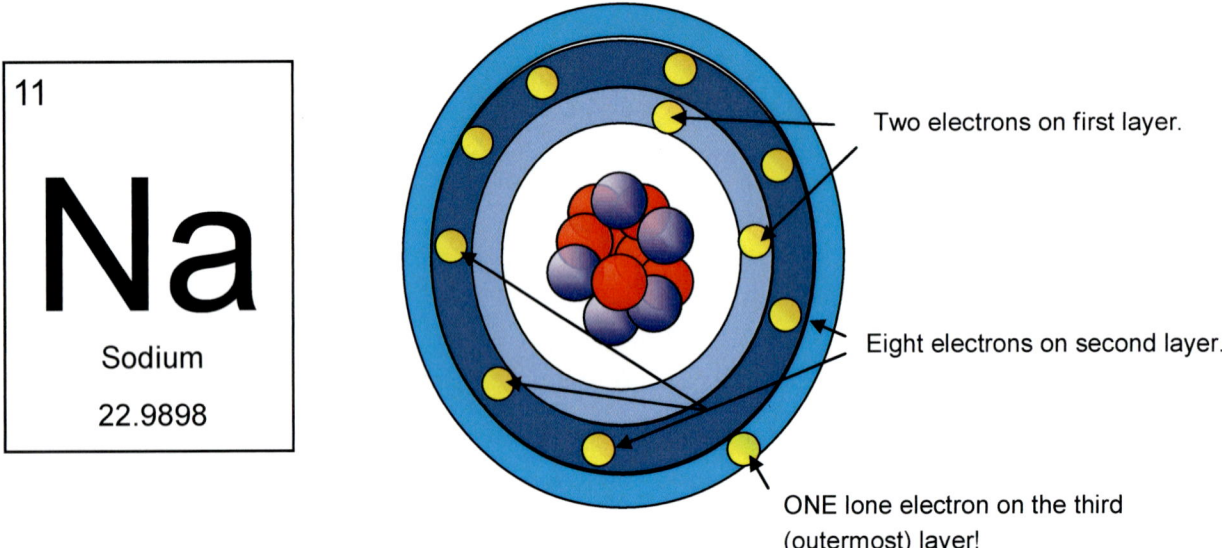

Two electrons on first layer.

Eight electrons on second layer.

ONE lone electron on the third (outermost) layer!

Now, what does this arrangement of electrons have to do with the behavior of atoms? Atoms that have their outermost orbit or shell completely filled with electrons are atoms that are the most non-reactive or stable. These atoms do not want to react with any other atoms of any other elements. The family of elements that is the epitomy of stability is the noble gas family found on the far right-hand side of the periodic table of elements (helium, neon, argon, krypton, xenon and radon). Each of the elements in this family have their outermost orbit completely filled. Another name for this family of elements is the inert gas family. Inert means no action or movement. These elements do not react very much whatsoever with any other elements. They are extremely stable elements. We might say they are "happy" the way they are.

The remaining elements on the periodic table, which do not have their outermost orbits or shells filled (this includes all the other elements but the noble gas family members), are not "happy" the way they are. Some are extremely "unhappy" and are, consequently, extremely reactive. To reduce this irritability, these elements often reorganize their electrons in such a way that it appears that their outermost orbits are now filled. Some of these irritable elements "borrow" electrons from other elements in an effort to make their outer layers of electrons appear full.

						2 He Helium
5 B Boron	6 C Carbon	7 N Nitrogen	8 O Oxygen	9 F Fluorine	10 Ne Neon
13 Al Aluminum	14 Si Silicon	15 P Phosphorus	16 S Sulfur	17 Cl Chlorine	18 Ar Argon
30 Zn Zinc	31 Ga Gallium	32 Ge Germanium	33 As Arsenic	34 Se Selenium	35 Br Bromine	36 Kr Krypton
48 Cd Cadmium	49 In Indium	50 Sn Tin	51 Sb Antimony	52 Te Tellurium	53 I Iodine	54 Xe Xenon
80 Hg Mercury	81 Tl Thallium	82 Pb Lead	83 Bi Bismuth	84 Po Polonium	85 At Astatine	86 Rn Radon

Noble gas family. All members have their outer layer filled with electrons. Another name for this family is the inert gas family. Inert means non-reactive.

Other elements find that if they can move one or two of the electrons found in their outermost layer, they, too, can appear to look like a member of the stable noble gas family. So, by gaining a few electrons or losing a few, atoms of these "irritable" elements can gain the stability of the noble gas family members. In many cases, atoms of elements that would like to get *rid* of electrons find they can easily do so when mixed with atoms of elements who desire to *gain* a few electrons. By allowing these atoms to come into contact with one another, the electrons can be moved accordingly and stability attained.

Noble gas family members are very stable. All other elements are "green" with envy and gain, lose or share electrons with other atoms to gain the appearance and stability of a noble gas family member.

This process of gaining or losing electrons is known as **ionization** and results in atoms forming "relationships" with other atoms to gain the stability "enjoyed" by the noble gas family members. These "relationships" are scientifically known as bonds which are links between various atoms. When electrons get transferred from one atom to another in the attempt to gain stability, the bonds that result are called **ionic bonds**.

A great example of this is the combination of sodium and chlorine atoms. Find the element sodium on the periodic table. Look for it on the left side, about half-way down with the symbol Na. Note that sodium has eleven total electrons. If we look at how these electrons are arranged, we find two on the first layer, eight on the second and then one on the third layer (for a total of eleven electrons). If the sodium atoms could lose the one electron on the third layer, they would "look like" neon, a noble gas family member. Note that these sodium atoms don't actually become neon, instead they just take on the stability of neon. Look now at the chlorine atoms.

Na (sodium)

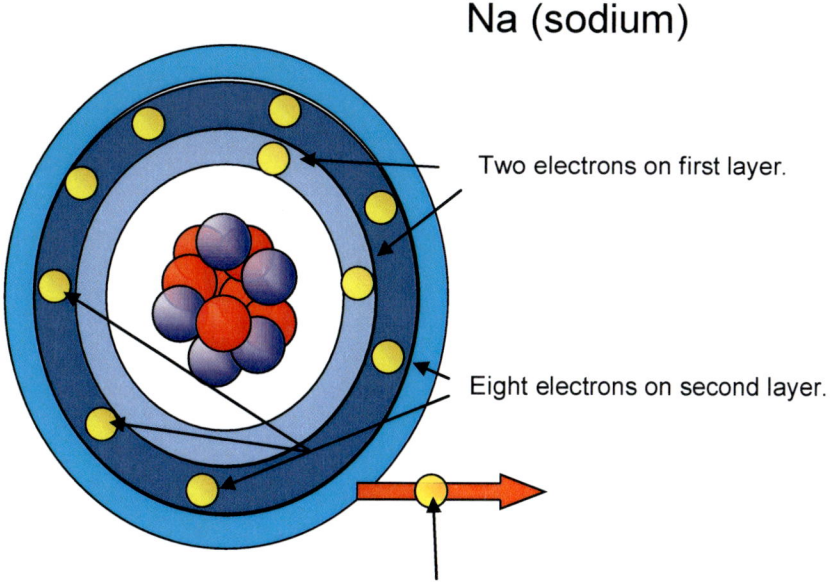

Two electrons on first layer.

Eight electrons on second layer.

ONE lone electron on the third (outermost) layer. If this electron could be given to another atom willing to receive it, this sodium atom could take-on the appearance of the element neon (with 10 electrons.) In doing so, it could attain the stability of neon and be "happy."

Find the element chlorine on the periodic table. It can be found over on the right-hand side in the family just next to the noble gas family. The element symbol for chlorine is Cl. The atomic number for chlorine is 17 meaning chlorine atoms have 17 electrons. If we look at how these electrons are arranged, we find there are two electrons on the first layer, eight on the second layer and then seven on the third layer. If chlorine atoms could each gain one electron from some other source, they could take-on the stability of argon (atomic number 18) and be very "happy."

Cl (chlorine)

Chlorine has seven electrons in its outermost layer. It will very willingly accept one more electron to have its outer layer filled with eight electrons. When it accepts one electron it will now have 18 total electrons and will attain the appearance and stability of argon and be very "happy."

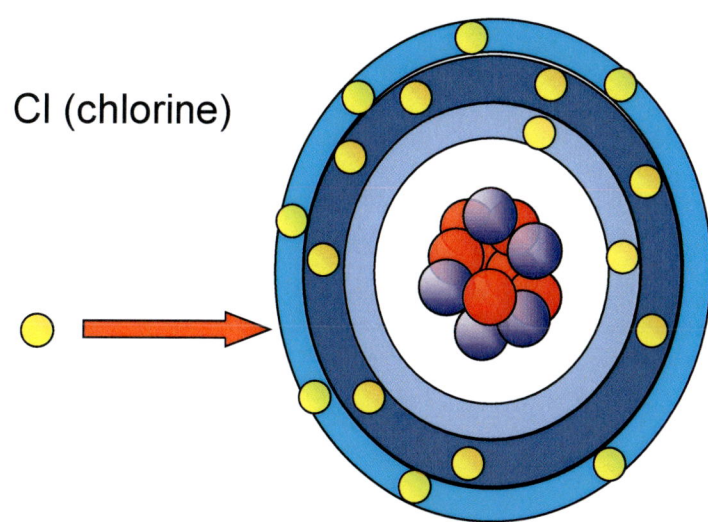

As you probably assume, if the sodium atoms get "mixed" with the chlorine atoms, the sodium atoms will readily "donate" their one extra electron to the willing chlorine atoms who are very happy to accept their donations. When this happens, ionic bonds are formed. The resulting combination of elements sodium and chlorine will produce the very stable compound sodium chloride, known more commonly as table salt! So, we see that elements from all across the periodic table can "work things out" by transferring electrons to form very stable compounds.

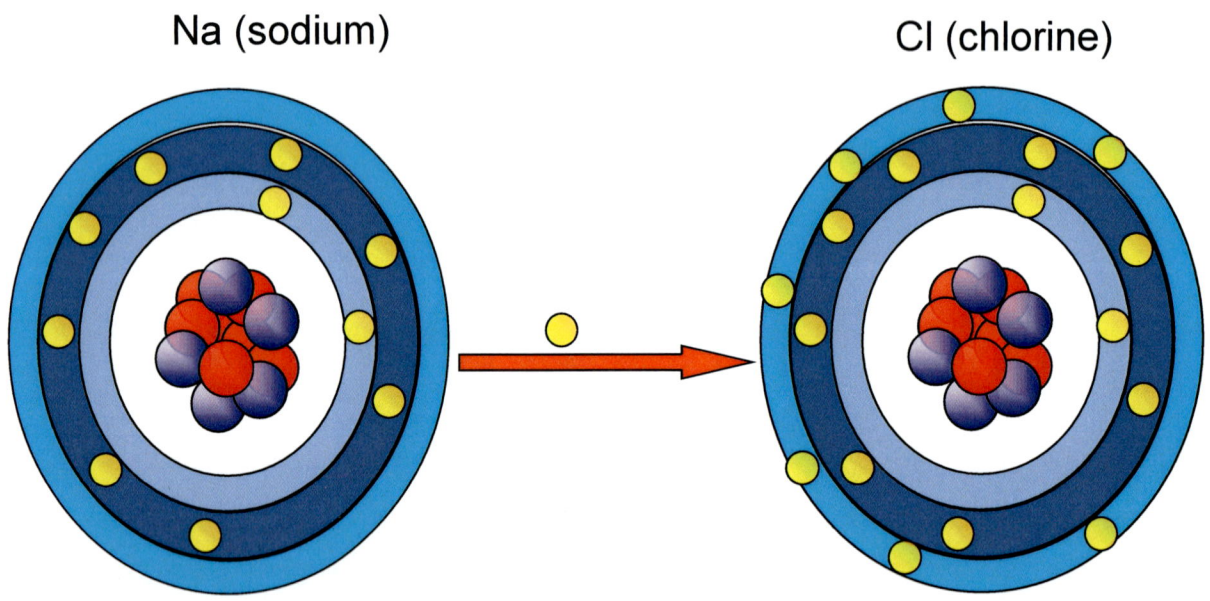

The sodium atom donates its outermost electron to the chlorine atom. In doing so, both atoms gain the appearance and stability of its nearest noble gas neighbor. The sodium atoms "looks" like neon and the chlorine atom, because it has gained one electron, "looks" like argon. This transfer of an electron results in the formation of an ionic bond between the two atoms.

In some cases, however, some elements find that sharing, rather than transferring electrons, is how they prefer to form bonds and, consequently, compounds. These elements are still attempting to fill their outermost layer of electrons and do so by allowing electrons to "be" at one atom for a short time and then moved back to another atom. We can liken this situation to the sharing of a toy. You have a ball (electron) which allows you to have a full toy box (outer layer of electrons). Your neighbor next door also likes the ball since it will fill up his toy box when it is at his house. Between the two of you, as long as you share the ball (electron) back and forth, each can have the stability the filled outer layer brings. This type of bonding which results from the sharing of electrons, rather than the transfer, is called **covalent bonding**. The name covalent comes from co– meaning with or among and –valent which refers to the valence electrons, those electrons found in the outermost layers. Covalent bonds are formed when electrons are shared between atoms.

Some elements form covalent bonds by sharing one electron and these bonds are called single covalent bonds. An example of this is chlorine gas. Chlorine gas consists of two atoms of chlorine each which share one electron. For part of the time, one of the atoms has a full outer layer and then the shared electron returns to the other atom allowing it to attain stability.

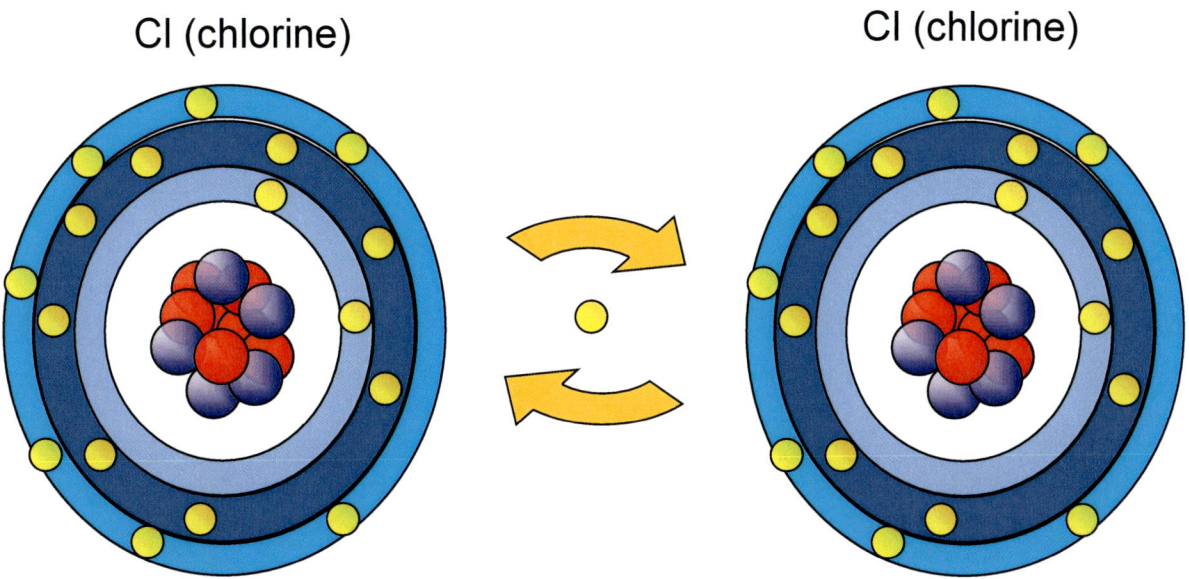

Some atoms like to share electrons, rather than transfer them. This sharing of electrons results in what is known as covalent bonds.

Other elements share two electrons and these bonds are called double covalent bonds. Oxygen gas is an example of a compound formed between two atoms of oxygen each sharing two electrons back and forth between them. Still other elements—nitrogen, for example— form stable compounds by sharing three electrons. These bonds that are formed when three electrons are shared are called triple covalent bonds.

> Ionic bonds form when atoms transfer electrons from one to another.
> Covalent bonds form when atoms share electrons.

If you look at the periodic table of elements, you can see that there are over 100 known elements and, obviously, many, many ways that the atoms of elements could combine to form compounds. However, as we study living things, we will concentrate on only a few very important elements which are vital to life. **Those important elements include carbon (C), hydrogen (H), oxygen (O) and nitrogen (N).** Note that there are many more elements that are required for living things to survive, such as sodium, potassium, calcium and magnesium, for example. We will discuss those elements later in this course. At this time, we will concentrate on carbon, hydrogen, oxygen and nitrogen.

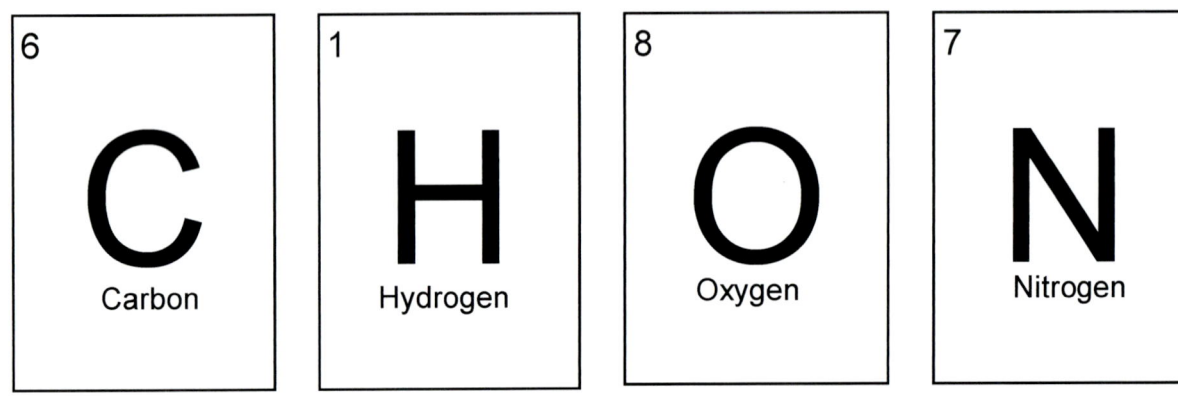

The compounds that these four elements form are necessary for life. Think about hydrogen and oxygen, for example. Can you think of any compound made from hydrogen and oxygen that is so important that we must consume some every day and that makes up over 70% of our bodies? Water! Yes, water is vitally important to the survival of all living things, albeit some to a lesser degree than others. The combination of hydrogen and oxygen atoms to form the compound water is extremely important!

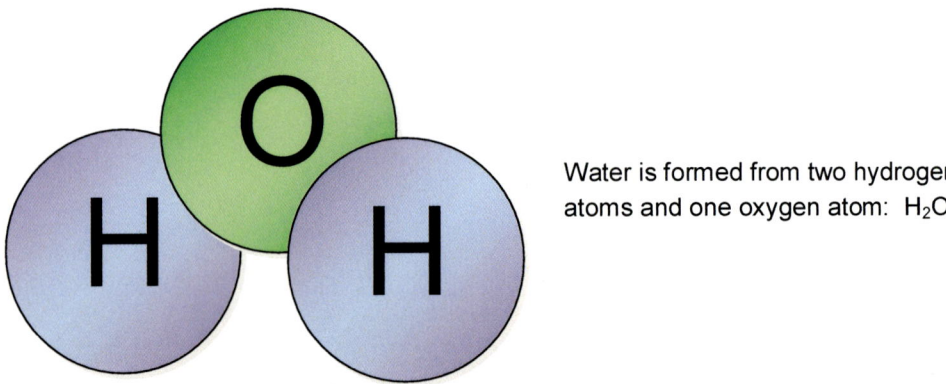

Water is formed from two hydrogen atoms and one oxygen atom: H_2O.

Before we end our discussion on basic chemistry, let's take a slightly closer look at the atomic structure of each of these elements. Recall that we said an atom of carbon has a total of six electrons (carbon is atomic number six.) This meant that in an atom of carbon, we'd find two electrons on the inner layer of electrons and then four on its outer layer. Because carbon would rather have some combination of electrons to make eight electrons on its out layer, we find that carbon has the desire to gain four electrons from neighboring elements. In other words, carbon atoms generally desire to form four bonds with other elements.

If we were to look at oxygen atoms, we'd find of the total of eight electrons, there are, again, two on the inner layer of electrons and six on the outer layer. Because oxygen would like to have eight on its outer layer, it has the desire to form two bonds with neighboring elements. So far we've said that carbon atoms desire to form four bonds and oxygen atoms desire to form two bonds.

Let's look at hydrogen atoms now. If we find hydrogen on the periodic table, we see that it is atomic number one indicating, for the most part, hydrogen atoms have one electron. Chemists have found that hydrogen atoms readily donate this lone electron to neighboring elements to gain their desired stability. Therefore, hydrogen atoms desire to form one bond with neighboring elements.

Finally, let's look at nitrogen. The nitrogen atom has a total of seven electrons. Two of these seven will be found on the inner layer of electrons while the remaining five will be on the outer layer. Knowing that all atoms desire to have eight on their outer layer, how many bonds do you guess nitrogen atoms like to form? If you said three, you're exactly right. Nitrogen atoms desire to form three bonds with neighboring elements.

Knowing these bonding arrangements of these four important elements will be very helpful in the upcoming lessons where we examine how these elements are used by living organisms.

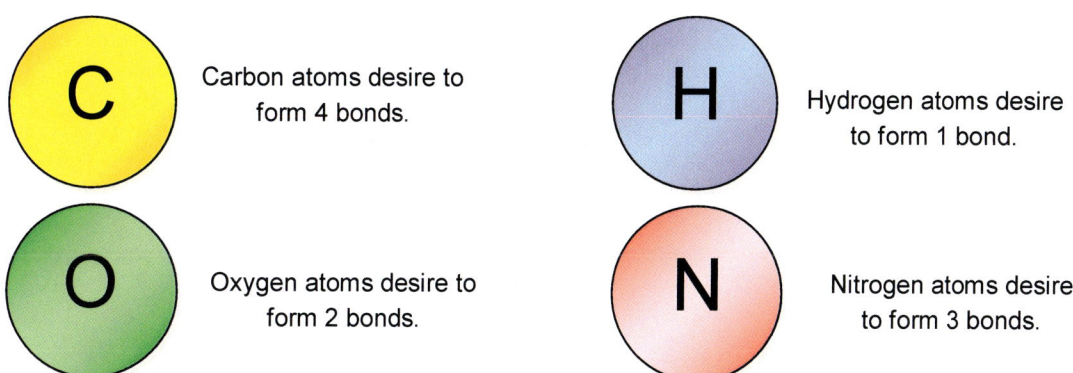

Lets pause and review now. In this lesson we've learned that:

- Everything in our world, living or non-living, is composed of matter. Theories tell us that matter is composed of very tiny bits known as atoms.

- Atoms consist of a central nucleus containing neutrons and protons. Circling around the nucleus are electrons.

- It is the arrangement of electrons in atoms that allows us to understand why some elements can be very reactive while others are very stable.

- Atoms that have their outermost layer of electrons totally filled are the most stable. The noble gas family members enjoy this stability.

- All other elements adjust the number of electrons in their outer layers by either transferring electrons from one to another and forming ionic bonds or by sharing electrons back-and-forth resulting in the formation of covalent bonds.

- As atoms join together in various ways to gain stability, compounds form.

- The elements carbon, hydrogen, oxygen and nitrogen are of vital importance to living organisms.

- Based upon their atomic structure, carbon atoms desire to form four bonds, oxygen atoms desire to form two bonds, hydrogen atoms desire to form one bond and nitrogen atoms desire to form three bonds with other elements.

Lesson 2 Lab Activity: In Quest of Carbon: The Marshmallow Inferno!

IMPORTANT: READ ALL OF THESE INSTRUCTIONS FIRST BEFORE BEGINNING!

The purpose of this lab activity is to allow you to isolate a sample of carbon from a marshmallow. Recall that the symbol for carbon is C and the atomic number for carbon is six. To isolate a sample of carbon, you will take a "raw" (uncooked marshmallow) and incinerate it over an open flame. Please read all of the instructions below before beginning this lab activity. To prepare for the marshmallow "roasting," you will need to collect the following supplies:

- an old, metal spoon
- candle in candlestick or holder
- tray or newspaper "placemat" in which to catch spills
- matches
- marshmallows - regular size, at least 4-5 in order to make observations

Safety considerations: If you are conducting this lab in a public facility or have a smoke-detector in your home, check to see if the smoke that you create will set it off or activate a smoke alarm system. If so, you may wish to re-locate to an outdoor table to complete this activity. Just as a precaution, you may want to locate the nearest fire extinguisher. Plan to have a bowl of water nearby should you need to extinguish any accidental fires or cool a burned finger. Note that a metal spoon will conduct heat and you may experience this phenomenon during the heating of the marshmallow. A kitchen mitt or pot holder may be used to protect your hand.

Procedure:

1. Begin the activity by taking out a "raw" marshmallow and make as many observations as you can from it. Don't forget to "dissect" the marshmallow to examine the inner parts of the marshmallow. Include observations made on sight, smell, touch, sound and of course, taste!

Write your observations here:

2. Next, think about from what ingredients are marshmallows made. Write what you think they are made from here and then check the ingredients label on the bag.

3. Continue the activity by placing a "raw" marshmallow on the metal spoon and heating it over the open candle flame. Allow the marshmallow to catch fire and allow it to cook until all that remains is a black, crunchy shell. This may take quite awhile to accomplish. The marshmallow may smoke for a while and then seem to quit and then begin to smoke again later with a different color smoke. Keep cooking until all smoking ceases. Write your observations here:

4. Allow the blackened marshmallow to cool and then make a new set of observations both inside and outside the marshmallow. Again, make observations of color, texture, smell and even taste! Know that the spoon and inside portions of the marshmallow may still be hot. Record these new observations and compare them with the initial set made of the "raw" marshmallows. Get another "raw" marshmallow, if necessary, to allow direct comparisons! Write your new observations here:

5. Finally, think about what remains of the marshmallow. Does it smell like ashes or charcoal? What is left now is a piece of carbon. The sugar, gelatin and starch compounds that make up marshmallows all contain carbon along with water and some other elements. The water that was originally found in these compounds has been cooked away. What is left is the carbon.

As an interesting side note, did you find that the underneath surface of the spoon you used to hold the candle turned black? What do you suppose this is? You may know that soot from a fire can turn things black like in a fireplace or inside a chimney. This soot, again, is another deposit of carbon.

In the case of the wax candle, depending upon which kind of candle you used, it could be made from a petroleum product or if you used a beeswax candle, from the product of bees. In either case, in the process of burning, not all of the wax gets burned and some carbon is actually found in the smoke. This carbon gets deposited on the underneath side of the spoon as soot.

In the next lessons in this course, we will examine in much greater detail how carbon along with water and other elements are vitally important to living creatures. Carbon is an essential ingredient!

To finish your lab, first make sure your candle has been extinguished and allowed to cool before moving to avoid being burned with the melted wax. Your spoon can be cleaned by allowing it to soak in soapy water overnight and then washed the next day. Clean up your work area and then make plans to use your remaining marshmallows for your next dessert or campout.

Finally, in the space below, write some things you learned in this lab.

I learned that...

Lesson 3: The Carbohydrates

In Lesson 2, we introduced the idea that there were four elements that were very important to all living things. Those elements were carbon, hydrogen, oxygen and nitrogen. We also discussed how the atoms of elements desire to form compounds in order to gain greater stability. These stable compounds, such as water, for example, are extremely important for the well-being of all living creatures.

Beginning with this lesson, we will explore four groups of compounds that are vital to living things. We will introduce each group of compounds first by looking at the elements which make up that group of compounds. Second, we'll look at the structure of those compounds. Then, finally, we will look at how they are used in living things. The first set of compounds we will look at are known as the **carbohydrates**, more commonly known as **sugars**.

Let's begin our discussion of carbohydrates by examining the elements from which carbohydrates are built. If we look at the name, carbohydrate, we can see two word parts: carbo– and –hydrates. The part carbo– tells us that the element carbon is a main component of carbohydrates. The second portion, —hydrate, is the root word of Greek origin for water. Water, as we discussed earlier, consists of hydrogen and oxygen atoms. So, carbohydrates consist of carbon, hydrogen and oxygen atoms. There are no nitrogen atoms found in carbohydrates.

> Carbo- refers to the element carbon.
>
> -Hydrate = water = H_2O

C + H₂O = a carbohydrate

There are many kinds of carbohydrates, but they all have a generic formula or recipe regarding how many atoms of carbon, hydrogen and oxygen each contains. This generic formula looks like this:

$$C_n(H_2O)_n$$

Recall, that the element symbol for carbon is C and water is H_2O. Note the little "n's" written as subscripts following the carbon and water parts of the carbohydrate formula. These subscripts tell us that the number of carbons and water parts can be adjusted, but the number of each is always equal. For example, if there are 6 atoms of carbon (n=6), then there would also be 6 parts of water. This carbohydrate would have the following formula:

$$C_6(H_2O)_6$$

This carbohydrate happens to be one the most important sugars of all! Its name is glucose, more commonly known as "blood sugar." Another way to write the formula for glucose is:

$$C_6H_{12}O_6$$

Note we just multiplied the water portion of the glucose molecule times six to get this result.

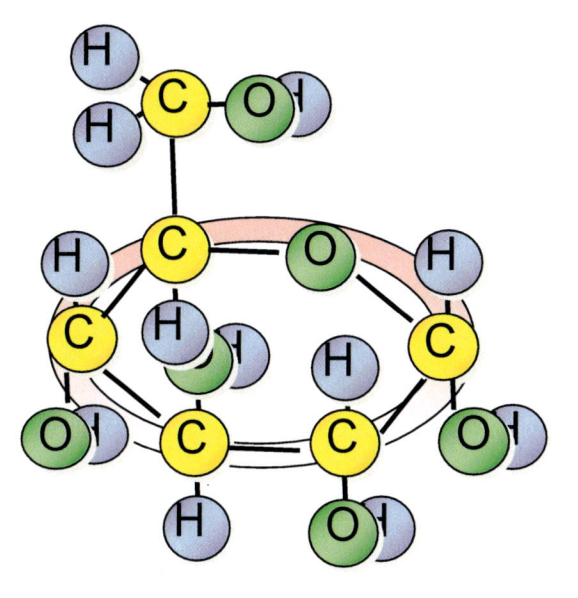

The glucose molecule is thought to exist as a circular, ring-shaped structure of carbon atoms. Note how each carbon atom has formed four bonds, each oxygen atom has formed two bonds and how each hydrogen atom has formed one bond.

Another carbohydrate that is found in many fruits is the sugar known as fructose. If we look at the formula for fructose, we see that the value for n is also six (n = 6.) The formula for fructose looks identical to that of glucose:

$$C_6(H_2O)_6$$

While glucose and fructose both have the same number of carbon atoms and water molecules in their formulas, the way these atoms are arranged in relation to each other results in differences between these two carbohydrates. These two different structures are known as **isomers.**

The natural sugar found in fruits is fructose.

To review, we've discussed that carbohydrates are composed of carbon atoms and molecules of water. The "generic" formula for a carbohydrate is $C_n(H_2O)_n$. Glucose, where n = 6, is one of the most important carbohydrates to living things. Glucose is more commonly known as blood sugar. A variation on the structure of glucose is fructose, a sugar found in fruits,/ and this variation is called an isomer.

Let's continue our exploration of carbohydrates. Both glucose and fructose are considered monosaccharides. The prefix for this term is mono– which means one. The suffix is –saccharide which means sugar. Monosaccharides are carbohydrates composed of one type of sugar.

> Mono = one
>
> Saccharide = sugar
>
> Monosaccharide = carbohydrate made up of one kind of sugar molecule.

Now consider disaccharides. The prefix di– means two so a disaccharide is a carbohydrate that consists of two types of sugars.

> Di = two
>
> Saccharide = sugar
>
> Disaccharide = carbohydrate made up of two kinds of sugar molecules.

Let's consider an example of a disaccharide. Sucrose, commonly known as table sugar, is a disaccharide. Sucrose is made up of one molecule of glucose and one molecule of fructose. These two molecules are bonded together to form this disaccharide.

Table sugar is sucrose.

Sucrose = glucose + fructose

Another very important disaccharide is lactose. Lactose is made up of one molecule of glucose and one molecule of galactose. Together, they form the disaccharide lactose. Lactose is the sugar naturally found in milk and, because of this, is the primary carbohydrate consumed by baby mammals as they nurse from their mothers. Obviously, lactose is a very important carbohydrate! Lactose can be found in all sorts of foods which contain dairy products such as milk, ice cream, yogurt and cheeses.

The natural sugar found in milk is lactose.

Lactose = glucose + galactose

Earlier we mentioned that the monosaccharide glucose is the most important carbohydrate for living things. Glucose is the primary fuel from which energy is produced in living organisms. It's like the gasoline that runs our cars. Glucose is the gasoline for our bodies! In the previous paragraphs, we noted that the carbohydrates sucrose and lactose were disaccharides, being composed of a combination of two carbohydrates. Note that glucose was found in both of those carbohydrates (sucrose = fructose + glucose and lactose = galactose + glucose) When we eat foods containing sucrose and lactose, our bodies have the capability of splitting these disaccharides back into monosaccharides thereby making glucose available for fuel. This splitting is accomplished through the use of chemicals known as enzymes.

Glucose is the "gasoline" for living things.

Enzymes are very interesting chemicals that do all sorts of amazing jobs in living things. Most of these jobs can be categorized into two main categories: matchmaking and cutting. Let's take a look at each of these jobs.

What is matchmaking? By definition, matchmaking is the process in which someone assists in the relationship of two other individuals. Think about this situation: suppose you were interested in meeting someone but were too timid or shy to just go up and introduce yourself. You might consider getting the assistance of a friend who could go to this person you would like meet and tell him or her you were interested in meeting. Your friend is functioning like a matchmaker. The friend was helping to speed-up the "reaction" between you and this new person you desired to meet, but the friend was not changed by the reaction. Enzymes work in a very similar fashion. Enzymes in living things work to speed-up reactions, but, in the process, they themselves do not get changed. This job of enzymes is essential when glucose is converted into energy for living things. We will discuss this idea in greater detail later in this course.

Enzymes can work like matchmakers.

Mulan

Fiddler on the Roof

The second job of enzymes is to cut. Enzymes can literally cut other compounds into pieces. A more scientific term for cutting is to lyse. Enzymes can lyse disaccharides into monosaccharides. Recall that sucrose is a disaccharide. The enzyme known as sucrase can lyse sucrose into its two components: fructose and glucose. Note that the ending of the word sucrose was changed to –ase in the enzyme name of sucrose.

Enzymes can also work like scissors to take things apart.

As we discussed previously, lactose is another disaccharide. The enzyme lactase can lyse lactose molecules into its two components: galactose and glucose. Note again how we changed the name of lactose to lactase when the enzyme was named. You may be familiar with someone who suffers from the digestive problems of being lactose intolerant. People who suffer from this problem generally have a deficiency in the enzyme lactase which prevents them from being able to digest lactose found in dairy products. They find that when they drink milk or eat ice cream or other dairy products they suffer a lot of digestive system discomfort. The lactose molecules are not lysed and bacteria that normally live in the digestive system begin having a "feast" on the new supply of lactose. As the bacteria begin consuming the lactose, they produce a byproduct of gas which causes the discomfort often associated with lactose intolerance. People who suffer from this problem often find it best to avoid eating dairy products, however, some find that taking lactase supplements helps to replace the low levels of naturally occurring lactase.

There are hundreds of other enzymes which work to speed-up reactions in our bodies as well as cut or lyse things. Enzymes are not particular only to carbohydrates. They work on all sorts of molecules and compounds in living things. As we move through this course we will discuss those enzymes and understand how very important they are in living things.

Let's go back to our discussion of various carbohydrates that are important to living things. We've talked about glucose (blood sugar), fructose (fruit sugar), sucrose (table sugar) and lactose (milk sugar). Maltose is another carbohydrate found in living things. Like lactose and sucrose, maltose is a disaccharide and is composed of two glucose molecules linked together. Maltose is found in seeds and grains of plants. Can you think of the enzyme name which can lyse maltose into its components? Yes, that enzyme is maltase.

We said maltose consists of two glucose units bound together. Plants have the ability to continue adding glucose units onto maltose sugars to create very large complex structures known as starches. Starches have a more scientific name which is amylose.

Because amylose consists of many, many glucose units bound together, amylose can be thought of as a storage mechanism of fuel for plants. You see, plants have the unique ability to produce their own glucose through the process known as photosynthesis. This process, for the most part, takes place in the plant's leaves. Other living creatures, like you and I, can't directly make our own glucose which is why we must continually consume food that contains glucose (or other carbohydrates that contain glucose). We will discuss photosynthesis later in greater detail. Amylose works like fuel that's been stored-up in a big tank outside your house. During times when the plant is actively growing and making glucose through photosynthesis, it continues to build amylose and store it up. In times of the year when the plant doesn't have leaves, it relies upon these stock-piles of amylose. Plants create these stockpiles in their seeds as well as their roots in the form of tubers like potatoes or bulbs.

Amylose is stored in a plant's seeds or roots.

Consider the seeds of plants which have these stockpiles of amylose. When seeds are planted, they are exposed to appropriate conditions of moisture and warmth in the hopes of getting them to sprout or germinate. The tiny baby plant inside the seed requires energy to begin growing. It will begin taking some of the amylose out of storage and turning it back into glucose for energy production. You might think of the situation like this: the "mother plant" which originally produced the seeds has kindly packed a "lunch" for each of its seeds in the form of amylose neatly packed within the seed. When conditions are right, the "lunch box" gets opened and the baby plant can use the fuel source until it grows its own leaves and can make its own supply of glucose.

Seeds, packed with amylose, are like packed lunch boxes.

If you were able to get your potato plant underway, by now you can see some shoots growing upward at the surface of the potting soil. The energy source for this growth came from the stored amylose in the potato. Because you placed the potato in optimum growing conditions, it responded by beginning to utilize the stored glucose in the potato. As soon as the shoots emerge from the soil surface they will grow leaves which will begin producing glucose for the plant's continued growth. With proper care, your potato plant will likely produce a large bushy plant which will produce lots and lots of glucose molecules which will be stored in "new" potatoes.

Can you guess the guess the name of the enzyme which breaks down this stored-up starch into individual glucose molecules? As you might imagine, the enzyme which breaks down amylose is amylase. Amylase is not only found in plants but in you and me, as well. Amylase is found in our saliva (spit) and works to begin digesting starches as soon as we begin chewing them in our mouths. Can you think of foods you eat that are high in starches? How about those foods made from grounds seeds, especially wheat seeds? Yes, flour, which is made from ground (milled) wheat seeds is just about all starch. Any food you eat which contains flour, such as breads, cookies, muffins or cakes, contains lots of amylose. As soon as you mix a bite of these foods with the saliva in our mouth, amylase enzymes go to work breaking them down into glucose molecules. The potatoes we discussed earlier are also loaded with amylose. As you eat a French fry or baked potato or any type of food made from the root of a plant (carrots, turnips, beets, etc.), amylase in your saliva begins to do its job of lysing amylose into glucose molecules.

Before we end our discussion of carbohydrates, there is one other carbohydrate that we need to consider. That carbohydrate is known as cellulose. Cellulose is not a sweet carbohydrate like the other carbohydrates we have discussed so far in this lesson. Nonetheless, cellulose is a carbohydrate made up of carbon, hydrogen and oxygen atoms. In fact, cellulose consists of glucose molecules which link themselves into very long and very strong chains. Cellulose is the carbohydrate which gives plants their strength. Cellulose is found readily in celery! Have you ever taken a piece of celery and pulled-off little "strings" of celery or maybe had the misfortune of getting one of those strings stuck between your teeth? Cellulose is the carbohydrate which makes these strings.

Cellulose gives the crunchiness to fresh vegetables. Cellulose is also the carbohydrate which makes wood in larger plants and trees. In fact, cellulose is what's left over in the cell walls of trees after the tree is cut down and made into boards. You might think of cellulose as being like the "bones" of plants.

Cellulose makes the crunch in celery and carrots.

Can you think of the enzyme which breaks down cellulose? Yes, cellulase is the name given to a group of enzymes which are able to lyse long strands of cellulose into smaller pieces. Most mammals, including humans, are limited in the amount of cellulase we have available to digest cellulose. However, animals like cows, sheep, deer and goats, known as ruminants, have the ability to breakdown cellulose and "harvest" this great supply of glucose. Actually, these ruminants also lack cellulase enzymes, however, they have an enormous supply of very tiny living creatures in their stomachs (microorganisms) which *do* have the ability to produce cellulase. These microorganisms digest the cellulose (grass or hay or leaves, etc.). Together, the ruminant and the microorganisms form a very special relationship where both of them benefit. The host ruminant provides the microorganisms with a continual supply of cellulose. In return, the microorganisms supply cellulase to break down the cellulose into useable glucose molecules which can be absorbed and utilized by the host animal. Not on-

Ruminants consume large amounts of cellulose to be broken down in their rumens.

ly do these ruminants have this capability, but termites do, too! Within the stomachs of termites, they, too, have microorganisms that assist them in digesting the wood products the termites consume. Again, a relationship exists which benefits both of the living creatures involved. Relationships between living creatures where each "partner" in the relationship may (or, in some cases, may not) benefit are called **symbiotic relationships**.

> A relationship between two living things where one partner may or may not benefit is called a symbiotic relationship.

In cases, where both "partners" benefit, a **mutual symbiotic relationship** is said to exist. This is the case between the microorganisms which live in the stomachs of ruminants and termites and their hosts. Both the ruminant (cow, sheep, deer, etc.) and termites and the microorganisms benefit from the relationship.

Other symbiotic relationships can be observed in living creatures where one member of the relationship benefits while the other member does not. A good example of this relationship involves the fleas and ticks which you may find on your pets. The fleas and ticks require a constant supply of blood (glucose) to survive and do so by extracting it from your dog or cat. This type of relationship where one benefits (the fleas and ticks) while the other suffers (your dog or cat) is referred to as **parasitic symbiosis** or **parasitism.**

Ticks, which suck blood from their hosts, are parasites.

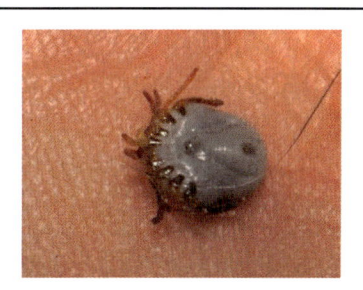

> A relationship where one member benefits to the detriment of the other is called parasitism.

Another type of symbiosis exists where one member of the relationship benefits while the other is not harmed. This type of relationship is termed **commensalism**. A good example of commensalism is the relationship between cattle egrets and cattle. Cattle egrets are large white birds which can often be found feeding on insects in tall grasses alongside cattle. As the cattle move along grazing, they stir up insects which are quickly consumed by the egrets. The egrets do no harm to the cattle, yet benefit from the movements of the cattle. In this case, the egrets benefit while the cattle are not affected.

Cattle egrets benefit from the movement of cattle in the grass. This is an example of commensalism.

A relationship where one member benefits while the other is unaffected is known as commensalism.

At this point, we will bring our introductory lesson on carbohydrates to a close. We will definitely visit the topic of carbohydrates again in Lesson 7 where we will discuss in greater detail how living things utilize carbohydrates (glucose in particular) to carry on life's processes.

Let's review:

- Carbohydrates are composed of carbon, hydrogen and oxygen atoms with the generic formula of $C_n(H2O)_n$.

- Glucose is the most important carbohydrate for living things in that glucose is the fuel on which living things "run." The formula for glucose is $C_6H_{12}O_6$.

- Some carbohydrates (sugars) are single molecules (monosaccharides) while others are composed of two types of carbohydrates (disaccharides). Enzymes work to join carbohydrates as well as cut or lyse carbohydrates into simpler units. Enzymes are named similarly to the carbohydrate on which they act; the ending of the carbohydrate name is changed to –ase.

- Fructose is the carbohydrate found in fruits and vegetables. Lactose is found in milk and milk products. Sucrose is table sugar. Maltose is found in grains.

- Plants can make their own glucose and store it in their seeds and roots for later use. This storage form of glucose in plants is known as amylose.

- Cellulose, while not sweet, is an important carbohydrate in that it provides strength and structure for plant cells.

- Interesting relationships, known at symbiotic relationships, exist between living creatures where one or both "partners" in the relationship benefits. Where both benefit, a mutualistic symbiotic relationship is said to exist. This relationship is important for ruminants who host microorganisms that help them break down the important carbohydrate, cellulose, into usable glucose.

Lesson 3 Lab Activity: Testing for the Presence of a Starch (Amylose)

PLEASE READ ALL OF THESE INSTRUCTIONS BEFORE BEGINNING THIS LAB ACTIVTY.

The purpose of this lab activity is to allow you to test various substances in your kitchen for the presence of starch (amylose.) Recall that amylose is formed by plants in an effort to store glucose to be used later.

Things you'll need for this lab include:

- Several small lids from plastic containers. Lids from yogurt cups, baby food, whipped topping work well. If you don't have these handy, you can use small Styrofoam plates or cups. The idea is to have a container with low sides on which you can place your powder for testing. If you use cups, plan on cutting down the sides to make them more shallow.

- Toothpicks or coffee stirrers. These will be used to stir your test powders with the testing solution. Plastic or regular household spoons can work, also.

- Eye dropper or syringe. These will be used to transfer your testing solution to your powders. We recommend you NOT use a drinking straw to transfer your test solution as your test solution is POISONOUS and should NOT be taken into your mouth or swallowed.

- Tincture of iodine. Tincture of iodine is a concentrated iodine solution which can be found in the first aid section of your grocery store or pharmacy. Note that iodine is POISONOUS and should **not be taken internally**. It can also stain one's skin or clothing. You will only need a very small amount for this lab so a small bottle is all you will need. Consider wearing an old shirt or lab coat to protect your clothes in this lab.

- Old newspapers on which to cover your work surface and a supply of paper towels or napkins to clean up spills.

- An assortment of powders in which to test. We recommend you test the following powders: white flour, whole wheat flower, granulated sugar, powdered sugar, corn starch, baking soda, baking powder, table salt, cornmeal and any other sugar products such as turbinado or sucanat, cake or biscuit or pancake mixes. **Do not use any sort of household detergents, cleaners or drain openers.**

- Foods you can test include; breads, muffins, potatoes, crackers, unpopped and popped popcorn, uncooked oatmeal and any sort of cold cereal. Fresh white potatoes or dehydrated potatoes are also recommended for testing.

Preparation of Iodine Testing Solution

Because tincture of iodine is highly concentrated, you'll need to make a dilute solution of it for this lab (a little goes a long way.) Make this solution by getting about 1/4 cup of tap water. Add to it a teaspoon or so of the tincture of iodine. The amount is not critical. Your goal is to create a solution which resembles brewed tea (light brown color.) Note that if you get the iodine on your hands or clothing it will stain. Stains on your hands are not permanent but may require a few days to wear off. It will not hurt you if you get it on your hands. Make sure to recap the container tightly and keep it out of reach of younger children. Iodine is used to cleanse the skin in the case of minor cuts and bruises as well as rid the skin of germs prior to surgery. It can also be used to kill skin fungi like ringworm. It works by creating "holes" in the coverings of germs and messing up their proteins inside.

Once you've prepared your iodine testing solution, label its container and set it aside. You might draw a "skull and crossbones" on the container just to remind you not to drink any!

Preparation of your Powders to be Tested

For each test powder, you will use one lid or prepared cup (see above for details.) With a pen or marker, label the lid with the powder you plan on placing in it. Then, with a clean spoon, transfer approximately 1 Tablespoon of each powder to each lid. It's important to use a clean spoon for each powder or clean and dry the one you are using between powders.

Testing Procedure

Begin testing your powders by testing corn starch first. Corn starch is basically pure amylose and will definitely give you a positive test result. Begin your test by getting some iodine solution in your dropper. Then, carefully add some drops to your corn starch powder. Using a toothpick or coffee stirrer, mix in the iodine with the powder. A positive test for amylose is an immediate color change of the iodine from the light brown color to an almost-black, purple color. Avoid touching the tip of your eyedropper to the powder which might contaminate your next test powder. If you should touch the powder, rinse the eyedropper in water and begin again.

To see a negative test result, test granulated table sugar next. Repeat the steps you followed when you tested the corn starch. You should find that granulated sugar does not create a color change in the iodine solution, but rather the iodine remains it original brown color. Recall that table sugar consists of sucrose which is obviously not amylose. Table sugar is usually made from two plant sources depending upon the location of where you live. If you live in the northern United States, table sugar generally is made from sugar beets whereas in the southern USA, table sugar is made from sugar cane. Both plants supply plant "juices" which are dried and processed into table sugar. Because sucrose has not yet been changed into amylose for storage, it will give you the negative starch test result.

On the next page you will find a data table in which to record your results as you move from powder to powder. Note in the first column, you will write the identify of the powder you are testing. In the second column, make a hypothesis (educated guess based on your previous experiences with the powder) as to the results of your test. Then, make your test and record your results.

When you test baking mixes, such as cake or biscuit mixes, think about the ingredients present in the mix. Check the label on the box and then make your hypothesis. You may find "interesting" results. One other thing to note is regarding powdered confectioners sugar. One might usually think that powdered sugar is just that: sugar that's been pulverized into a fine sugar. However, you might check the label on the bag or box and find that many times corn starch is added to help the powder flow more freely. Because of this, the negative result you'd normally expect may not occur due to the presence of the corn starch.

Testing the popped popcorn is fun, also! Try as many different foods as you can.

One other fun thing to do is have your teacher prepare some unknown powders for you to test. Based on your observations, see if you can determine the identify of these powders. Remember to only use powders that are safe for you to make observations of (no chemical or powdered cleaners.)

Finally, clean-up your work area when you are finished. The iodine solution is safe to pour down the drain followed by a good flush of water. Dispose of your powders in a trash container and wash any of the lids or utensils for future use.

Powder	Hypothesis	Observed Results	Positive or Negative for Amylose

Lesson 4: The Lipids

In Lesson 3, we discussed the first group of important compounds found in living things: carbohydrates. In this lesson, we will explore a second very important group of compounds known as lipids. A more commonly used term for lipids is fats. As we did in Lesson 3, first we will look at the structure of lipids. Then, we'll examine how lipids behave and, finally, we'll discuss how lipids are used in living organisms.

Like carbohydrates, lipids are composed of the elements carbon, hydrogen and oxygen. Lipids do not have any nitrogen atoms present. However, unlike carbohydrates, lipids are large molecules. Their large size gives us insight to one of their very important uses in living things. So, first, let's look at the structure of lipids.

All lipids have two basic parts to their structure: the glycerol portion and the fatty acid portion. Let's look at the glycerol portion first. The glycerol portion consists of three carbon atoms, each bonded to each other and hydrogen atoms to make up the four bonds that carbon desires to form.

Friendly Biology

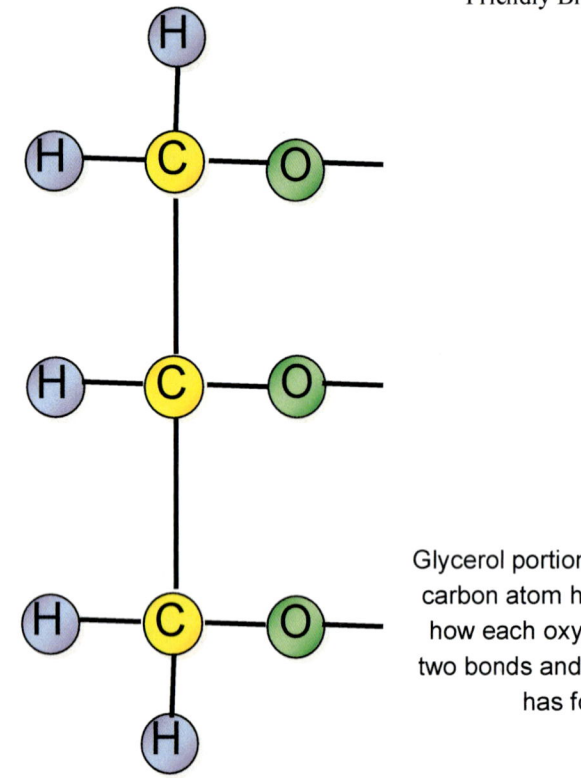

Glycerol portion of a lipid. Note how each carbon atom has formed four bonds and how each oxygen atom desires to form two bonds and how each hydrogen atom has formed one bond.

Note that on one side of the glycerol portion of a lipid, there are three oxygen atoms. The glycerol portion can be found in all lipids. Let's look at the fatty acid portion of lipids next.

Fatty acids or fatty acid chains, as they are often called, are just that: chains of carbon atoms ranging in length from 4 to 24 carbon atoms long. They have various names according to their lengths (and possible presence of other elements such as nitrogen and sulfur). For this course, we will limit our discussion to understanding the basic structure of a fatty acid chain.

As we mentioned above, fatty acid chains range in length from 4 to 24 carbons. Bound to these carbon atoms are hydrogen atoms. Look at the diagram below to see a simple fatty acid chain.

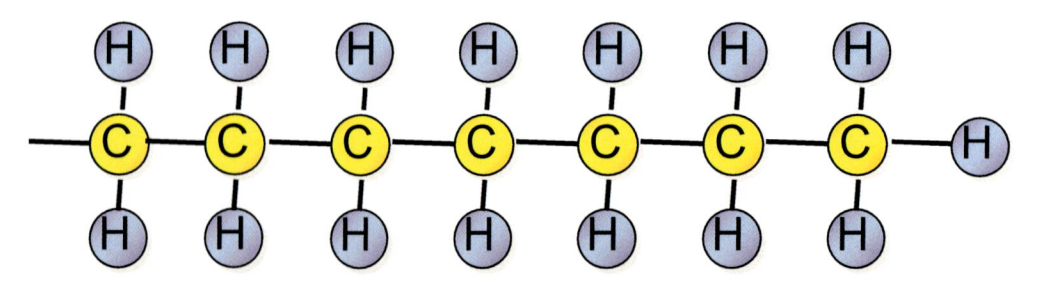

Fatty acid chain portion of a lipid. Again, look at the number of bonds formed by each atom.

Fatty acid chains are named according to their structure. Examples of fatty acids include butyric acid found in butter and arachidonic acid found in peanuts. Others include stearic acid and palmitic acid.

How are these fatty acid chains connected to the glycerol portion of a lipid? These fatty acid chains are linked to the oxygen atoms in the glycerol portion of the lipid. Because there are three oxygen atoms present in glycerol, there can be up to three fatty acid chains linked there (one chain to each oxygen atom). If one fatty acid chain is present, the lipid is called a monoglyceride, where the prefix mono– means one and glyceride refers to lipid or fat. Look at the diagram below of a monoglyceride. Note that there are only hydrogen atoms bound to the other two potential bonding sites on the glycerol molecule.

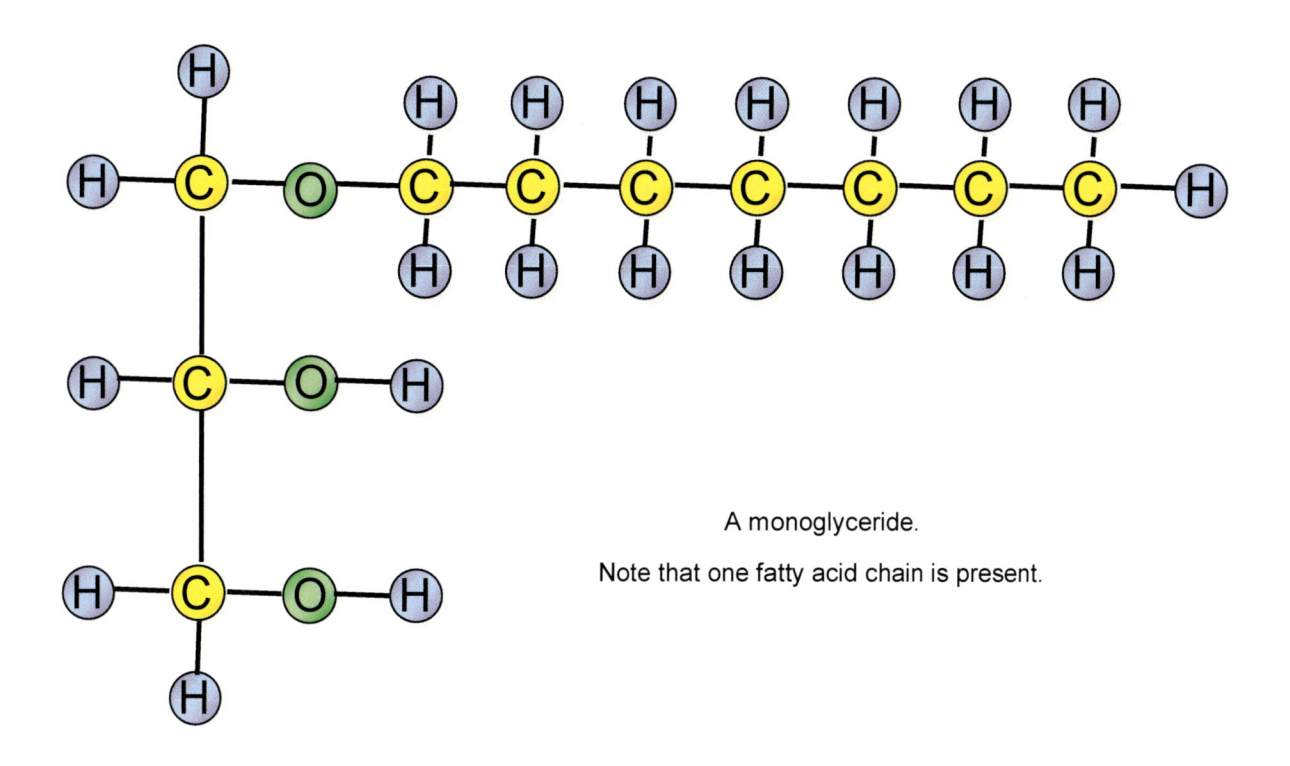

A monoglyceride.
Note that one fatty acid chain is present.

If two fatty acid chains are bound to the oxygen atoms of the glycerol unit, this lipid is called a diglyceride, where the prefix di– means two. Here is a diagram of a diglyceride.

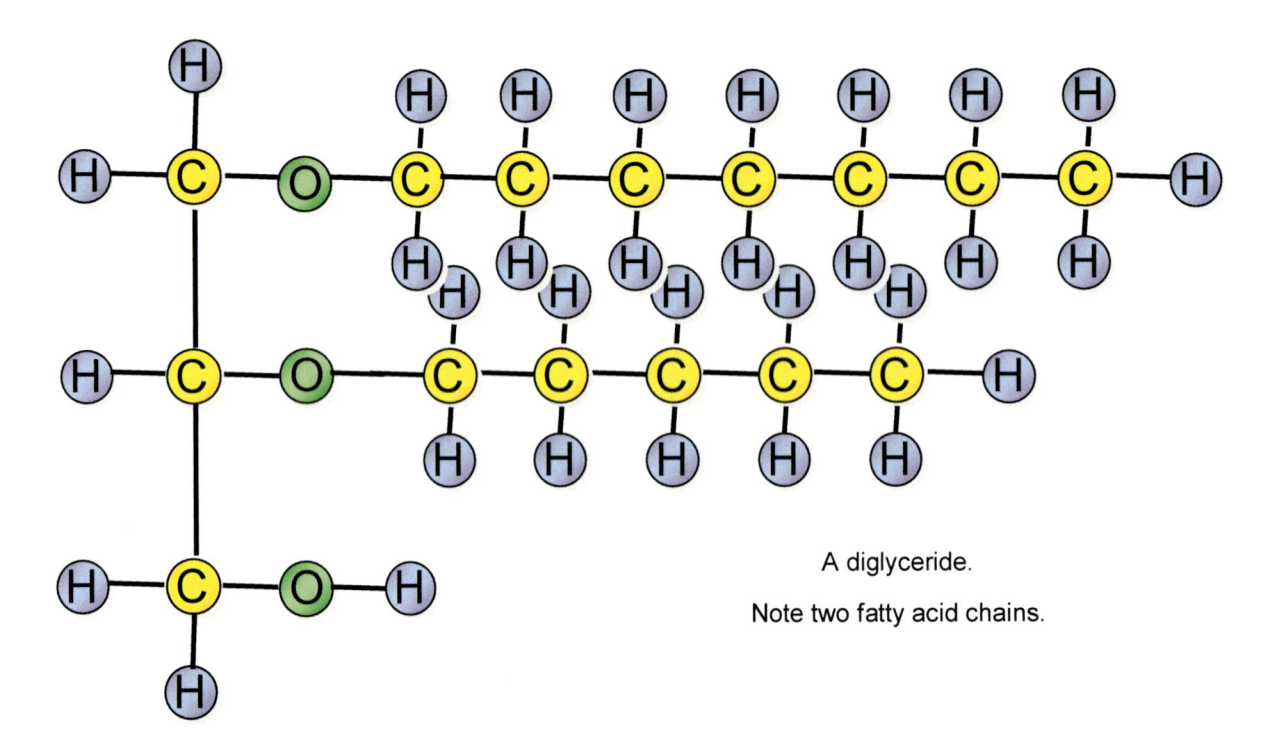

A diglyceride.
Note two fatty acid chains.

If all three oxygen atoms have a fatty acid chain present, the lipid is called a triglyceride. The prefix tri– means three. Here is a diagram of a triglyceride.

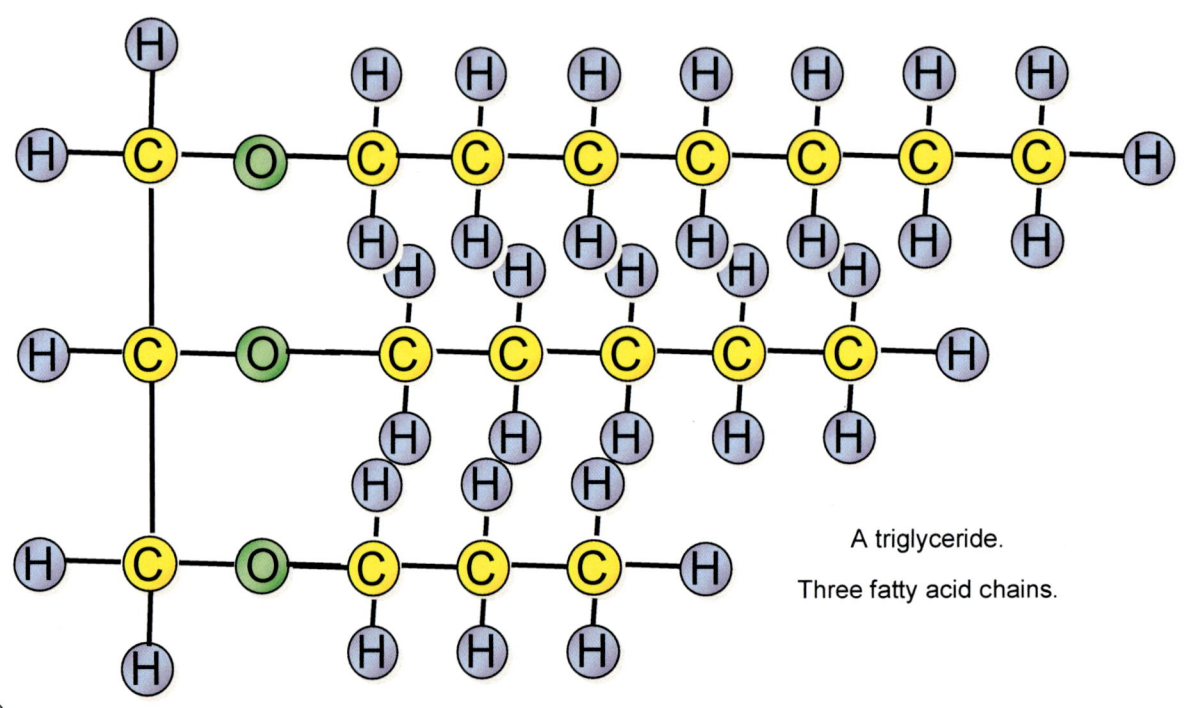

A triglyceride.
Three fatty acid chains.

Let's pause and review. In this lesson we've introduced a second set of compounds which are extremely important to living things. They are known as the lipids or fats. We learned that lipids contain carbon, hydrogen and oxygen atoms and that lipids are much larger molecules than carbohydrates. We learned that lipids consist of two main structures: the glycerol portion and the fatty acid chains. We learned that there can be up to three fatty acid chains bound to the glycerol portion of a fat. A monoglyceride has one fatty acid chain, a diglyceride has two and a triglyceride has three fatty acid chains attached to the glycerol molecule. Let's continue now.

Let's take a closer look at fatty acids. Fatty acids can be classified into two main sub-groups: the saturated fatty acids and the unsaturated fatty acids. This division is made according to the presence of double bonds found in the chain of carbon atoms. When a double bond is found between two consecutive carbon atoms in a fatty acid chain, those carbon atoms require fewer hydrogen atoms to be bonded to them in order for the carbon atoms to acquire their four desired bonds. Because, at this point, there are fewer hydrogen atoms present, this type of fatty acid is said to be unsaturated.

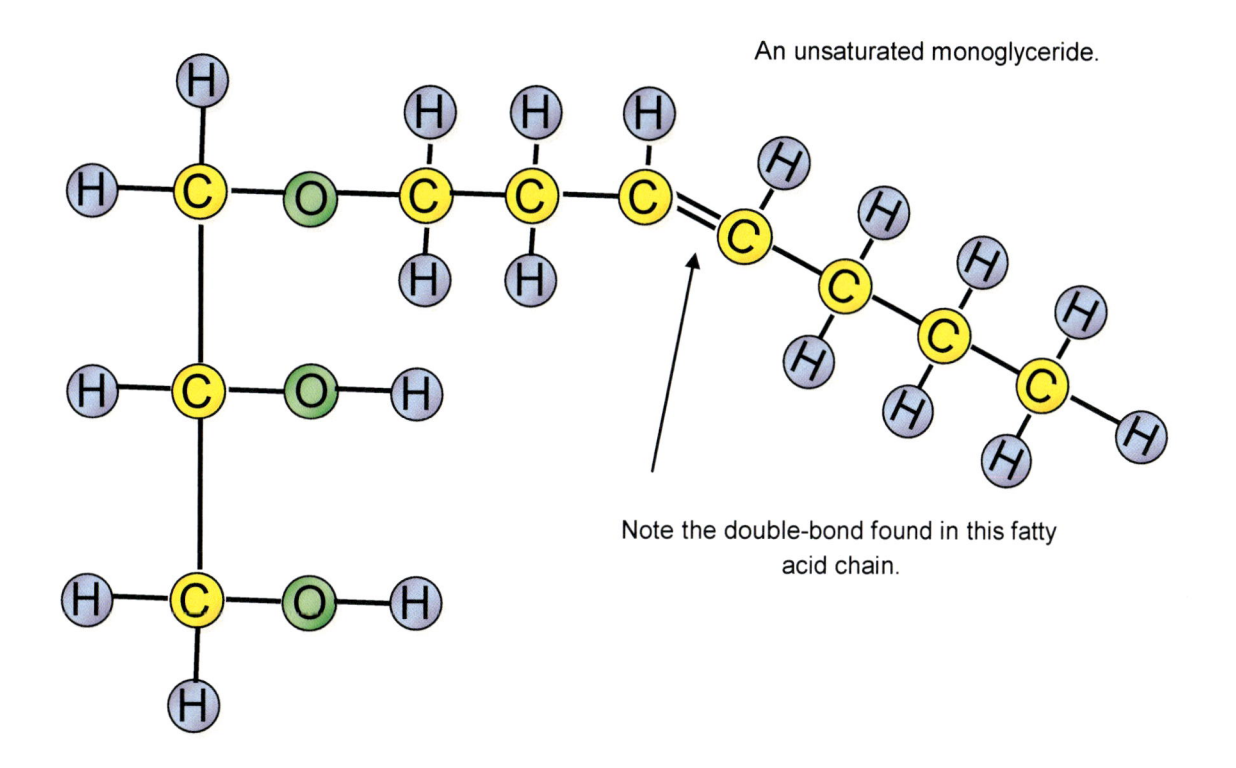

An unsaturated monoglyceride.

Note the double-bond found in this fatty acid chain.

If a fatty acid chain has only single bonded carbon atoms throughout its length, each carbon atom has its desired number of four bonds fulfilled by hydrogen atoms. We can say that the fatty acid is "filled to the brim" with hydrogen atoms and is therefore considered saturated.

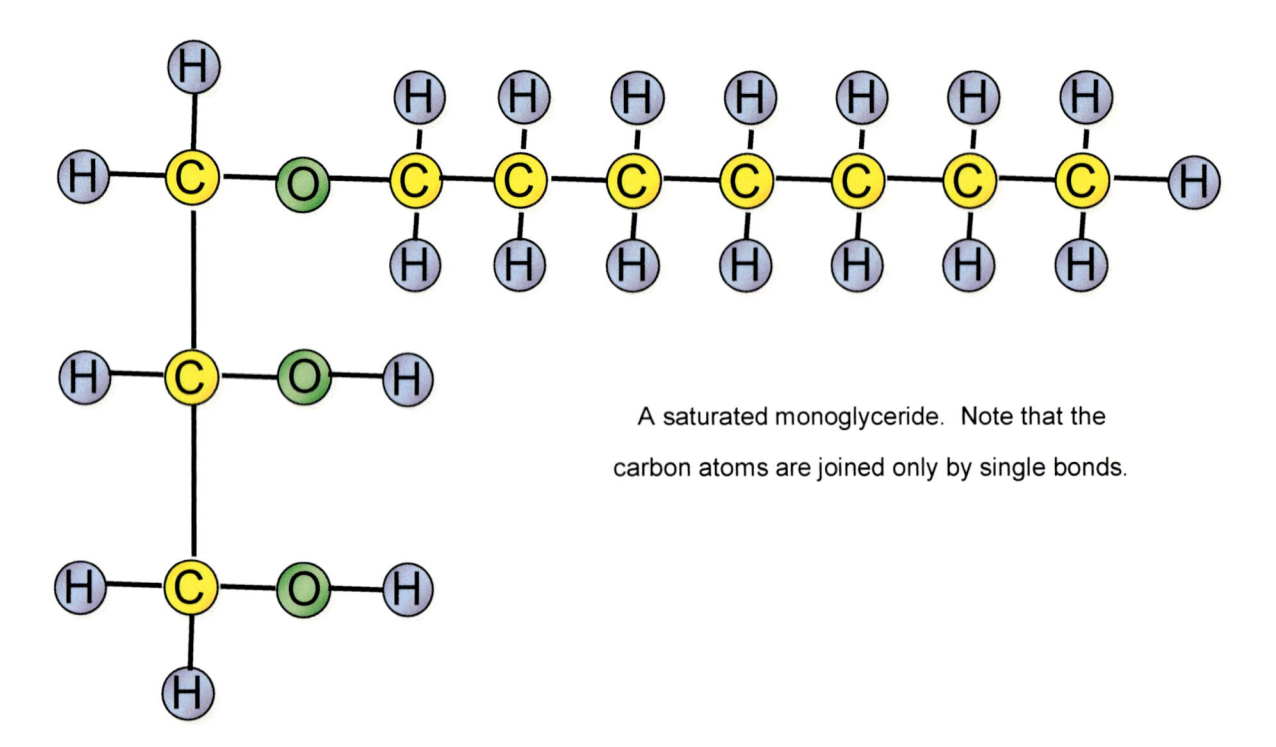

A saturated monoglyceride. Note that the carbon atoms are joined only by single bonds.

When we investigate the sources of naturally occurring fats, we find that saturated fats come from animal sources. Fats like lard, which comes from hogs, is a saturated fat. Tallow is the fat which comes from beef animals and is also considered to be a saturated fat. Fats from animal sources are solids at room temperature. Consider a piece of uncooked bacon. Note the meaty portion and the fatty portion. Before the bacon is cooked, the fatty portion is a solid at room temperature.

Saturated lipids are usually solids at room temperature and are derived from animal sources.

On the other hand, fats which originate from plant sources are unsaturated fats. There are many of these fats that are used in your kitchen. Salad oil is made from fats extracted from corn. Soybean oil, which comes from soybeans, and canola oil, with comes from seeds of the canola plant, are two other commonly used oils in cooking. Cookies purchased from the grocery store are often prepared with palm oil made from seeds of palm trees. Cottonseed oil is also used in many commercially prepared baked items. Another oil often used in the preparation of foods is olive oil which, obviously, comes from olive trees. All of these fats originate from plants. Are they solids or liquids at room temperature? If you said, *liquids*, you are correct. Unsaturated fats which originate from plants are liquids at room tem-

Unsaturated lipids are usually liquids at room temperature and are derived from plant sources.

Let's pause and review, once again, before we continue our discussion of lipids. We said that:

- Lipids are composed primarily of carbon, hydrogen and oxygen atoms.

- There are two main portions of lipid molecules: the glycerol portion and the fatty acid portion.

- The glycerol portion consists of three carbon atoms each with an attached oxygen atom and the necessary hydrogen atoms to make carbon's desired four bonds.

- Fatty acids are chains of carbon atoms ranging from 4 to 24 carbons in length.

- The lipid is classified as being a monoglyceride, diglyceride or triglyceride according to the number of fatty acid chains present in the lipid.

- Saturated fatty acids are solids at room temperature and originate from animal sources.

- Unsaturated fatty acids are liquids at room temperature and originate from plant sources.

Let's continue our discussion of lipids. The way that the lipid molecule is designed allows it not to mix with watery substances. Try this lab activity.

Lesson 4 Lab Activity 1: Behavior of Lipids

Fill a clear, one quart canning jar half-full of water. To the water, add a cup of salad (corn) oil. Any other cooking oil will do. Do not mix the two liquids. What do you observe?

Does the water mix with the oil?

Which liquid stays on top of the other?

Gently stir the two liquids. What happens?

Will the oil mix with the water? (Don't shake the jar just yet as we'll use it again shortly.)

Lipids are said to be insoluble in water. They do not mix with water. Have you ever gotten a fat on your hands, like butter or margarine? You probably know that you can't wash these fats off of your hands with water alone. Water will not mix with the fats to carry them away from your hands and down the drain of the sink. This inability to mix with water is due to the way the carbon, oxygen and hydrogen atoms are arranged in lipid molecules. Substances which do mix or dissolve in water have their atoms arranged in a way similar to that of water. These substances are said to be water soluble.

Now, take some food coloring and drop a few drops into the oily top layer. Watch closely what happens. Does the food coloring mix with the oil? Where does the food coloring go? What happens when the food coloring meets the water? Continue to make observations while you gently stir the oil and water mixture. What happens? Does the oil change color? What does this tell you about the solubility of the food coloring? You should find that the food coloring is not soluble in the oil and therefore does not color the oil. On the other hand, the food coloring is water soluble and does mix or color the water in the jar.

Lipids are not soluble in water. They are hydrophobic.

Knowing that lipids are insoluble in water is important because lipids are used in living organisms to keep water in or out of certain places You might think of lipids as being like raincoats which can keep water off of you! Lipids are used like barriers or walls to keep water in or out of the cells of living things. We will discuss this capability in much greater detail in a later lesson. For now, knowing that lipids are insoluble in water is important.

Before we leave the discussion of how lipids and water behave, we need to learn two other pairs of terms which are used to describe this behavior. In addition to describing lipids as being insoluble in water, we can also say that lipids are immiscible in water. Substances that do mix with water are said to be miscible. For example, we can say that butter is immiscible in water (does not mix) while something like pancake syrup is miscible (it does mix.)

The second set of terms used to describe lipid and water interactions are the terms hydrophobic and hydrophilic. Recall from Lesson 3 when we discussed carbohydrates. We said that hydro– is the root word for water. In this case, we have the term hydrophobic. Phobic or phobia means fearing. You may be familiar with the term arachnophobia which means fear of spiders. Hydrophobic, therefore, literally mean water-fearing. Lipids are hydrophobic: they fear (or, maybe more accurately, repel or do not mix with water). On the other hand, substances that mix with water, like the pancake syrup referred to above, are considered to be hydrophilic. Again, hydro– refers to water while –philic translates to loving: water-loving! Pancake syrup or anything else that dissolves or mixes well with water is hydrophilic.

Hydro = water
Phobic = resisting or fearing
Philic = attracting or loving

Water soluble = miscible = hydrophilic
Water insoluble = immiscible = hydrophobic

In this lesson so far, we've explored the structure of lipids and how they behave with other substances (mainly water) and how that's important to living organisms. There is another concept regarding lipids that is very important when it comes to living organisms.

Take a look at a major similarity between the structure of a lipid and the structure of glucose. Both glucose and lipids are composed of the same elements: carbon, hydrogen and oxygen. Recall that the chemical formula for glucose was $C_6H_{12}O_6$.

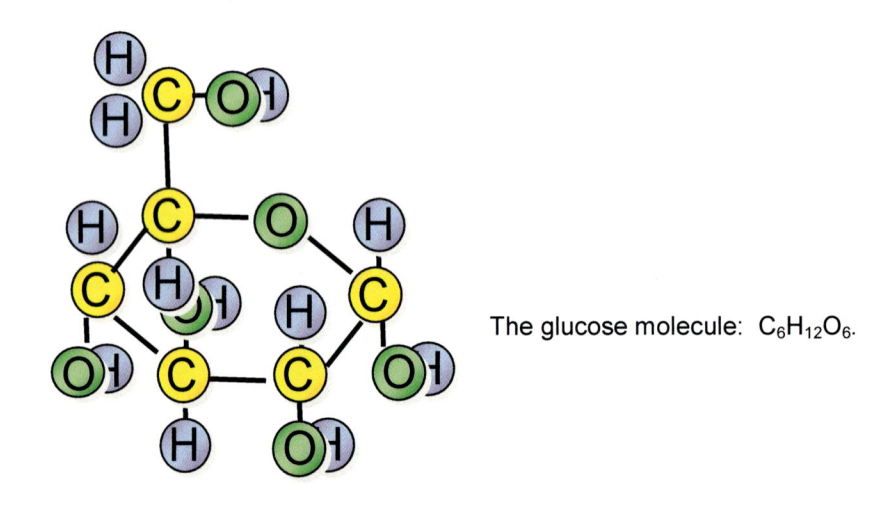

The glucose molecule: $C_6H_{12}O_6$.

Recall, also, that glucose is the "gasoline" on which our bodies run. Lipids, if "cut into pieces" can be "turned into" glucose molecules! Think about all the carbon, oxygen and hydrogen atoms present in one lipid molecule! You can imagine quite a number of glucose molecules which can be generated from one lipid molecule. In essence, lipid molecules work like storage locations or stockpiles of the elements necessary to make glucose when conditions require the living thing do so. When glucose supplies get low in a living organism, lipid molecules are taken out of storage and converted into glucose.

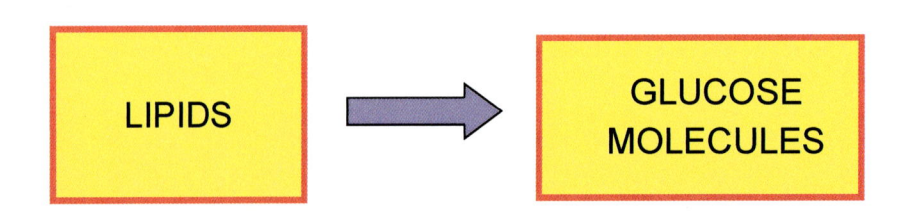

LIPIDS → GLUCOSE MOLECULES

This "cutting-up" of lipid molecules is accomplished by enzymes in living organisms. Can you think of the name given to enzymes which "cut-up" lipids? Recall from Lesson 3 where we first introduced enzymes, that we said that enzymes have two primary functions: matchmakers or "cutters" of other compounds in the living organism. We also learned that enzyme names usually ended in –ase. Therefore, an enzyme which breaks down lipid molecules is called…can you guess it? Yes, a lipase enzyme. There are many types of lipase enzymes which together accomplish the task of breaking down the complex lipid molecules and transforming them into glucose molecules to be used to fuel the living organism.

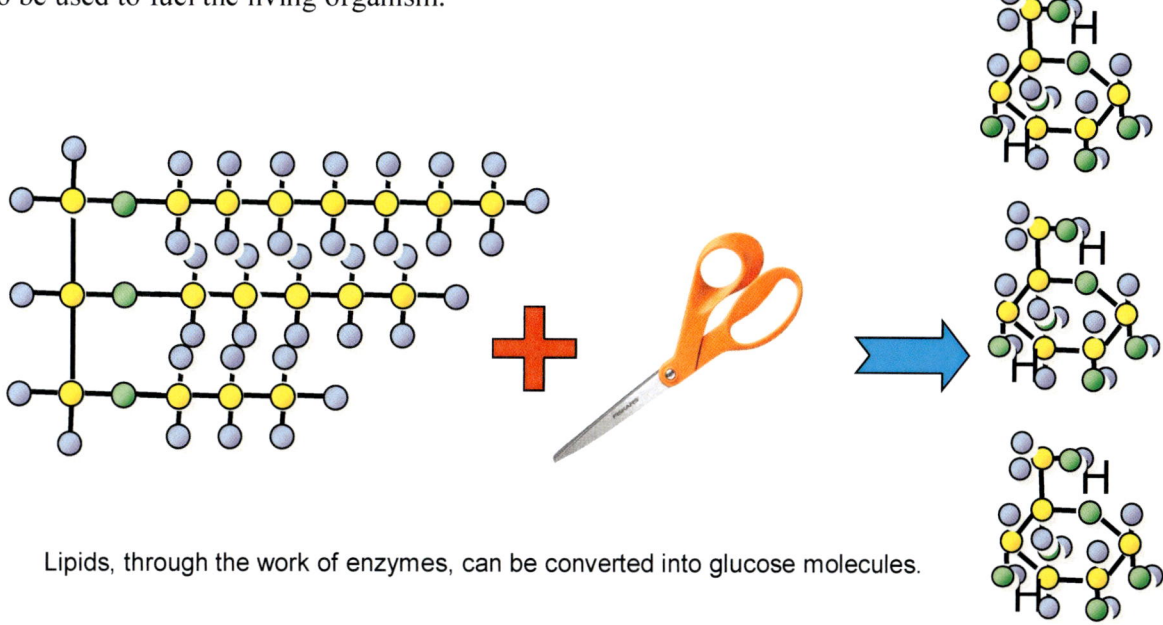

Lipids, through the work of enzymes, can be converted into glucose molecules.

An interesting "tidbit" of information regarding the amount of stored-up energy inside lipid molecules is the fact that fats hold about nine times the potential amount of energy found in glucose. In other words, if you compared a cup of glucose to a cup of a fat, the fat would have the potential to produce about nine times the amount of energy as that of the cup of glucose. This explains why animals that live in cold, Artic areas can survive for long periods of time utilizing their stores of fat on their bodies! Lipids work like storage locations for vast amounts of potential energy.

The energy stored in one cup of a lipid is roughly equal to NINE cups of glucose!

Lesson 4 Lab Activity 2: Making Butter

Let's do a fun lab activity next: making butter! The purpose of this lab will be to make some butter, obviously, but in the process you will be able to observe first-hand how lipids are hydrophobic. This hydrophobia, in the case of preparing butter, is very useful in the process. Let's begin.

Items to gather:

- Clean pint jar with lid.
- Heavy whipping cream (this works better than table cream). A pint is enough for one student. (Note that this is whipping cream, not whipped topping or the aerosol squirt topping.)
- Table spoon or wooden spoon.
- Source of running water.

Procedure:

1. Pour 1 pint of heavy whipping cream into the clean jar. Tightly screw on the lid.

2. Gently shake. Gently agitate. Gently rock back and forth. Continue doing so for several minutes. After a few minutes, you may sense that the whipping cream is no longer a liquid. It doesn't seem to splash around inside the jar as it once did. Carefully open the jar and take a peek inside. At this point the cream is more like whipped cream. It's not butter yet! Replace the lid on the jar and continue shaking.

3. After 15-20 minutes of shaking, suddenly you'll feel another change in the liquid inside the jar. If you look through the side of the jar, you'll see a lump of butter bouncing back and forth within the watery buttermilk. Continue to agitate the butter 1-2 more minutes. If you've been agitating your cream longer than 20 minutes with no sign of butter forming, sometimes the heat of your hand delays the butter formation. Try running cold water over the jar for a few minutes and then shake some more. Usually, the butter will readily form.

4. After you see your lump of butter in the jar, carefully open the lid and observe. The butter is fat which has clumped together to form one large "glob." The buttermilk surrounding the butter is the water that was once part of the cream.

Let's pause here and discuss some things about milk and cream and cows. Cows, like all mammals, produce milk for their offspring. Because milk provides all of the nutrition for the baby mammal, it includes a beautiful assortment of carbohydrates (remember the lactose we discussed in Lesson 3), proteins, vitamins, minerals *and fats!* As the cow is milked, all of these milk components are evenly dispersed in the milk. If the freshly milked milk is allowed to sit undisturbed, the fat portion will rise to the surface of the milk as cream. (Think back to the water and oil lab activity we completed earlier in this lesson.) If all of the cream that rises to the top is skimmed-off, the milk that remains is called skim milk. The fat content of skim milk is 0% (all of the fat has been removed.)

Milk straight from the cow can range in fat content from 3-6%.

You may be familiar with 1%, 2% and whole milk that you buy at the grocery store. Each of these percentages reflects the amount of fat in the milk. Milk straight from the cow ranges from 3-6%, with Holstein cows giving milk at 3-4% fat and Jersey cows giving milk at 5-6% fat. Within these breeds, the amount of fat produced by individual cows also varies. In other words, some Holstein cows may produce milkfat at the 4-5% level while other Holstein cows may produce milkfat at the 3-4% level.

At the milk processing plant, *all* of the fat (cream) is separated from the milk by machines. If you or your parents or grandparents grew up on a farm, you may be familiar with machines called cream separators. Cream separators were used on the farm to separate the cream from the fresh milk before the milk was sold to the milk plant. The cream could then be sold or used to make butter. Before cream separators were invented, farmers simply allowed the milk to set overnight to allow the cream to rise to the top. It was then skimmed off using ladles.

Once the cream is removed from the milk, then, according to needs of the customers of the dairy plant, the fat is returned to the milk at the specified levels of 1%, 2% or whole milk which is usually 3% fat. And then, of course, you can buy skim milk with 0% fat. In this way, you can be assured the milk you purchase at the grocery store has a consistent amount of fat present.

The milk you buy at the grocery store varies in how much fat is in the milk. Whole milk is 3%. You can also purchase, depending upon where you live, 2%, 1% and 0% or skim milk.

You might wonder why milk (other than skim milk) you buy from the grocery store does not separate with the fat rising to the top once again. At the processing plant, not only is the fat put back into the milk at specified levels, the milk is also homogenized. The prefix homo– means the same and the meaning of the root word –genize (from genesis) is to make or generate. Homogenization means to make the milk the same throughout and in this case, the fat is evenly dispersed throughout in the watery portion of the milk. The fat is made into very tiny droplets which remain suspended in the milk and do separate. Homogenization results in a milk product where the fat no longer separates from the watery portion of the milk.

> Homo = same
>
> Genize (genesis) = to make or create, generate
>
> Homogenize = to make the same throughout

The heavy whipping cream that you began your butter with is usually produced at 36% fat. So even though it has quite a bit more fat in it compared to whole milk (3%), it still has a good bit of water present. So, as the fatty components within the milk clumped together to form the butter in your jar, the liquid portion that remained is this watery portion of the milk. This watery portion is known as buttermilk.

Before you can eat your butter, your next job will be to "wash" your butter. You may be wondering, "How do I *wash* butter?" Washing your butter means you need to wash away all of the buttermilk. By removing the buttermilk your butter will remain fresher for a longer period of time and not spoil as quickly.

5. Because the buttermilk is water soluble, you can use water to wash it away. To wash your butter, first drain-off the buttermilk. Open the lid to your jar and then carefully pour-off the liquid portion of the milk. Use the lid of the jar to keep the butter from also pouring out.

Then add fresh cold water to your butter in your jar and, using your spoon, smash your "ball" of butter over and over again to push out any remaining milk. The water will get cloudy as the buttermilk leaves the butter and moves into it. Pour off this "dirty" water and add more fresh water. Be careful not to let your butter escape your jar. Add more cold water and smash the butter again to remove more milk. Pour off and repeat the process until the water you pour off is crystal clear. Pour off any remaining water and you can now taste your butter. You may add a sprinkle of salt if you desire. A slice of bread or soda cracker is a great tool for testing your butter. Enjoy your fresh butter!

As we end this lesson, let's recall what we've learned about lipids:

- Lipids are composed primarily of carbon, hydrogen and oxygen atoms.

- There are two main portions of lipid molecules: the glycerol portion and the fatty acid portion.

- The glycerol portion consists of three carbon atoms each with an attached oxygen atom and the necessary hydrogen atoms to make carbon's desired four bonds.

- Fatty acids are chains of carbon atoms ranging from 4 to 24 carbons in length which can potentially attach to the oxygen atoms of the glycerol portion of the lipid molecule.

- The lipid is classified as being a monoglyceride, diglyceride or triglyceride depending upon the number of fatty acid chains present in the lipid.

- Saturated fatty acids are solids at room temperature and originate from animal sources. Examples include bacon or meat fats.

- Unsaturated fatty acids are liquids at room temperature and originate from plant sources. Examples include corn, soybean or palm oils.

- Because lipids are made from many, many carbon, hydrogen and oxygen atoms, they are used in living things like stock-piles of raw materials to be converted into glucose, the fuel for living things.

- Lipids are broken down by enzymes known as lipases.

- Lipids are insoluble in water. We can also say that lipids are immiscible in water and are hydrophobic. Substances which are soluble in water are said to be miscible or hydrophilic.

- Milk produced from cows varies in fat content. At milk processing plants, all milk fat is removed from milk and then reintroduced at specified levels. Milk with no fat reintroduced is called skim milk. The fat portion of milk is what is used to make butter.

Lesson 5: The Proteins

In Lessons 3 and 4 we began our discussion of substances which are very important to living organisms. In Lesson 3 we learned about carbohydrates and in Lesson 4 we discussed lipids. In this lesson we will explore proteins. As we did in Lessons 3 and 4, we will first look at the structure of proteins, then, how they behave in relation to other substances and, finally, we'll examine their role in the lives of living creatures.

Like carbohydrates and lipids, proteins are made of carbon, hydrogen and oxygen atoms. However, proteins have another element: nitrogen. Together, these four elements make up a wide assortment of very important substances in living organisms. To understand how proteins

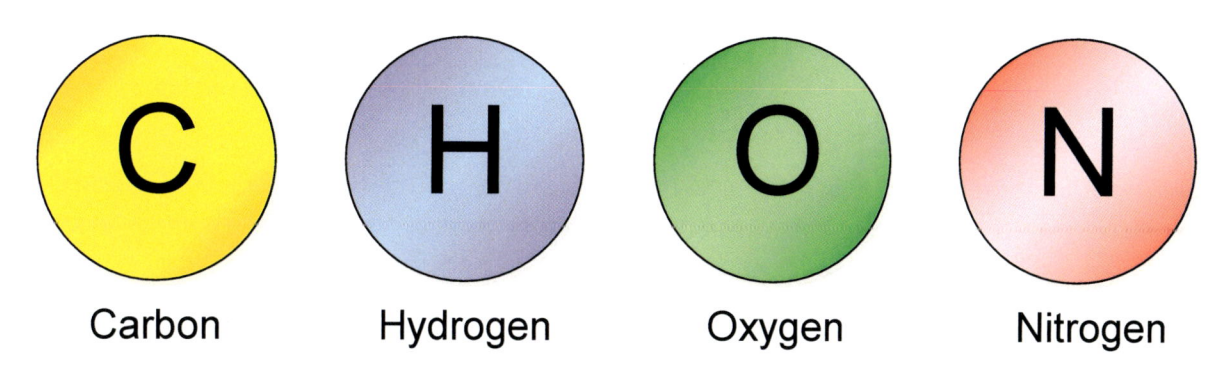

C Carbon   **H** Hydrogen   **O** Oxygen   **N** Nitrogen

Friendly Biology

are built, we need first to look at a group of substances known as amino acids. Proteins are built from amino acids. Let's look closer at these amino acids.

Amino acids consist of carbon, hydrogen, oxygen and nitrogen atoms. Like lipids, they are formed in two parts: a part common to all amino acids, known as the common part, and then a part that varies and makes amino acids unique. This part that varies in structure is called the R-group. Let's look at the common part first.

The common part consists of two carbon atoms bonded together with an assortment of hydrogen, nitrogen and oxygen atoms bonded to them. Look at the diagram below.

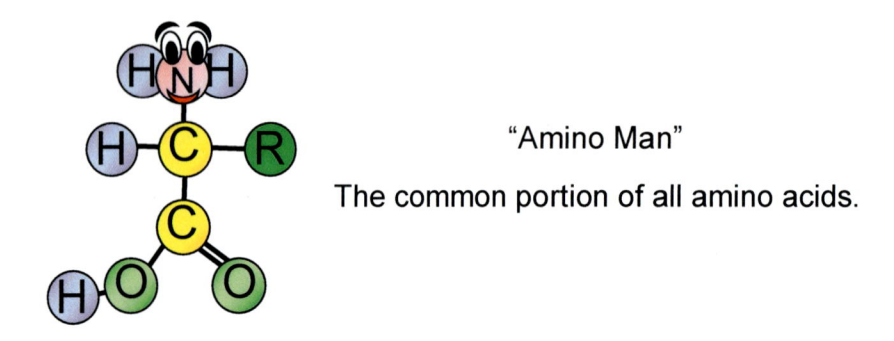

"Amino Man"
The common portion of all amino acids.

We like to think that this common portion of an amino acid resembles a little person known as "Amino Man." The portion of Amino Man from his "waist" down, is known as the carboxyl group.

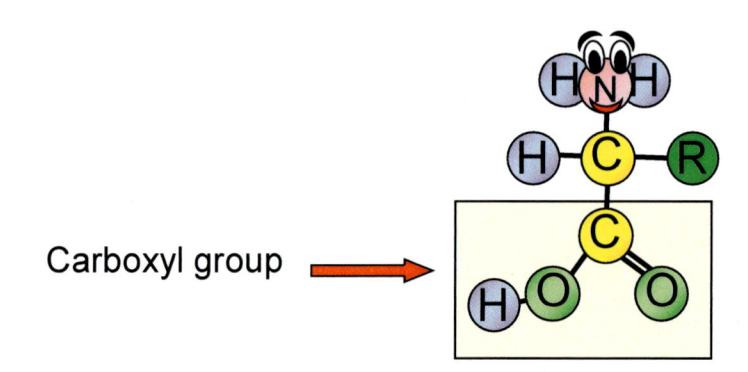

Carboxyl group →

64

Note that Amino Man's "feet" are made of oxygen atoms and on his right foot, he has one "toe," a hydrogen atom. This oxygen/hydrogen atom combination on his right foot is called a hydroxyl group.

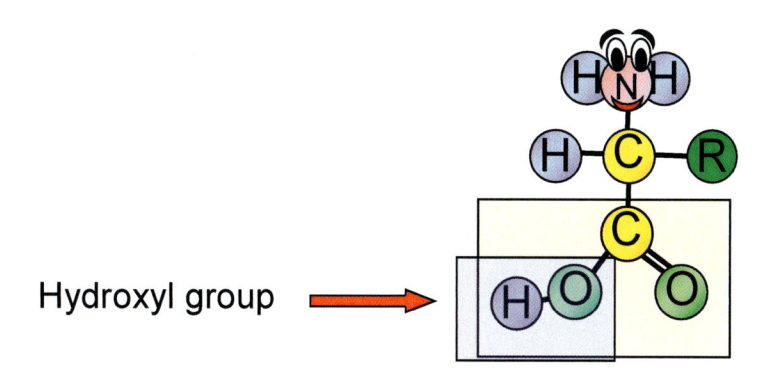

Hydroxyl group →

Amino Man's "body" is made of carbon atoms. Note that his "head" is a nitrogen atom. He has two "ears" which are hydrogen atoms. This combination of one nitrogen atom and two hydrogen atoms (NH_2) is called an amine group. Did you already figure out that amino acids are named for this amine group? Take a closer look at the nitrogen atom in Amino Man's head. How many bonds are present? Recall from Lesson 2 where we learned that nitrogen atoms desire to form three bonds. Here we can see the three bonds formed by the nitrogen atom: two are made with each "ear" of hydrogen and one form the "neck" down to the first carbon atom in Amino Man's "body."

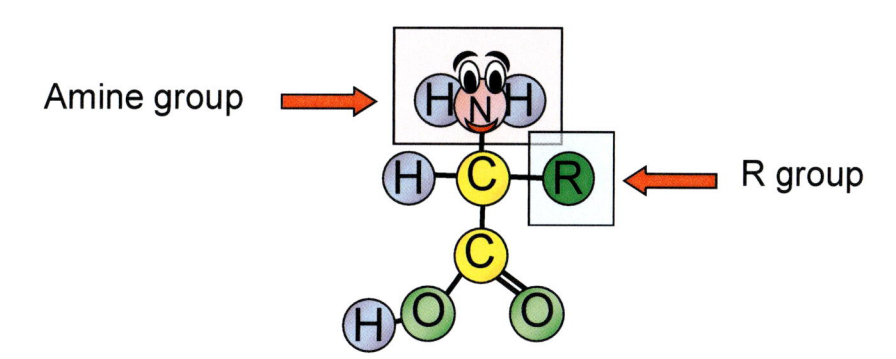

Amine group →   ← R group

Note in the amino acid diagram above that one of Amino Man's arms is a hydrogen atom. His opposite arm is noted by the letter "R" and, as we mentioned earlier, this is the part of the amino acid that varies. There are specific R-groups which allow us to differentiate between the amino acids. Let's look at some of these R-groups.

The simplest R-group is a single hydrogen atom (H.) This amino acid is named glycine. Look at the diagram of glycine below.

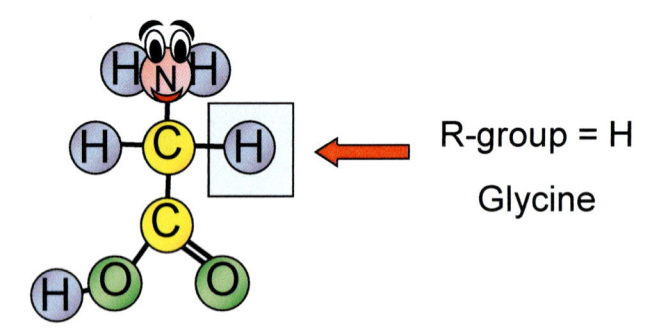

R-group = H
Glycine

A slightly more complicated amino acid is alanine. The R-group for the amino acid alanine is a carbon atom with three attached hydrogen atoms. Look at the diagram of alanine below.

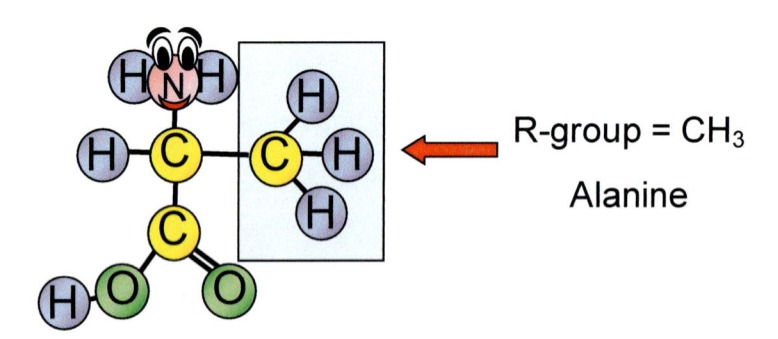

R-group = CH_3
Alanine

As you can imagine, the complexity of the R-group increases with each new amino acid. Some of these amino acids incorporate other elements into their R-groups besides carbon, hydrogen and oxygen atoms. Look at the diagrams of twenty of the 50 or so identified amino acids on the next pages. Note that in addition to carbon, hydrogen, oxygen and nitrogen, we find sulfur to be another element found in amino acids.

Valine

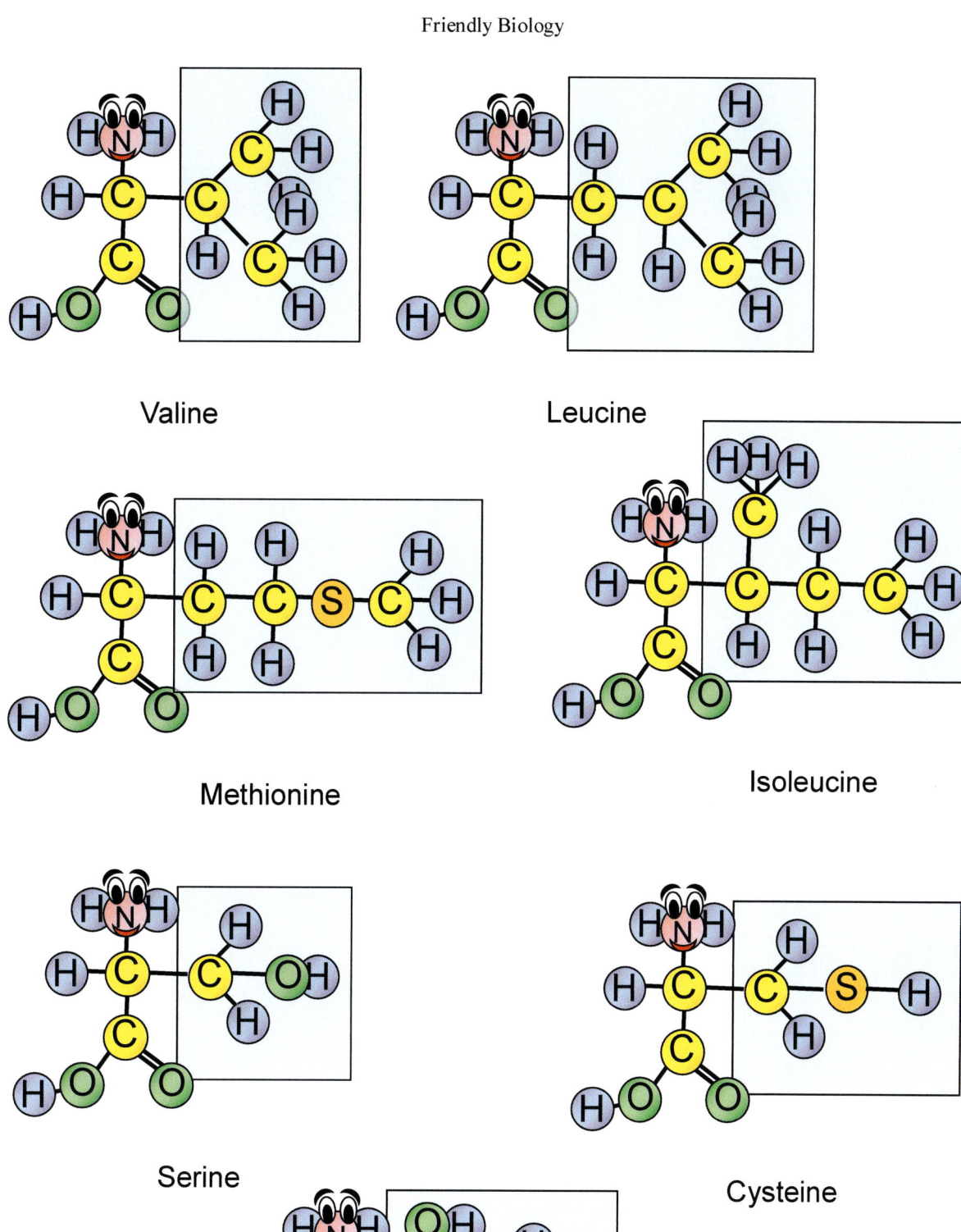

Leucine

Methionine

Isoleucine

Serine

Cysteine

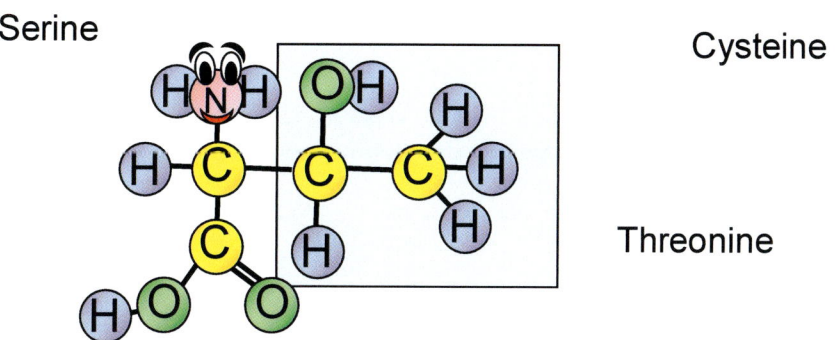

Threonine

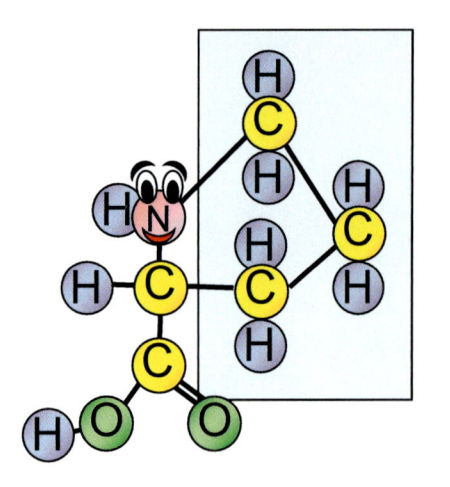

Friendly Biology

Proline

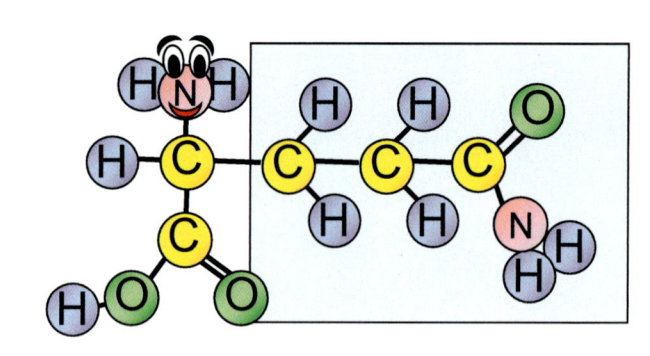

Glutamine

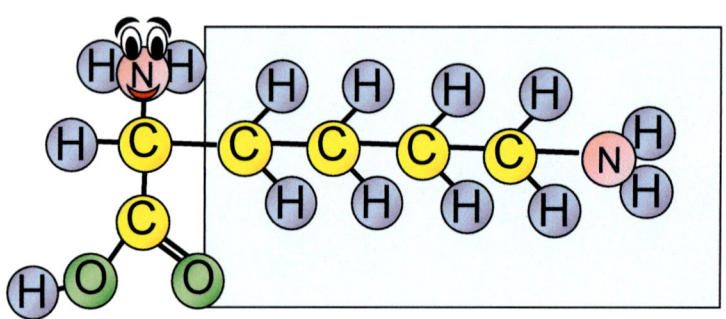

Lysine

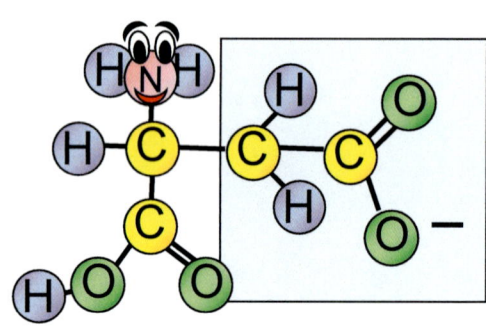

Aspartic Acid

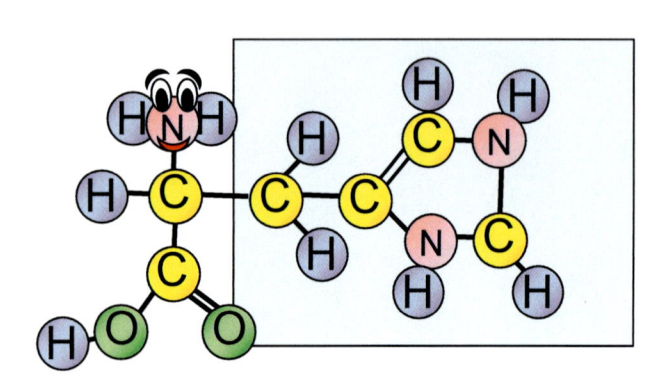

Histidine

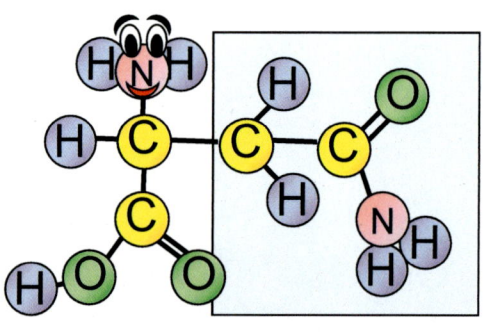

Asparagine

68

Arginine

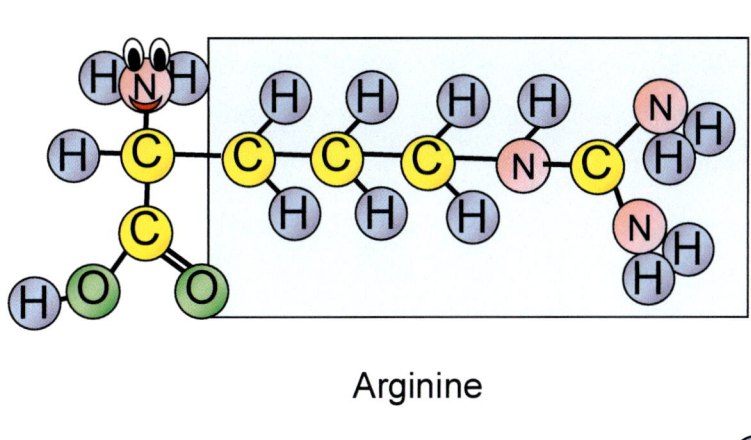

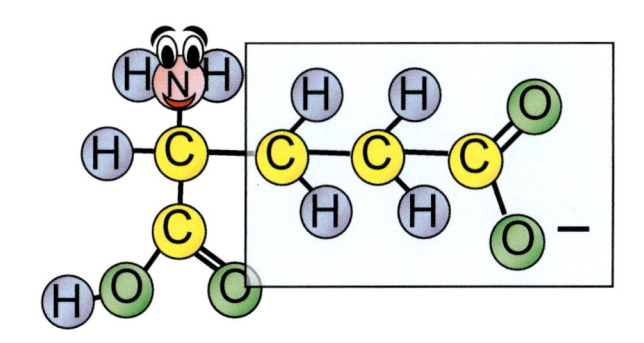

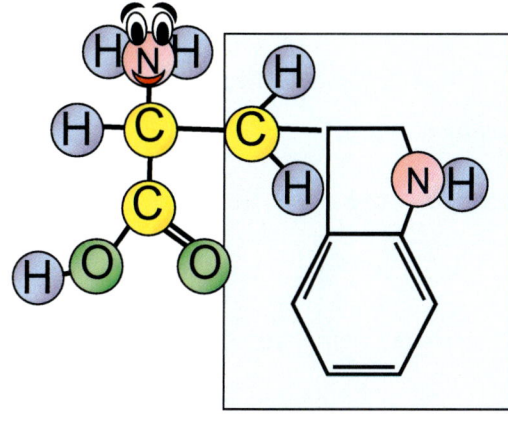

Glutamic Acid

Tryptophan

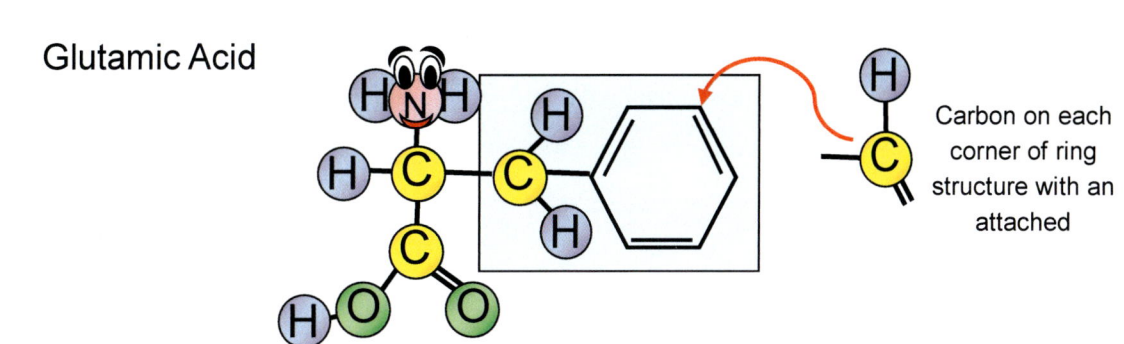

Phenylalanine

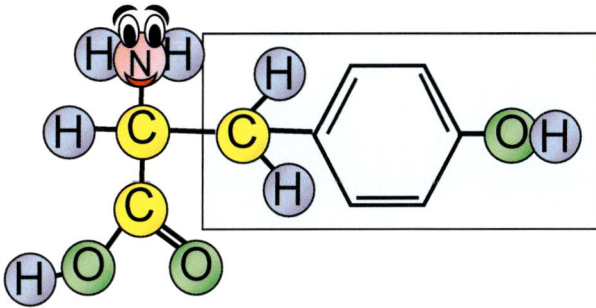

Carbon on each corner of ring structure with an attached

Tyrosine

The twenty amino acids shown on the previous pages are a very special group in that they are the ones that are necessary for human beings to live. Of those twenty, eight must be taken in through the foods we eat. The remaining twelve can be made in our body from other raw materials which, obviously, are carbon, hydrogen, oxygen and nitrogen atoms or even other amino acids. The eight that we cannot make ourselves are called the eight essential amino acids. These include: valine, iso-leucine, leucine, phenylalanine, threonine, tryptophan, methionine, and lysine.

Before we continue, let's pause and review what we have discussed so far:

- Proteins are made of the elements carbon, hydrogen, oxygen and nitrogen (with some having the element sulfur).

- Proteins are built of smaller sub-units called amino acids.

- Amino acids have a common portion and then a varying R-group.

- There are twenty amino acids which human beings require for life. Of those, eight are considered the essential amino acids in that we must consume them in our diet. We cannot make them in our bodies.

Now that you have an idea of what amino acids are, let's explore how they are used to create proteins. First, we need to understand that there are many, many different proteins in our bodies. All of these proteins are unique and all are made from amino acids that have been linked together in long chains. The amino acids could be likened to beads in a necklace. It is the sequence of these amino acids which determines the protein that is made. By changing the order of the amino acids, our body can build all sorts of necessary proteins.

Amino acids are linked together like beads on a necklace to form a protein.

A useful analogy of this amazing system to consider is our alphabet. We have 26 letters in our alphabet. We can rearrange those 26 letters in many, many ways to make thousands and thousands of words. Similarly, our bodies use the twenty amino acids it requires to make the many, many proteins our bodies need to survive. Instead of letters of the alphabet, our bodies use amino acids!

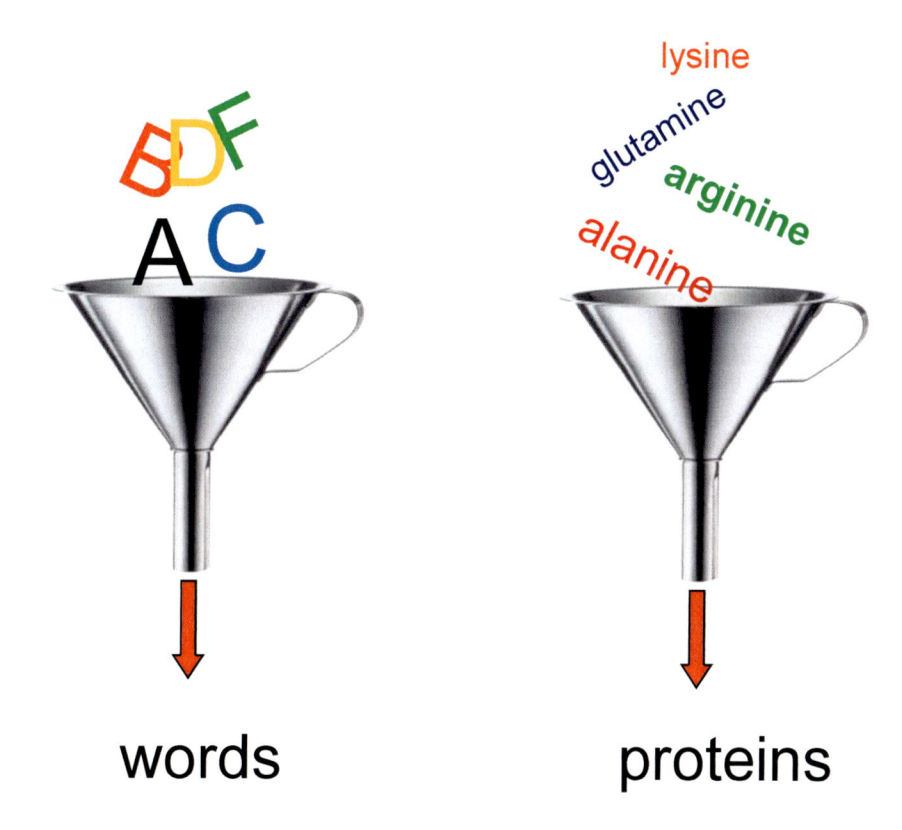

words

proteins

Let's look a little closer at how these amino acids link themselves together. Recall the Amino Man structure of the amino acids. Note how on Amino Man's head there was the amine group (NH_2). Then, recall how on his right foot there was a hydroxyl group. These two groups are instrumental in forming the link between amino acids as they form proteins. By using "equipment" in our bodies we can position each amino acid where the amine group is brought near the hydroxyl group of a nearby amino acid. As they link together, the amine group donates one of its hydrogen atoms while the hydroxyl group is released from the neighboring amino acid. This results in the formation of a molecule of water (H-O-H or H_2O). In doing so, the two amino acids become joined together and the molecule of water is "released." While we have simplified this linking process, it is quite complex and is accomplished through the work of many enzymes.

Recall from our earlier discussion in Lesson 3 that we introduced the concept of enzymes. We said that enzymes work to lyse or cut molecules apart. We also learned that enzymes can work like matchmakers in that they can speed up reactions without themselves being changed in the process. The linking of amino acids to build proteins is a great example of how enzymes speed up reactions. Enzymes are totally necessary for proteins to be built in living things.

This process of building proteins is called dehydration synthesis. These terms may sound complicated, so let's break them into pieces to make them easier to understand. When we learned about carbohydrates, we learned that the root word hydro– refers to water. When we put the prefix de– in front of it, it means to reduce or remove. Dehydrate means to remove or reduce water. This may be a word you are familiar with when thinking about drinking enough water in hot weather to prevent dehydration. Synthesis means to make or produce. So, dehydration synthesis means to make or produce by removing water. As we learned earlier, the linking of amino acids involves the removal of water, hence the name dehydration synthesis. Enzymes facilitate this process of dehydration synthesis.

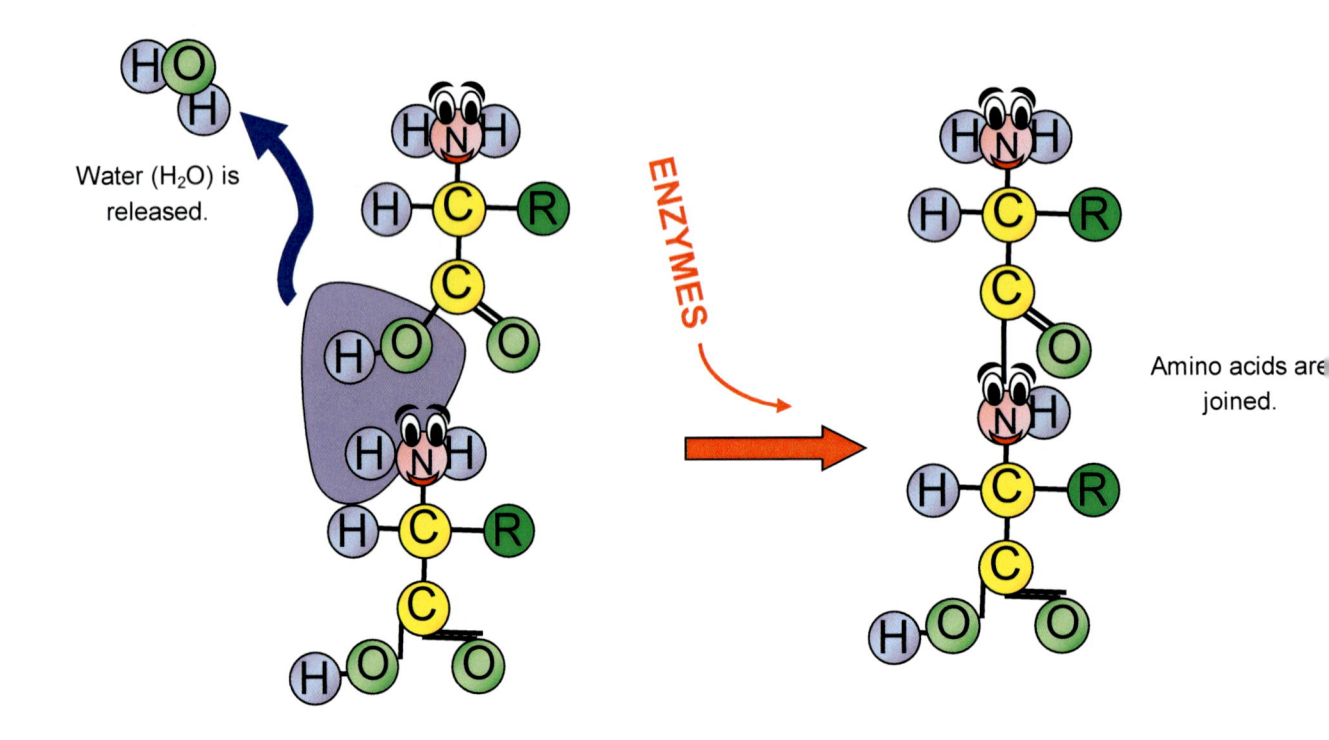

Water (H_2O) is released.

ENZYMES

Amino acids are joined.

Through the action of enzymes, amino acids are linking together to form proteins. In the process, water is released. Hence the name, dehydration synthesis.

We now need to learn two more new terms when we discuss the linking of amino acids to build proteins. The first term is peptide. A peptide is a group of amino acids that have been linked together in the early stages of building a protein. When two amino acids are linked, the resulting molecule is called a dipeptide. Remember from our discussion of lipids (for example, diglycerides) that the prefix di– means two. Hence, a dipeptide consists of two linked amino acids. When three amino acids are linked through dehydration synthesis, a tri-peptide is formed. Recall that the prefix tri– indicates three. When several are linked together, a polypeptide is said to have formed. The prefix poly– refers to many or several.

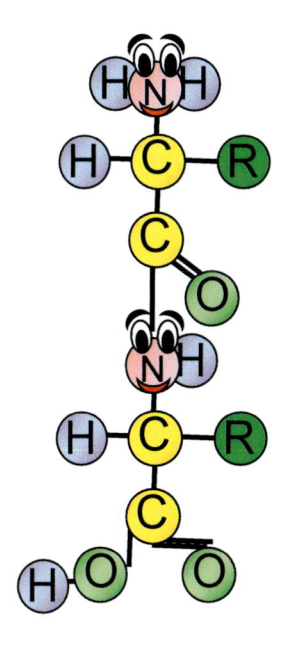

Two linked amino acids form a dipeptide.

Three linked amino acids form a tripeptide.

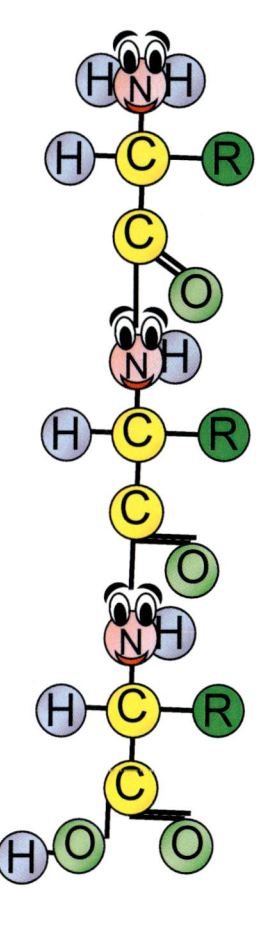

Several linked amino acids form a polypeptide.

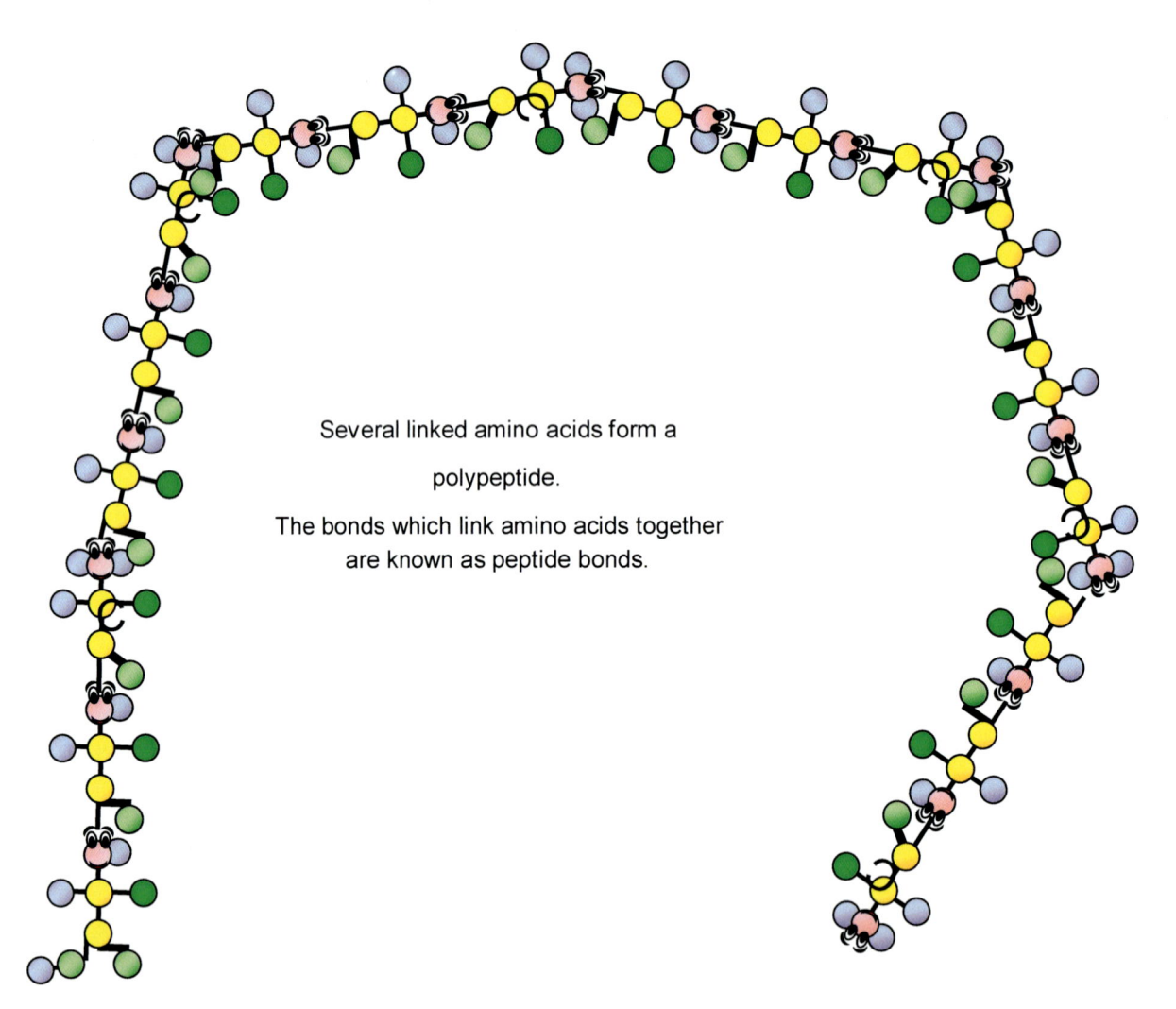

The bonds which link amino acids together are known as peptide bonds.

The second term we will add to our discussion is peptide bond. When two amino acids are acted upon by enzymes and become linked together through dehydration synthesis, the resulting bond that now forms between the two amino acids is called a peptide bond. Peptide bonds are the links between the amino acids in proteins.

Speaking again of enzymes, can you think of the name, in general, which is given to the enzymes that work to assemble amino acids into peptides? Peptidases are enzymes which do this work. While we are discussing enzymes, let's consider yet another job that enzymes have when it comes to dealing with proteins.

Earlier in our discussion of amino acids, we talked about a special set of amino acids known as essential amino acids. We said these essential amino acids are those that we cannot make ourselves and, therefore, we must get those from the foods we eat. However, these amino acids are not usually found as separate, free amino acids in our foods. They, too, are found bound into proteins that were once part of the animal or plant product from which our food was made. In order for our bodies to get those essential amino acids, we must first break down those proteins into free amino acids. This process is known as hydrolysis. If we take the word hydrolysis apart, we find the prefix hydro– which, as we already know, refers to water. Then we see the root word -lysis which we know means to cut. Hydrolysis means to cut with water. Through the use of enzymes and water, we are able to lyse the proteins we eat to get the amino acids our bodies need. Essentially, we are able to take animal or plant proteins, cut them into small "bites" of amino acids, rearrange those amino acids and rebuild proteins our bodies need. These enzymes that work with water to break down proteins are known as proteolytic enzymes.

> De = to remove
> Hydrate = water
> Dehydrate = removal of water
> Syntheses = to make or produce
> Dehydration syntheses = the make through the removal of water.

> Hydro = water
> Lysis = to cut or take apart
> Hydrolysis = cutting or taking apart using the addition of water.

Now that we've discussed how proteins are made, let's learn the jobs that proteins do for living things. Proteins are used for structural components in living things. Proteins are analogous to boards, nails, bricks, mortar, screws, tape, glue, steel beams, concrete and any other building product. Proteins make up the walls and membranes of cells of living things as well as components of all parts of living things. Proteins are found in just about all parts of living things. While carbohydrates and lipids are mainly involved in energy production, it's the proteins which make up the structure of the living organism.

The proteins that make up the food we eat are not the proteins we need for *our* bodies. We must first dismantle our food proteins through hydrolysis and then rebuild them through dehydration synthesis to make proteins our bodies need.

Like boards and bricks and nails compose a house, proteins do the same to make the structure of living things.

One interesting protein that you are likely familiar with is called albumen. When you crack open an egg, the egg white is primarily made of the protein albumin. Albumen is also a very important protein in the blood of living things in that it helps blood maintain the correct amount of water within the blood. Another protein found in blood is known as fibrin. Fibrin is the substance that, when triggered, makes your blood clot, which is very important when you suffer a cut or injury. Without the protein fibrin, your blood would not clot and you might suffer greatly from the loss of blood.

Another set of proteins which is very important for humans and animals are those known as the immunoglobulins. The prefix immuno– in this word refers to the immune system. The immune system is the body system which helps defend us from invaders which cause disease or illness. The root word –globulin refers to protein. Immunoglobulins are proteins that work to assist the body in fighting disease.

When baby mammals are born, the first milk produced by their mothers, known as colostrum, isn't really much like milk at all. Instead, colostrum is a megadose of naturally made immunoglobulins designed especially for the baby. From her exposure to her environment, the mother has produced this set of defense proteins in an effort to give her babies a sort of jump-start to fighting diseases. In domestic animals like cattle, sheep, goats and horses, livestock owners make a concerted effort to make sure the calves, lambs, kids and foals receive plenty of colostrum in the first 1-2 days of life. This greatly enhances their chance for survival. Eventually, the level of colostrum is reduced and replaced by milk to continue nourishing the newborns.

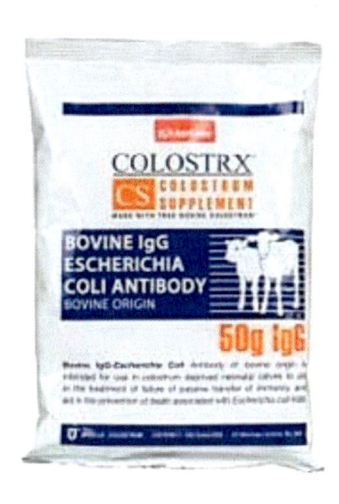

Colostrum is the first milk made by mammals for their offspring (babies.) It is so important for survival of livestock that dried or frozen colostrum is made available for newborns who are unable to nurse or have been orphaned. Note on the label the notation, "IgG." This indicates that immunoglobulin G is the main protein found in this replacement product.

We have one last concept to consider regarding proteins in living things. It is important to know that proteins are very large molecules. They are much larger than carbohydrates and fats. The linkages of amino acids become quite lengthy requiring the molecule to fold and twist upon itself. Because of their size, proteins often have difficulties when it comes to moving in and out of living organisms on the cellular level. This concept will be discussed in greater detail in Lesson 8 where we introduce parts of cells.

Let's pause now and review what we've learned so far about proteins. We said that:

- Proteins are made up of carbon, hydrogen, oxygen and nitrogen atoms and sometimes sulfur and other elements.

- Proteins are built from smaller units known as amino acids.

- The structure of an amino acid consists of a common portion and an R-group. It is the R-group that makes amino acids unique.

- Amino acids are linked together to form peptides. The bonds formed between amino acids are known as peptide bonds.

- The process of linking amino acids into peptides is known as dehydration synthesis. Water is released in the process.

- Some amino acids cannot be made by our bodies and we must regularly eat foods which contain those amino acids. Our bodies utilize enzymes through a process known as hydrolysis to cut apart proteins containing those amino acids. This process allows us to utilize those amino acids for our own needs.

- Proteins are used primarily as structural components in the cells of living organisms. They also serve as vital components within the blood and immune systems of human beings.

- And finally, proteins are generally very large molecules.

Friendly Biology

Lesson 5 Lab Activity: Proteins and Enzymes

PLEASE READ THESE ENTIRE INSTRUCTIONS BEFORE BEGINNING THIS LAB.

The purpose of this lab activity is observe how proteins are affected by lysing enzymes. Recall that enzymes can work as matchmakers to speed-up reactions or like scissors to cut or lyse chemical substances. In this lab, you'll take some prepared proteins and apply an enzyme to them and observe the results.

Materials you'll need to gather include:

- Three packages of Jello[R], any flavor will do, but get the same flavor for all three. Or you may substitute unflavored gelatin packets.

- Fresh pineapple. Peel and core the pineapple. Cut the pieces into bite-size chunks.

- Canned chunked pineapple.

- Container of powdered meat tenderizer.

Procedure:

1. Prepare one package of Jello[R] by dissolving the powder into heated water according to the package directions. Save the box from the Jello[R]. Pour the mixture into a glass bowl, cover with a lid or plastic wrap and place in the refrigerator to cool.

2. Prepare a second package of Jello[R] by dissolving the powder in heated water according to the package directions. Pour the mixture into a glass bowl and add the fresh chunks of pineapple. Cover the bowl with a lid or plastic wrap and place in the refrigerator to cool.

3. Prepare the third package of Jello[R] as you did with the two previous packages. To this batch of jello, add the pineapple chunks which came from the can. Cover with a lid or plastic wrap and place in the refrigerator to cool.

4. Allow all three containers to set at least four hours in the refrigerator.

5. After four hours, remove all three containers from the refrigerator and remove the covers. What do you observe? Write your observations here:

Jello[R] with no added pineapple:

Jello^R with fresh pineapple:

Jello^R with canned pineapple:

If you're reading this lab for the first time, it's okay to stop here and begin conducting the lab. The remainder of what has been written below will "spoil" the fun of finding the results of this activity. So, turn back to the previous page and begin the lab. Then, continue with the discussion below.

Why do you think you are seeing these results? Just in case you're not sure your results are correct, you should set well-set Jello^R in the first bowl where you only have Jello^R, un-set Jello^R in the second bowl where you've added the fresh pineapple and then, well-set Jello^R in the third bowl where you've used the canned pineapple.

Write some ideas here as to why you're seeing these observations:

Here is an explanation of what is happening in this lab activity. First, look at the box from the Jello^R to determine what ingredients are used to make Jello^R. Write them here:

The main ingredient here that we are interested in is the gelatin. Do you know where gelatin comes from? Gelatin is a protein which comes from processed tendons and bones from cattle and hogs. Processing consists mainly of boiling these tendons and bones to remove collagen which is a very strong protein. The heating causes the collagen molecules to break apart into smaller peptides which become gelatin. This gelatin is further processed into the powder found in the Jello packages. If you've ever baked a turkey or chicken or a roast, you may have observed a layer of gelatin which forms on the juices in the roasting dish after it is a cooled. The warmed gelatin, whether it be from a Jello^R package or juices from a baked turkey, remain liquid or runny. Once cooled, however, bonds form between the protein molecules causing the liquid to move to the semi-solid state of gelatin.

Obviously, with our experiment here, the observation that we have some *set* Jello^R and some *non-set* Jello has something to do with the pineapple, right? And then, there must be something happening with the fresh versus canned pineapple that makes a difference.

Fresh pineapple contains a chemical known as bromelain. Bromelain consists of two enzymes known as proteases. Notice the ending of the term proteases? Recall that the ending –ase indicates an enzyme (ie. lactase, amylase, etc.) These two enzymes lyse or cut the bonds forming between the setting Jello^R resulting in a bowl of non-set Jello^R.

Now, what about the Jello^R made with the canned pineapple? Think about the canning process. If you canned any fruits or vegetables at home, you know that part of the process is getting the food you are canning to a high enough temperature to kill any microorganisms present so the food won't spoil or cause you to get food poisoning. At the pineapple canning factory, the pineapple is heated to a temperature high enough to kill any microorganisms and it also makes the enzymes dysfunctional. The enzymes are said be denatured. So because the enzymes in the canned pineapple are denatured, they no longer have an affect upon the gelatin proteins ability to form bonds with itself. The result is a bowlful of *set* Jello^R.

Other fruits which contain enzymes that prevent the proteins of gelatin to set include kiwis, figs, papayas and pawpaws. Again, if these fruits are heated and their enzymes denatured, one can use them in Jello^R desserts without problems.

So, what about the meat tenderizer you got for this lab activity? Take a look at the label to determine the main ingredient. The usual ingredient found in most meat tenderizers is bromelain! That chemical sounds familiar. So, how do you think meat tenderizer works? Again, the enzymes here in the bromelain break down the proteins in the muscle fibers which make up the meat. By applying bromelain, the meat fibers become softer and therefore tenderized.

Let's finish the lab by sprinkling some of the meat tenderizer on your first bowl of prepared Jello^R (the one with no added pineapple.) What do you think will happen? Try it and see. Complete your lab by eating your Jello^R and then washing all your dishes and utensils.

Lesson 6: pH

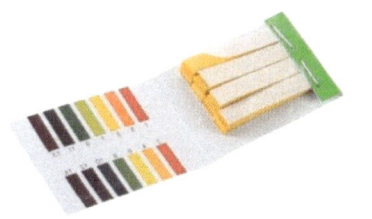

In the last three lessons, we explored the structure of and roles played by carbohydrates, lipids and proteins in living organisms. Throughout each of these lessons, we also discussed how the utilization of these three groups of molecules was dependent upon the action of enzymes. In this lesson we will investigate how pH affects an enzyme's ability to complete its job. We will also look at other ways the pH level is important to the survival of living organisms.

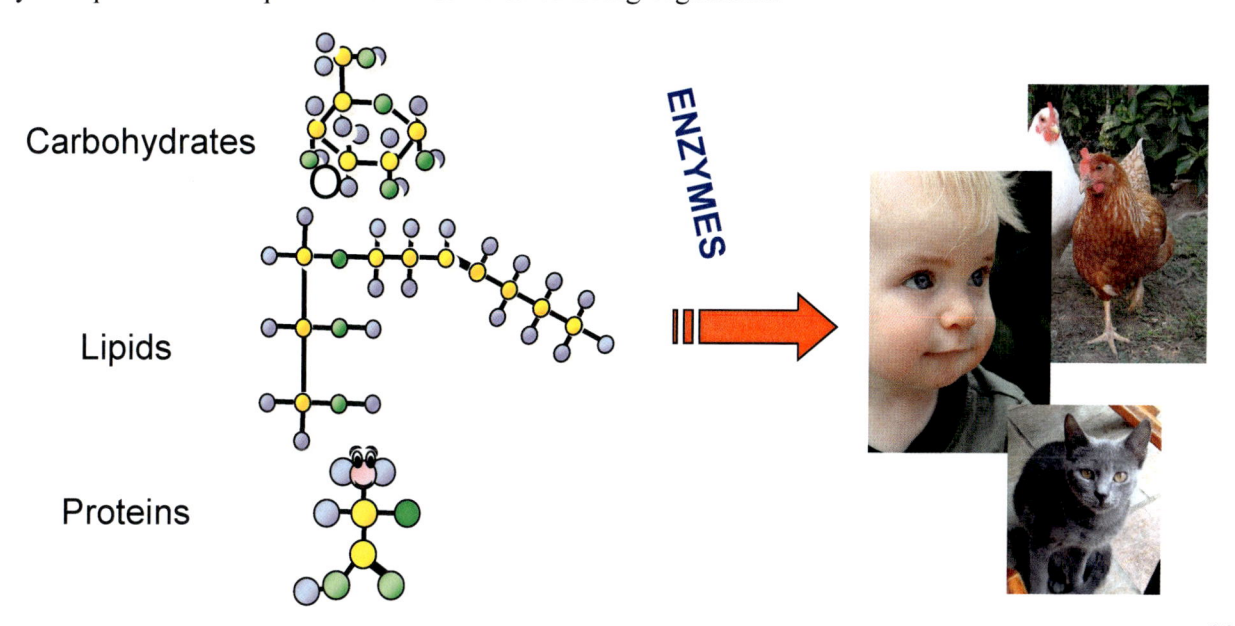

Carbohydrates

Lipids

Proteins

ENZYMES

First, let's discuss what pH is and why the name pH is written with a lower case "p" followed by an uppercase "H." While it is unclear exactly what the p stands for in the term pH (some say it stands for power, some say potential), most biologists agree that the p means percent hydrogen ion.

The upper case H is the symbol for hydrogen ion. So pH is a measure of the level of hydrogen ions in a substance. We have discussed in our earlier lessons that hydrogen atoms are found in carbohydrates, lipids and proteins. Hydrogen atoms are also present in many other substances. One group of those substances is called acids. Acids are chemicals which have hydrogen atoms in their structure that they would really like to get rid of in order to gain better stability. The hydrogen atoms that leave the acids are called hydrogen ions. Some acids have greater "desires" than other acids to be rid of these hydrogen ions. They are called strong acids. Other acids that have a lesser "desire" to be rid of these hydrogen ions are called weak acids. pH is a measure of this "desire" to get rid of hydrogen ions.

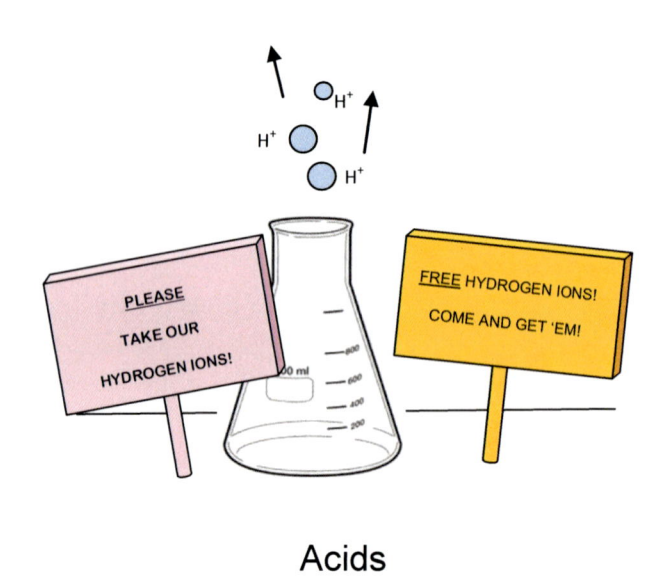

Acids

Let's look at some examples of acids. A very strong acid is known as hydrochloric acid. This is the acid that is found in our stomachs and is also known as gastric acid. Hydrochloric acid is responsible for breaking down foods that we eat. You may be familiar with the term gastric acid reflux or heartburn. These conditions result when hydrochloric acid gets outside the confines of the stomach where protective mechanisms are not in place. If we examine hydrochloric acid, we find that it has an enormous desire to get rid of hydrogen ions and is therefore classified as being a strong acid.

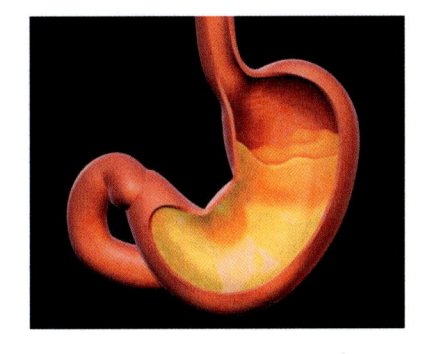

Hydrochloric acid, also known as gastric acid, is a very strong acid which works with enzymes in our stomach to break down the food we eat.

Another acid you may be familiar with is acetic acid. Now, you might stop and say, "I've never heard of acetic acid before!" The common name for acetic acid is vinegar. Vinegar, which is often used in cooking, is considered a weak acid. Its desire to get rid of hydrogen ions is much less than hydrochloric acid's, but it does have the desire to do so.

The common name for acetic acid is vinegar. Acetic acid is considered to be a weak acid.

The acid found in automobile batteries is known as sulfuric acid. The desire of sulfuric acid to get rid of hydrogen ions is near that of hydrochloric acid. Like hydrochloric acid, sulfuric acid is also considered to be a strong acid.

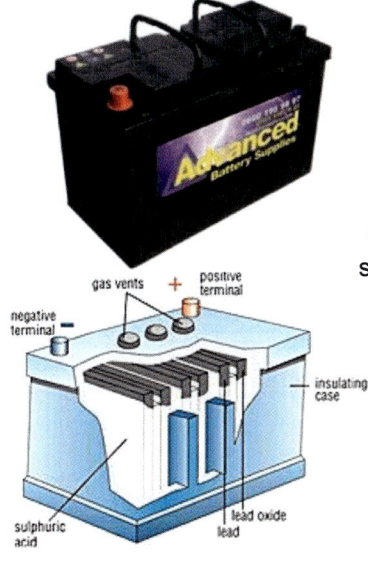

Sulfuric acid, a strong acid, is used in car batteries. The sulfuric acid reacts with plates of lead inside the battery to create a current of electricity used to start the car.

To keep track of these degrees of desire to get rid of hydrogen ions, a scale has been developed known as the pH scale. Numbers on the scale tell us the relative degree of desire to get rid of hydrogen ions.

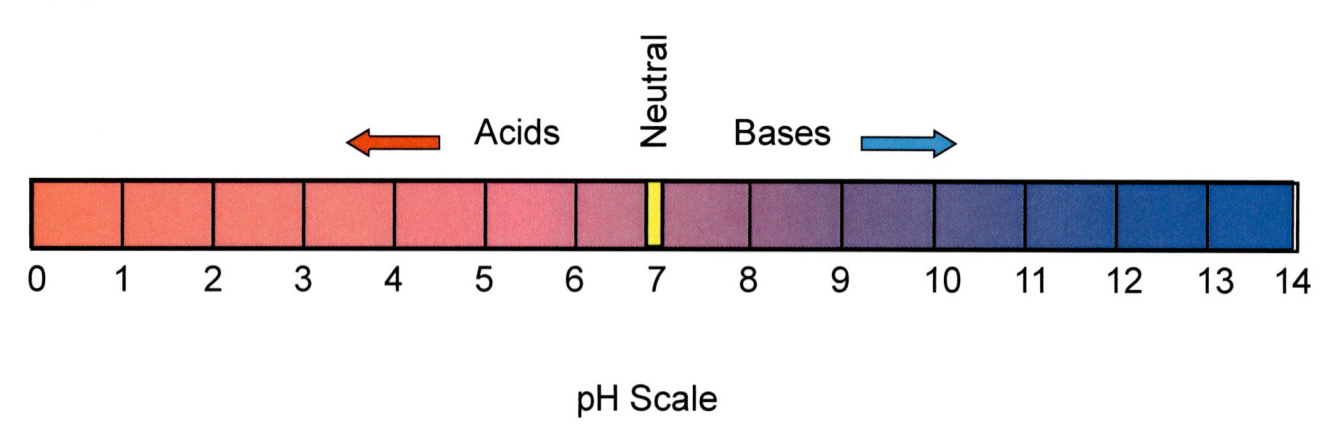

pH Scale

The pH scale's numbers range from 0-14. However, the acids we've been discussing so far in this lesson only occupy half of the scale: 0-6.9. The remainder of the scale (7.1-14) are used by another group of substances which are known as bases. Bases are substances which have desires to accept hydrogen ions. They, like acids, have varying degrees of desires to accept hydrogen ions. Therefore, like the acids, we have substances known as weak bases, moderate bases and strong bases.

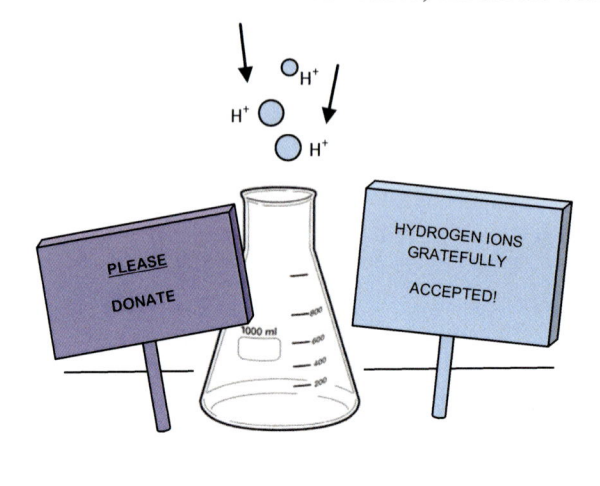

Bases

Take a look at the pH scale below. Note that below the value of 7 on the scale you find the acids. Above the value of 7 you find the bases. The farther away from seven you go, either below or above, the stronger the acid or base becomes. The strongest acids are nearest zero on the pH scale. The strongest bases are nearest 14 on the pH scale.

Friendly Biology

← Acids | Bases →

0 1 2 3 4 5 6 7 8 9 10 11 12 13 14

strong acids   weak acids   weak bases   strong bases

On this pH scale, you can see various acidic and basic substances and where they fall on the pH scale. Note that pure water falls directly on 7.0. On the pH scale 7.0 is identified as being neutral, neither acidic nor basic. Any substance with a pH of less than 7.0 is said to be acidic. Substances greater than 7.0 are said to be basic.

Note that the pH of human blood is 7.4. However, not every part of the human body has a pH of 7.4: some areas have a lower pH (like we discussed earlier in the stomach) and some areas have a higher pH. For the most part, the body functions best at a slightly basic pH. This is extremely important to the function of enzymes in our bodies and for all living creatures. Enzymes work best at specific pH values. If the pH values rise above the optimum or best pH value or, on the other hand, fall below the optimum or best pH value, the work of the enzyme is affected.

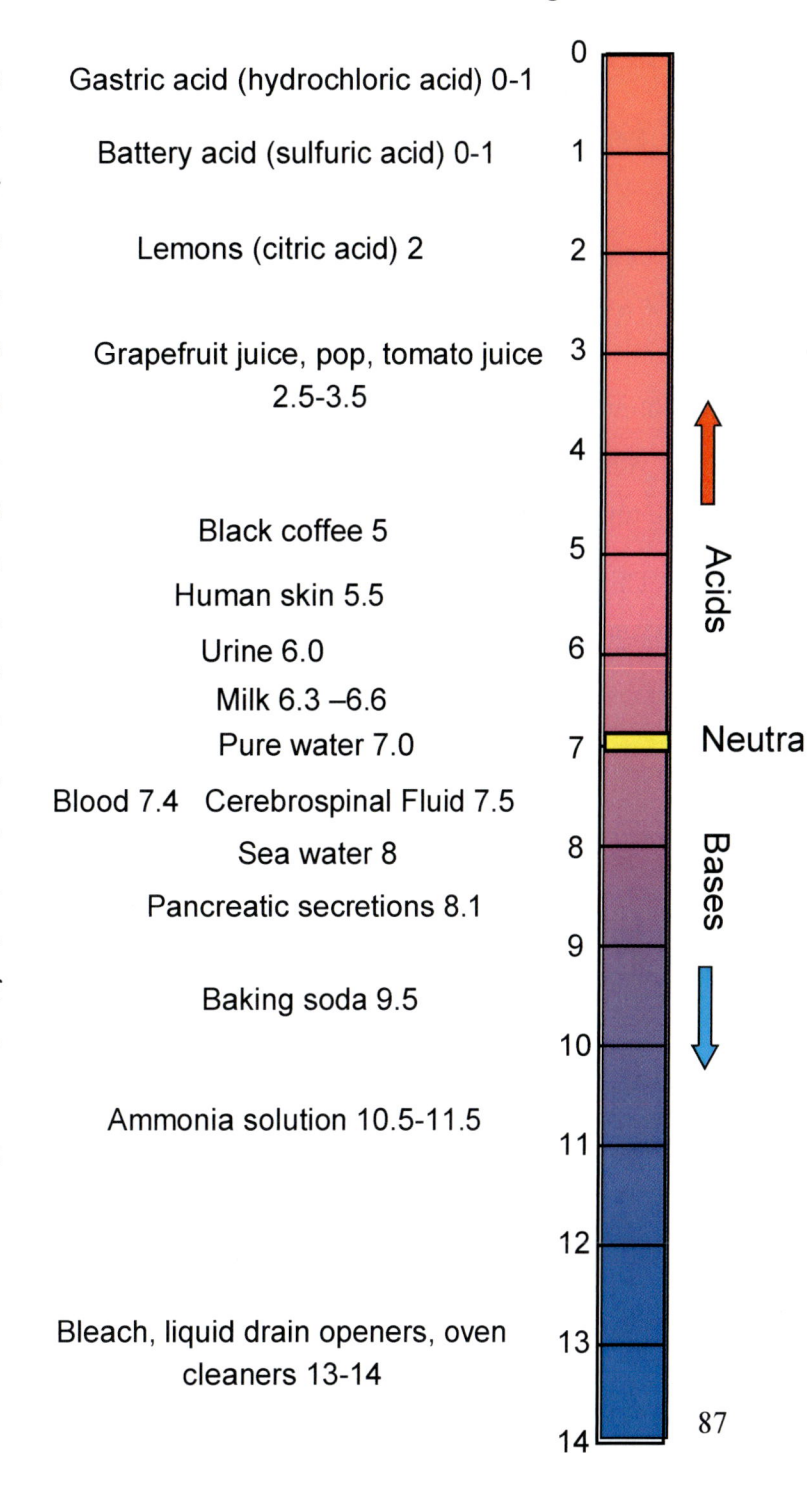

Gastric acid (hydrochloric acid) 0-1

Battery acid (sulfuric acid) 0-1

Lemons (citric acid) 2

Grapefruit juice, pop, tomato juice 2.5-3.5

Black coffee 5
Human skin 5.5
Urine 6.0
Milk 6.3 –6.6
Pure water 7.0
Blood 7.4   Cerebrospinal Fluid 7.5
Sea water 8
Pancreatic secretions 8.1

Baking soda 9.5

Ammonia solution 10.5-11.5

Bleach, liquid drain openers, oven cleaners 13-14

Acids ↑ Neutral ↓ Bases

One important thing to understand about the pH scale is that it is a logarithmic scale. This means that as you move from value to value, the strength of the acid or base does not change by one unit of strength but rather ten times! An acid with a pH of 3 will have ten times the strength of an acid at a pH of 4. The same holds true for the basic side of the pH scale. A base with a pH of 9 is ten times stronger than a base with a pH of 8. Likewise, two "jumps" on the pH scale indicate ten times ten or one hundred times the strength.

Because pH levels are so very important to living things, from time to time it becomes necessary to check these levels. This can be done by using an electronic pH testing device or through the use of pH indicators. These pH indicators are substances which change color in the presence of acidic or basic substances. A widely used pH indicator is litmus powder which is often affixed to small strips of paper and called litmus paper. Litmus paper comes in two forms: red litmus paper and blue litmus paper. Red litmus paper (which is actually a light reddish color) will turn blue in the presence of a base. In other words, if you take a piece of red litmus paper and dip it into a solution of baking powder (a base), it will turn blue.

On the other hand, blue litmus paper will turn red in the presence of an acid. If you were to take a piece of blue litmus paper and dipped into a container of vinegar (acetic acid), it would turn from blue to red.

Litmus paper is helpful to tell us if we have either an acid or base present, but it's limited to telling us that much information. There are other pH indicators which can tell us by color changes a more precise pH value. These indicators are also placed onto strips of test paper and a commonly used brand name is Hydrion[R] paper. By dipping the test strip into the unknown solution, one can match the color that the paper changes to with a set of standard colors on the paper's container. The associated pH value can then be determined. To get very precise pH readings, electronic pH meters are available and are widely used in laboratory settings.

Electronic pH meters can give precise pH measurements.

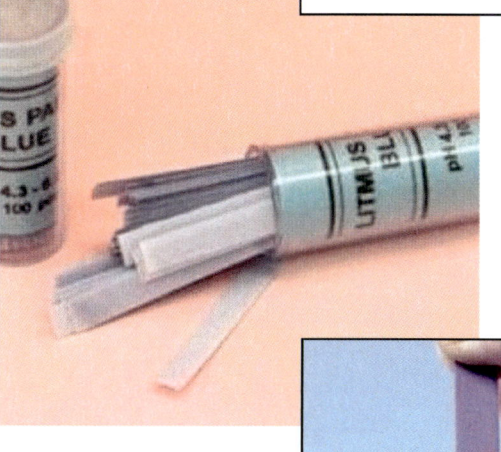

Litmus paper comes in either red or blue. Blue litmus paper will turn red in the presence of an acid. Red litmus paper will turn blue in the presence of a base. Here we see blue paper turning red after being dipped into a clear liquid. Based on the results, we can say that the clear liquid must be acidic.

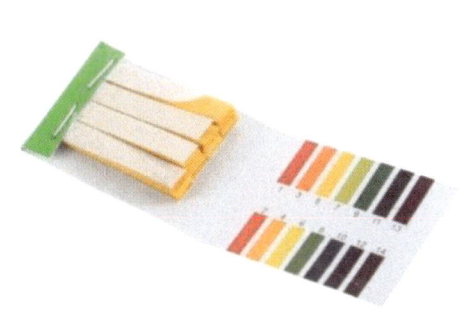

There are test strips available which can determine pH values at various ranges of the pH scale. Some can test the full range of 0-14.

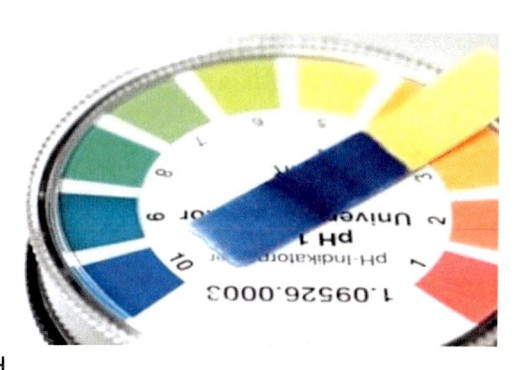

Friendly Biology

Lesson 6 Lab Activity 1: Determining pH using Red Cabbage Juice

READ ALL INSTRUCTIONS BEFORE BEGINNING THIS LAB ACTIVITY!

To get some hands-on experience at checking pH values, we will use yet another pH indicator which is very simple and easy to acquire. There are certain plant pigments which change color in the presence of acids or bases. A pigment is a substance which imparts color to a living thing. Green leaves have chlorophyll pigments within them which gives them a green color. We have pigments in our skin which impart various shades of skin color. For our pH indicator, we will use purple pigments found in red cabbage, grape juice or blueberries. Any of these pigments work equally well as pH indicators.

To prepare the indicator solution, we will first need to remove the pigments from the plant tissue. In the case of grape juice, this process has already happened in the processing of the juice. Grape juice can be used as a pH indicator just as you would buy it from the grocery store. If you choose to use red cabbage as your pH indicator source, you will need to rupture the cells of the cabbage leaves in order to release the pigments. The easiest way to do this is by boiling the cabbage. Here's how it can be accomplished:

1. Purchase a small head of red cabbage (green cabbage will not work as it doesn't have the red pigments). You won't need very much cabbage at all to get a good supply of cabbage juice.

2. Chop the cabbage into pieces and place into a sauce pan. Add enough water to cover the cabbage.

3. Bring to a boil and reduce the heat to a simmer for about 5 minutes or until the cabbage leaves lose their purple color.

4. Turn off the heat and allow the cabbage and juice to cool. Pour the cabbage through a colander or strainer to remove the cabbage leaves. Catch the juice in a jar or bowl. Be careful handling any of the hot juice. The cabbage juice should be a dark purple color.

After cooling, the juice is ready to be used as a pH indicator. Begin by gathering various household substances to test. An ideal acid to test is vinegar (acetic acid). In a small jar or test tube (baby food jars work very well), pour a small amount of vinegar. To this vinegar, add a few drops of the cabbage juice. What changes to do you observe?

Another readily acquired acid in your kitchen is citric acid. Citric acid is the acid of citrus fruits like oranges, limes and lemons. It is also used in home canning certain fruits and can be purchased under the brand name of Fresh Fruit[R] in the canning section of the grocery store. Lemon-lime carbonated drinks often use citric acid as a flavoring component. Try testing a sample of lemon juice or orange juice with your cabbage juice. Do you see similar results to that of the vinegar?

Next, try some milk. What do you think the pH of milk might be?

To see how cabbage juices react in the presence of a basic solution, mix up a solution of baking soda. A teaspoon in a cup of water should work fine. Add a few drops of the cabbage juice. What do you observe? Try the same procedure with a stronger base, ammonia. Ammonia can be purchased in the cleaning solutions section of your grocery store. Note that ammonia readily evaporates and can be irritating to your nose and mouth. Be cautious when working with ammonia. If you have experienced asthma or allergies to things you breathe in, you may want to avoid testing ammonia.

How does the cabbage juice change when you test the ammonia? What does this tell you about ammonia? Another extremely strong base is chlorine bleach (Clorox[R]). We recommend you do not test chlorine bleach! You will see the expected color change only briefly as the properties of bleach cause the color to rapidly disappear. Chlorine bleach can cause skin irritation and ruin clothing. We do not recommend you test chlorine bleach.

Substance Tested	Color Observed	I conclude this substance must be an acid/base.

One final topic to discuss regarding acids and bases is how they taste. Taste can readily help us identify substances as being acidic or basic. First, think about vinegar (acetic acid). Vinegar is often used to preserve cucumbers into pickles. Using vinegar works because the bacteria and molds that spoil cucumbers cannot survive at this low pH level. How do pickles taste? Yes, pickles do indeed taste sour! What about the citric acid of the juice of lemons and limes? Yes, they, too, taste sour. Acids, in general, taste sour.

On the other hand, how do basic substances taste? Have you ever had the misfortune of tasting baking soda? Sometimes if baking soda is not adequately mixed into the batter of cookies, we may get a taste of it. Baking soda has a bitter taste as do other basic substances. So, as a general rule, acids taste sour and bases taste bitter!

Let's review what we've learning regarding pH in this lesson:

- pH is a measure of how basic or acidic a substance may be;
- The pH scale ranges from 0-14 with values less than 7 being acidic and values greater than 7 being basic. The value of seven is said to be neutral. The pH scale is a logarithmic scale;
- Enzymes in living organisms are dependent upon specific pH levels to work properly;
- The pH of blood is 7.4;
- The pH of a substance can be obtained through the use of an electronic pH testing instrument or by pH indicators which turn colors according to the pH. Litmus paper is often used to determine the presence of an acid or base whereas Hydrion^R paper can help determine a more specific pH value.

Lesson 6 Lab Activity 2: Making Yogurt

READ ALL INSTRUCTIONS BEFORE BEGINNING THIS LAB ACTIVITY!

In this lab activity you will explore how milk can be preserved as yogurt. Because we will be working with milk that is placed in warm temperatures, it will be very important to keep everything as clean as possible. You will be purposefully adding a known bacteria to the milk to make the yogurt. You don't want to accidentally add a "wild" bacteria which might result in something other than yogurt. Keeping your hands, containers and utensils clean will help keep your yogurt safe to eat.

Items you will need include:

1. Clean pint canning jar with lid.

2. Milk—any degree of fat content will work, but we've found whole milk to give best results.

3. Store bought yogurt with live cultures. You will be using the bacterial cultures already at work in the purchased yogurt to change your milk to yogurt. Look on the label of the yogurt you purchase. It must say made with active or live cultures. If it has been pasteurized following preparation, the bacterial cultures will have been killed and will not work to make more yogurt. It does not matter if the yogurt you buy has been flavored or is plain yogurt. Plain is ideal but fruit-flavored will work. You will add your own flavorings after you make your yogurt. The key is to have live cultures in the yogurt you begin with.

4. Wooden or plastic kitchen spoon.

5. Heat source. In order for your milk to become yogurt, the bacteria that we you intentionally add to the milk must be exposed to an ideal growing temperature. The ideal growing temperature is about 110 degree F. There are several ways to accomplish this temperature. You can create an incubator with a light bulb, but unless the light bulb is regulated by a thermostat (switch which turns off the light bulb when hot enough and then back on when cool), you may find the temperature goes above 110 degrees and your bacteria are killed. Another way to get the correct amount of heat is to create a hot water bottle system. This can be accomplished by filling some empty pop bottles or other container with very hot water (hotter than 110 degrees). Be very careful as water this hot may cause burns to your skin. Do not use boiling water. Place lids over these water bottles. By wrapping 1 or 2 of these hot water bottles next to

your bottle of milk, you can keep your milk at a suitable temperature. Yet another way to accomplish the need for a heat source is to turn on your kitchen oven to 275 degrees F. Allow it to heat and then turn the oven OFF. With the oven off, it will cool over time but yet provide enough heat to keep your bacteria happy. If you have a gas oven which has a pilot light that is always burning, you can place your jar in the oven, without turning on the oven to get a suitable yogurt-making environment.

Steps to Make your Yogurt:

1. Thoroughly wash your glass jar and lid with hot soapy water. Allow them to air dry.

2. If you have access to litmus paper or Hydrion[R] strips, test the pH of your milk. Write down your observations.

3. Fill your jar approximately 3/4 full with milk. Heat the milk in a microwave oven until the milk is similar to the temperature of bath water. If you do not have a microwave oven, you may place your jar of milk (with lid in place) in a bowl of hot water (not boiling!). Allow the jar to sit in the hot water until the milk is the same temperature as the water. This method may require that you put in a fresh supply of "hot" water every so often. The idea here is to warm the milk yet not overheat it. We want the bacteria that we introduce to find themselves in a warm, suitable growing environment.

4. Open your store-bought supply of yogurt. With your clean spoon, scoop out about 1/4 cup of yogurt. If your yogurt has fruit in it, it's not necessary to stir-up the yogurt. Just take your 1/4 cup scoop right off the top. Add this yogurt to your warmed milk. Stir it in and replace the jar lid. The process of adding a culture of bacteria to another supply of "food" for the bacteria is called inoculating. You inoculated the milk with yogurt-forming bacteria.

5. Now, you will need to incubate your inoculated milk. Place your jar in your warm location for at least 24 hours. After this time has passed, observe your jar. If conditions were right, the bacteria will have utilized the lactose in the milk as their main carbohydrate source and multiplied many times. The "new" bacteria will also consume more of the lactose in the milk. In doing so, they produce an acid byproduct which changes the pH of the milk and causes the milk proteins to cur-

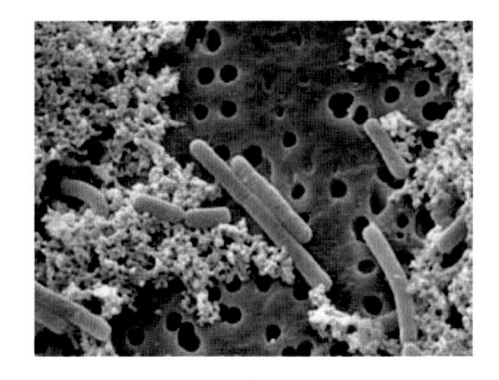

Lactobacillus bacteria consume lactose and produce lactic acid byproducts. The pH is lowered and the milk turns into yogurt.

dle. If things have gone "well," and the yogurt has formed, you should see a semi-solid layer of curdled milk in the jar with a thin, clear watery layer above it. If you don't see this and only see what appears to be milk, allow the jar to incubate another 12 hours. Make sure it is still warm for your jar.

6. Once the yogurt forms, it's ready to be eaten! You may find the yogurt to be not quite like the store-bought yogurt you may be accustomed to eating. Before you dig in, take another pH reading of the yogurt. How might you expect the pH to have changed? Then, taste a sample of your yogurt. How does it taste? Does the taste of your yogurt agree with what you learned earlier about acidic substances? Because fresh yogurt is rather tangy in taste, you may find it necessary to add some sweetener, maybe some sucrose and then maybe some fresh fruit.

7. If you enjoyed your yogurt and would like some more, reserve a 1/4 cup sample of your yogurt to inoculate another batch of warmed milk. As long as you use clean hands, jars, lids and spoons, your yogurt-making bacteria should continue to multiply. If you find some sort of product that you didn't expect, such as very foul smelling or off-colored curds when you check your yogurt, pour all of it out, wash all of your equipment very thoroughly and begin again.

Lesson 6 Lab Activity 3: Making Fresh Cheese (Queso Blanco)

READ ALL INSTRUCTIONS BEFORE BEGINNING THIS LAB ACTIVITY!

This lab activity combines concepts you've just learned about pH with concepts you learned in the last lesson about proteins! The purpose of this lab is to see how lowering the pH of a protein can cause it to permanently change its shape. This permanent change in shape is known as coagulation of a protein. You may be familiar with the term coagulation when you think about how blood coagulates or clots. In the case of blood coagulation, the trigger that makes blood clot is the release of a chemical by cells which begins the clotting reaction.

Coagulation can also be caused by heating. When you cook an egg, the heat you apply to the pan gets transferred to the albumin (protein) of the egg white which causes it to stiffen and turn white. The proteins in the yolk of the egg also coagulate. In this lab, we will examine yet another means of causing proteins to coagulate and that is by lowering the pH of the protein.

Materials you will need for this lab include:

- One gallon of whole milk
- 1/4 cup vinegar
- Metal sauce pan or stock pot capable of holding 1 gallon of milk
- Metal spoon
- Food thermometer which measures to at least 185 degrees F.
- Clean cloth and colander to strain cheese curd from whey
- Cheese press (optional)
- Chopped spices or peppers to flavor cheese (optional)

Procedure:

1. Transfer the milk to the sauce pan or stock pot and begin slowly heating. Stir frequently to avoid scorching the milk. Heat to 185 degrees F.

2. Once milk is at 185 degrees F turn off the heat and pour in vinegar. Pour the vinegar in three

equal additions (ie. pour in one-third of the total 1/4 cup, wait a minute or so, pour in the second third, wait and finally pour the remaining third of the 1/4 cup of vinegar). Stir continually as you pour in the three "doses" of vinegar. Recall that vinegar is a weak acid and has the scientific name of _____ (check back a few pages if you've forgotten this). The vinegar will lower the pH of the milk and coagulate the milk proteins. These proteins are known as casein. You should begin to see the curds forming which look like white clumps that separate from the watery part of the milk known as whey. The whey appears like a yellowy green liquid. Gently continue to stir for another five minutes and then allow to rest for five minutes.

3. Place the clean cloth in a colander over a bowl large enough to catch the whey. Carefully pour the curd/whey mixture into the colander. Allow the whey to drain through while catching the curds in the colander.

4. If you'd like to flavor your cheese, add chopped fresh spices or peppers at this point. You can also add salt if you'd like. Gently mix the curds with the spices. If you desire a drier cheese, you can hang it in a bag over a bowl or sink to allow additional whey to drain from the curd. Otherwise, you can eat it fresh at this point or transfer it to a cheese press.

5. Your cheese press can be a simple cylinder cut from a plastic container or section of clean PVC pipe. Place the cylinder over a porous surface or cloth inside a pan. Add the curd and then insert a can or other round object which can fit down inside the cylinder. Place a weight on top of the can to press the cheese. Start with about 10 pounds of weight at first (large sack of flour or sugar) for about 20 minutes. Increase to about 25 pounds for 2-5 hours. Your cheese can then be removed from the press, wrapped in wax paper and refrigerated. Note that this type of cheese will not melt when cooked.

For additional recipes and instructions on making cheeses of all kinds, go to the New England Cheese Making Supply website: www.cheesemaking.com.

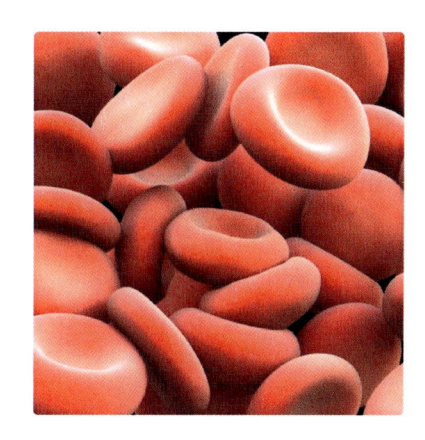

Lesson 7: The Cell (Part 1)

In this lesson we will begin to explore cells. Living things are composed of small units of structure known as cells. These small units of structure can be likened to a vast array of small rooms where activities take place that allow a living thing to live! It is within the cells of living things that the life processes we've discussed in our earlier lessons take place: carbohydrates get used as fuel, fats are stored, more structural components are built from proteins, pH values are adjusted and enzymes are hard at work making all these events happen. These processes occur in a very organized fashion in these small "rooms" of life known as cells.

Cells were discovered by a Dutch scientist named Antonie van Leeuwenhoek in the 1600's. Leeuwenhoek was one of the inventors of the first light microscope and the first cells he saw were cells that made up the bark of a cork tree. While looking at these cells he thought, it has been said, that they resembled jail cells in a prison and that's how cells received their name.

The scientific term for the study of cells is cytology. The prefix cyto– means cells and the root

> Cyto = cell
>
> -ology = the study of
>
> Cytology = the study of cells

word –ology means the study of, hence cytology means the study of cells.

In our study of cells, we will look at the many parts which make up most cells. Then, as we learn about each part, we will also learn about the job of each part. Understand from the beginning that not all cells have all of these components. Initially, we'll talk about all of the parts and then later we'll point out which cells have these various parts. These parts of a cell are known as the organelles of a cell because he parts of cells function like "little organs" within cells. Let's begin.

On the outer surface of cells we find a skin-like structure known as the cell membrane or plasma membrane. In most cells, this outer "skin" is made of two layers of molecules known as phospholipids. A phospholipid is a lipid molecule with an attached phosphate group. The lipid portion of the phospholipid positions itself so that the lipid portion is exposed on the inner surface and outer surface of the membrane. The phosphate group is within the membrane. The cell membrane is often referred to as being composed of a phospholipid bilayer.

Diagram of an Animal Cell

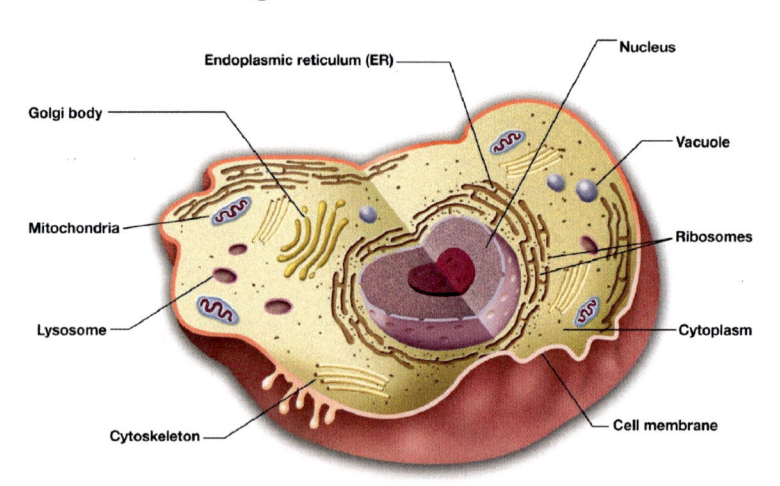

- Endoplasmic reticulum (ER)
- Nucleus
- Golgi body
- Vacuole
- Mitochondria
- Ribosomes
- Lysosome
- Cytoplasm
- Cytoskeleton
- Cell membrane

The purple balls represents the phosphate group of the phospholipid molecule. The yellow ribbons below represent fatty acid chains extending towards the center of the cell membrane.

The phosphate groups are hydrophilic and attract water to the surface of the cell. The fatty acid chains are hydrophobic and create a barrier for water to pass through the membrane.

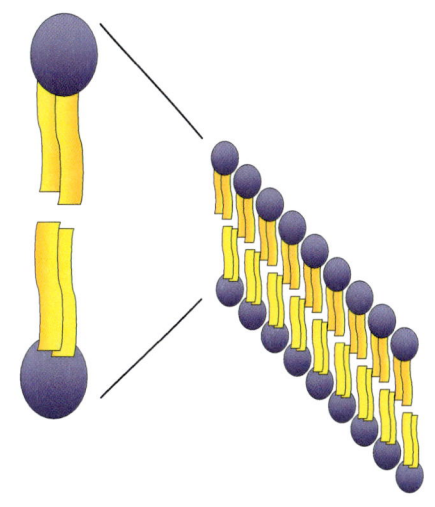

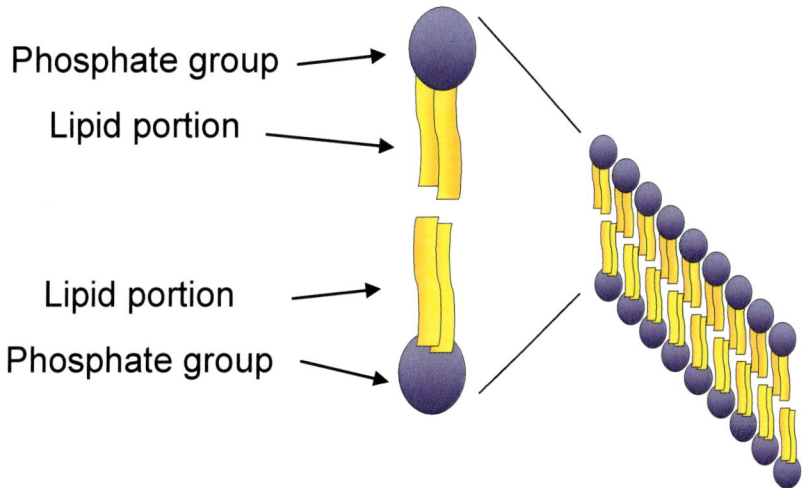

Phosphate group
Lipid portion
Lipid portion
Phosphate group

Recall from Lesson 4 that lipids are insoluble in watery solutions. Because of this, water finds it difficult to pass directly across cell membranes, much like water finds it difficult to penetrate *our* skin!

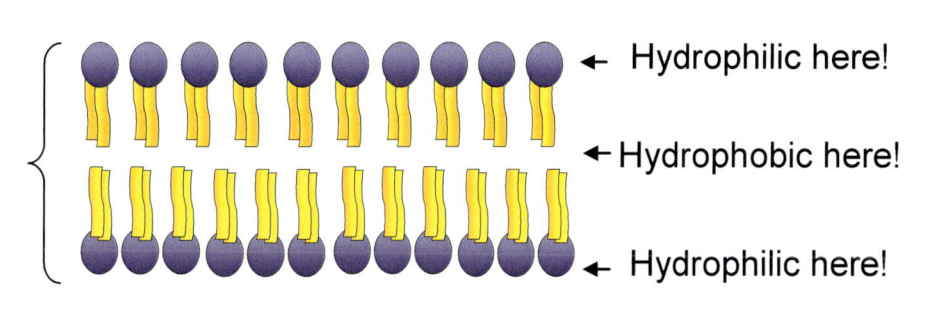

Phospholipid bilayer

← Hydrophilic here!
← Hydrophobic here!
← Hydrophilic here!

However, water must move in and out of cells in order for them to survive, so cell membranes have particular "doors" or "gates" which open and close to allow substances, including water, to readily pass. Therefore, cell membranes are said to be semi-permeable. This means cell membranes have the capability of allowing certain things to enter and leave the cell. Look at the diagram below to see how cell membranes are designed. Note the passageways which allow things to move in and out of the cell.

The cell membrane works like a fence with gates. At certain locations, gates allow certain substances to pass. Other substances are not allowed to pass. In this way, cells are capable of keeping unwanted substances outside their boundaries, yet keep needed substances within. Wastes can also be removed and needed nutrients brought in.

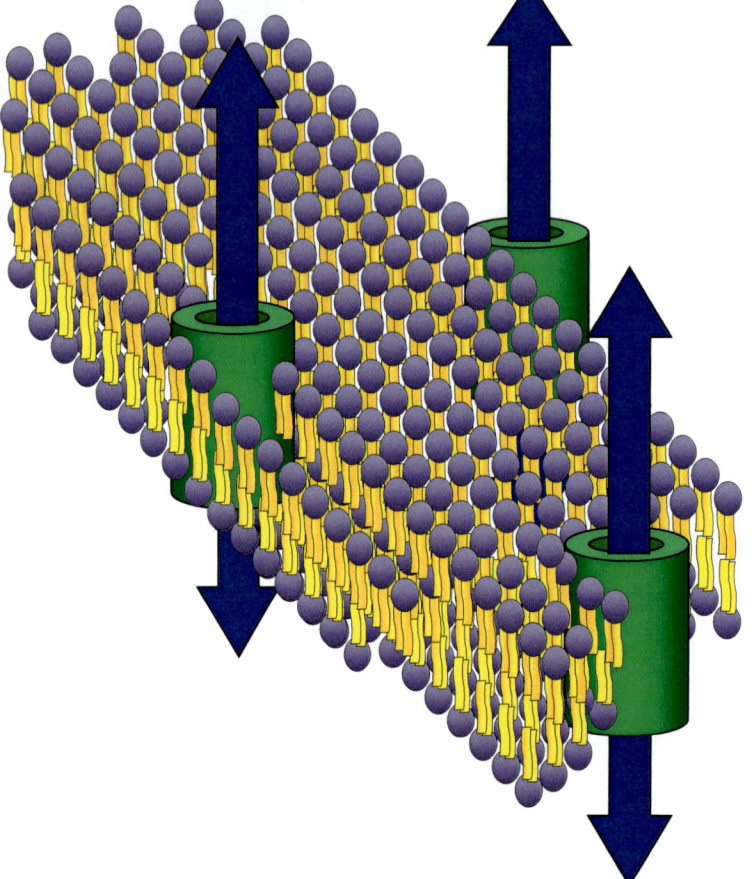

Before we move deeper into the cell, we need to examine one more protective layer which surrounds cells of certain living things. In this particular case, these living things are plants. Plants have another layer of protection surrounding each cell known as the cell wall. The cell wall in plants can be found just outside the cell membrane. It serves as a skeleton for the plant cell in that it provides structure and form, much like our skeleton does for our bodies. These "skeletons" make up the walls of the cells that Leeuwenhoek saw when he was viewing cells for the first time.

The material that makes up the cell wall in plants is cellulose. Recall from Lesson 3 that cellulose is a carbohydrate. Cellulose is made up of long strings or chains of glucose molecules that have been linked together. If you've ever broken a stalk of celery and found the long strings it can make, you're witnessing chains of cellulose molecules! The cell wall, composed of cellulose, makes the walls, roof and floor of a plant cell. It gives plant cells great strength as evidenced by the fact that when a plant dies, it leaves behind its cell walls which, in the case of a tree, become wood!

Diagram of a Plant Cell

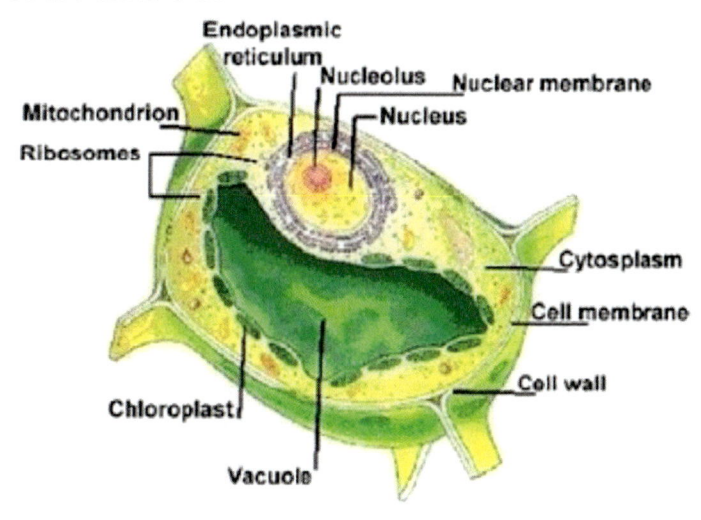

Labels: Endoplasmic reticulum, Nucleolus, Nuclear membrane, Mitochondrion, Nucleus, Ribosomes, Cytosplasm, Cell membrane, Cell wall, Chloroplast, Vacuole

The crunchiness of celery is due to the cellulose found in the cell walls of the celery plant.

Recall, also, from Lesson 3, that the enzyme cellulase is utilized to break down cellulose. Now you know that the enzymes that are used with the breakdown of cellulose are working on the cell walls of plants. The microorganisms found inside the stomachs of cows and termites have cell walls for their breakfast, lunch and dinner!

If you've ever built something made with wood or studied trees, you may be aware that wood can generally be categorized into two groups: softwoods and hardwoods. Softwoods come from fast-growing trees like pines and firs and make up most of the lumber used to build a house. Hardwoods come from slower growing trees like oaks, maples, ash and walnut trees. Hardwoods are used to make furniture and hardwood flooring due to its great strength and ability to resist scratching and denting.

Can you think of what might be the major difference between these two types of woods on a cellular level? It's the thickness of the cell wall which determines whether a wood is considered a softwood or hardwood. The cell walls in pines and fir trees are thinner which results in wood that is soft compared to the wood of oak, maple, ash and walnut trees. Hard woods, with their thick cell walls make strong floors and furniture.

Wood made from fast-growing trees like pines, firs and spruces is known as softwood. Their cell walls are thinner.

Wood from slower growing trees, like oaks, maples and walnuts is known as hardwood. The cell walls in the cells of these trees are thick!

People who burn wood to heat their homes know that choosing hardwoods over softwoods yields much more heat per log. Softwoods can still be used, but hardwoods, because of all those layers and layers of cellulose, provide a greater supply of fuel. By breaking the bonds that hold the molecules of cellulose together (burning the wood) lots of energy in the form of heat can be released to heat our homes or roast our hotdogs.

Let's pause and review. We've said:

- Cells are the structural units of living things;

- Within cells, there are various "little organs" known as organelles;

- Cells have an outer covering known as the cell membrane or plasma membrane. The cell membrane functions like a skin holding certain things within the cell and keeping control over what enters and leaves the cell. We learned that cell membranes are said to be semipermeable;

- Plant cells have an extra layer of protection which functions like a skeleton for the plant and is called the cell wall which consists of cellulose.

Let's continue on with our exploration of cell organelles. The organelle of the cell which serves as the "brain" of the cell is known as the nucleus. Just like our brain, the nucleus of the cell works to control all activities of the cell. You might think of the nucleus as a cookbook or owner's manual which the cell utilizes minute-by-minute to find out what needs to happen next in the cell. Generally, the nucleus of a cell is spherical, like a ball, within the cell but not always. In some cases the nucleus may be oval or, sometimes, like in the case of adipose cells (fat tissue cells), the nucleus is smashed-up next to the cell membrane making it appear almost flat.

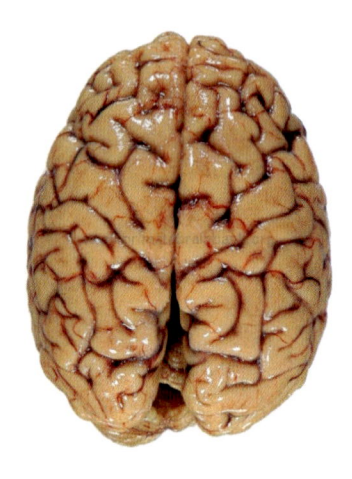

The nucleus of the cell functions like the control center (or brain) of the cell. It can also be likened to a cookbook which holds all the "recipes" for the cell to function.

In more complex living things like animals and plants, the nucleus, like the cell itself, is enclosed in a membrane. This membrane is known as the nuclear membrane. Another name for the nuclear membrane is the nuclear envelope. In simpler living things, like bacteria, for example, there is no nuclear membrane present.

Organisms whose cells have a nuclear membrane present are referred to as being eukaryotic or eukaryotes. Organisms whose cells do not have a nuclear membrane are referred to being prokaryotic or prokaryotes. The prefix eu– in eukaryote means good while the root word karyote refers to kernel or nut (cell in this case). The prefix pro– in prokaryote refers to before. Some biologists believe that prokaryotes existed earlier in time than the eukaryotes due to their more primitive formation (no nuclear membrane). Regardless of the timing of these groups of organisms, eukaryotes have nuclear membranes (the eu– DO have nuclear membranes) while the prokaryotes have no nuclear membrane (the pro– have NO nuclear membranes).

> Eukaryotes do have a nuclear membrane. (Eu = Do)
>
> Prokaryotes have no nuclear membrane. (Pro = No)

Within the nucleus of the cell are structures that hold the information regarding the "goings-on" of the cell. If we refer back to our analogy that the nucleus is like a cookbook or owner's manual, the structures within the nucleus that hold this information could be compared to the chapters within the cookbook: the vegetable chapter, the meats chapter, the breads chapter, etc. In the cell, these "chapters" would be known as the chromosomes. The word chromosome is derived from the prefix chromo– which means colored while the root word –some refers to body. Together, chromosome literally means colored body. Historically, when biologists were first studying the nuclei (plural form of nucleus), they noted worm-like structures within the nucleus. As they prepared microscope slides of these cells, they used colored stains to better visualize the cell organelles. They noted that the worm-like structures tended to "take-up" stain well and so gave them the name chromosome (colored bodies.)

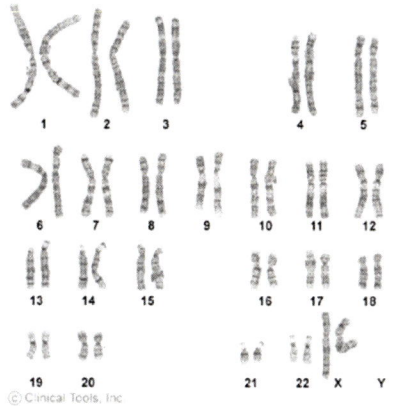

Chromosomes of the human are found in pairs.

> Chromo = colored   -somes = bodies
>
> Chromosomes = colored bodies

Let's return to our discussion of chromosomes as being like the chapters in a cookbook. As the chapters in a cookbook focus on certain types of foods, the information found on chromosomes is specific for certain parts of the cell as well as certain cell functions in regard to the whole organism. For example, certain chromosomes carry information for body size or shape, while other chromosomes carry information for eye, hair or skin color. The chromosomes carry all of the cell's information.

The number of chromosomes living things have in the nuclei of their cells varies. Some organisms have several chromosomes while simpler living things may have only one chromosome. As humans, we have 23 pairs of chromosomes for a total of 46 chromosomes in the nuclei of each of our cells. This is not exactly true for *all* of our cells and we will discuss this in greater detail later. Simpler organisms have fewer chromosomes. Bacteria may have one only chromosome. All other living creatures have a precise number of chromosomes, each with specific information. It's for this reason that organisms can only reproduce with other organisms having the same number of chromosomes and more specifically, the same information found on each chromosome.

Let's continue by exploring the structure of a chromosome in greater detail. To do so, let's return to our cookbook analogy. We said that the chromosomes were like the chapters in a cook book. What do you find within each chapter of a cookbook? You find recipes! A recipe gives a specific plan for completing a food dish. Likewise, the chromosomes within the nuclei of cells have separate sections which provide specific directions for the cell to carry out its job. These "recipes" in cells are called genes. The word gene comes from Latin which means to generate or make.

Nucleus	Chapters	Recipes
Cookbook	Chromosomes	Genes

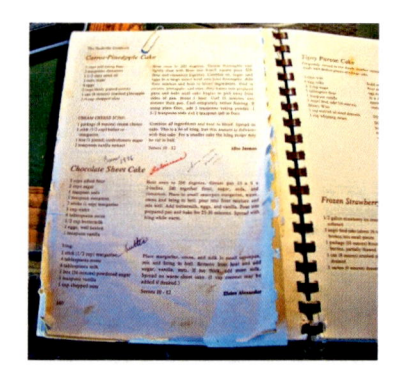

Genes work like recipes to accomplish specific tasks for the cell. For example, a specific gene in humans carries the information to code for eye color or hair color or skin color or face shape. Another gene is responsible for other parts of the body. All parts, and we mean ALL parts, within our bodies are coded for in our genes. Humans have an estimated 23,000-25,000 genes! That's a LOT of "recipes"! Together, they work to keep all of our cells working properly, which is a truly amazing feat!

However, like in the kitchen, sometimes the recipe gets fouled-up and the dish we had hoped for doesn't turn out as planned. In a similar fashion, if a gene gets messed up for some reason, the structure or function that was prescribed by the gene doesn't happen correctly. The results of these "mistakes" may be minor and not be a problem for the organism. Sometimes, however, these alterations may cause the organism great problems or even cause death. We'll come back to our study of genes in Lessons 12 and 13.

Before we move on, let's review. We have said that:

- The nucleus is the cell organelle responsible for controlling the activities of the cell;

- The nucleus may or may not be bound by a membrane. Living creatures which do have a membrane bound nucleus are known as eukaryotes while organisms without a nuclear membrane are known a prokaryotes (Eu = do and pro = no);

- Within the nucleus one can find chromosomes which carry all of the information as to how the cell appears and functions;

- The number of chromosomes and information on each chromosome is specific for each living thing;

- Chromosomes consist of smaller subunits known as genes. Genes consist of specific codes for specific parts of the organism, both for the way that part is made as well as how it works.

Let's move on now to another very important organelle found in cells: the mitochondrion. If the cell nucleus is analogous to a brain or cookbook of recipes, the mitochondrion is analogous to a power plant. The mitochondrion (singular) is the place where raw materials are utilized to produce energy just like a power plant takes raw material (usually coal) and through a series of steps, produces electricity.

The mitochondrion is like a power plant for the cell.

Instead of coal, can you think what raw material the mitochondrion might use? Do you remember how in Lesson 3 we learned how glucose is considered to be the fuel of our bodies? Yes, it is in the mitochondrion that glucose, with the assistance of many, many enzymes, is converted to energy-containing molecules which are then utilized to make energy. This process of converting glucose into energy is known as respiration. Breathing is also referred to as respiration. This is the same word, but these are two very different meanings.

Glucose → Mitochondrion → ENERGY!

The process of converting glucose into energy is known as respiration.

When viewed under a microscope, mitochondria (plural) appear like rounded beans. They have a convex surface and a concave surface. The concave surface refers to the side that is "pushed inward." The convex surface is the opposite surface that bulges outward. Within the mitochondrion, there is a folded membrane known as the cristae. It is on this membrane surface that the enzymes do their work with glucose (respiration.)

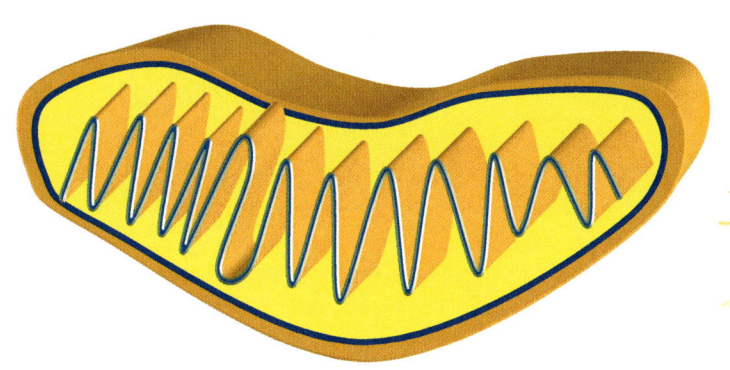

On the interior of the mitochondrion, one finds a folded membrane known as cristae where respiration takes place. Respiration is the conversion of glucose into energy-containing molecules known as ATP's.

The energy-containing molecules which result from the respiration process are known as adenosine triphosphate molecules or ATPs for short. For every one glucose molecule that gets processed through respiration, 38 ATPs are formed. The ATPs are then broken down and energy is released which can be used by the cell to accomplish things. Biochemists believe there to be three main sets of reactions that are utilized to convert glucose into energy: glycolysis, Kreb's cycle and the electron transport chain. Within each of these three sets of reactions there are multitudes of individual reactions all controlled by a multitude of specific enzymes.

1 Glucose Molecule → 38 ATPs → Movement, Body heat, etc.

Besides the main ingredient of glucose for these reactions, there is one other essential ingredient: oxygen! Oxygen, found in each breath we take, is a vital ingredient used to convert glucose into energy! We need to point out, however, that some living creatures (primarily bacteria) produce the energy they require with very little or no oxygen at all. While these organisms can get the job done, making this conversion is much less efficient. Instead of the 38 ATPs produced per glucose molecule in the presence of oxygen, they only produce 2 ATPs without oxygen. Creating energy from glucose under little or no oxygen levels is known as anaerobic respiration. The prefix an– refers to without and the root word aerobe refers to oxygen. Creating energy in the presence of oxygen is known as aerobic respiration. We, as humans, obviously employ aerobic respiration in our cells.

An important byproduct or waste material of respiration is carbon dioxide (CO_2). As humans, carbon dioxide is exhaled each time we take a breath. We'll take another look at the importance of carbon dioxide in our next lesson.

One use of the energy available in the ATPs is to contract muscles which results in movement. In warm-blooded creatures, the energy produced in the form of heat is used to maintain appropriate body temperature.

Let's review. We've said that:

- The mitochondrion is the organelle of the cell which is responsible for generating energy for the cell;

- Mitochondria are bean-shaped having within a folded membrane (cristae) where the energy production takes place;

- Glucose is the fuel used by the mitochondrion to produce energy. This process is called respiration. Oxygen in the environment provides for very efficient respiration;

- The product of respiration is adenosine triphosphate (ATP). It is when these ATPs are broken down that energy is released to be used by the cell. Carbon dioxide is a byproduct of the reaction;

- Many living things require oxygen in order to convert glucose into energy. Some do not, however the efficiency is greatly reduced.

Friendly Biology

Lesson 7 Lab Activity 1: Osmosis with Eggs (Part 1)

PLEASE READ ALL INSTRUCTIONS BEFORE BEGINNING THIS LAB.

The purpose of this lab will be to explore how the cell membrane of a cell can allow various substances to pass through while limiting others. This capability is referred to as selective permeability. Instead of looking at tiny, individual cells in this lab, we will utilize fresh eggs which have a membrane just inside the shell that functions very similarly to cells. This lab requires several days to complete so if you plan to have it completed within one week, getting started promptly is important.

Materials you will need for this lab include:

- At least 4 fresh eggs (uncooked). The lab really requires only 1-2 eggs, but having spares on hand is very helpful.

- 5 glass jars (quart-sized canning jars work very well). It's best if they are clear containers.

- 4 disposable Styrofoam cups, 8 ounces or more.

- Wax pencil or marker which can write on glass or plastic containers. It needs to be water-proof. A dark-colored crayon will work.

- Graduated cylinder or large syringe which measures in milliliters or cubic centimeters (cc's).

- Vinegar. It's best to get 1 gallon of vinegar because running out can cause delays in the lab.

- White sugar. At least five pounds.

- White corn syrup (cheap store brand works great).

- Access to running water. This lab is a very "wet" lab so plan on using a kitchen counter or table over a floor that can be mopped dry. We suggest using a layer of newspapers over the table surface or counter to make clean-up easier.

Procedure:

1. The first thing you'll need to do is label your eggs. Using a permanent marker or crayon label them with a letter, number or name. It doesn't really matter what you label them, just something to help you remember which egg is which.

2. Next, you'll need to find the volume of each egg. Because we can't readily use a math formu-

la to find the volume of an egg due to its irregular shape, we'll find the volume using the displacement method.

Here's how the displacement method works: suppose you fill your bathtub about three-fourths full of water. When the water is still, you mark the level of the surface of the water on the side of the tub. Then, when you jump into the tub, the water will rise, right? If you totally submerge yourself under the level of the water, the volume of you will equal the amount the water rose on the side of the tub. In other words, the amount of water you displaced by jumping in the tub equals the amount of volume of you! We'll use this same technique to find the volume of your eggs.

So, we'll first prepare your volume-measuring container. You'll use it several times throughout this lab. Begin by taking one of your glass jars and label it "volume-measuring container." Fill it about three-fourths full of water. Set the container on a level surface and on the outside of the jar, using your wax pencil or crayon, mark the level of the surface of the water. Gently, lower your first egg into the water. (Do not allow any water to splash out of the container. If you do splash water out, you'll need to start over.) Mark the level of the water again. Carefully, with one hand over the opening of the jar, pour out the water and then remove the egg, too, and set it aside.

Now, take your empty jar and begin adding known increments of water until you reach the first mark you made. For example, if you have a 30 cc syringe, add 30 cc increments of water until you are close to the mark. Then add portions until you reach the mark adding up the amounts you've added as you go. The value you end-up with will become your "starting value." Write it down here: _____.

Next, continue adding water until you reach the mark you made after you put the egg into the jar. This value will be the "total volume." Write it here: _____. If you subtract the "total volume" from the "starting volume," the answer you get will be the amount of water displaced by the egg which equals the volume of the egg. Do your math work here for this first egg: Total volume - starting volume = _____

_____ - _____ = _____

Repeat these steps for all four of your eggs. Take your time as this step is important. Record your results below.

Egg ID	Total Volume	Starting Volume	Egg Volume

3. The next thing you'll have to do is remove the shell from your eggs. You'll do this by submerging them in straight vinegar (acetic acid) for 1-2 days. The vinegar reacts with the calcium carbonate of the shell to dissolve it leaving the membranes below exposed. Because your egg label was placed onto the shell, it, too, will be dissolved. Because of this, it's important to label the cups into which you place your eggs. So, take your four disposable cups and label them for each of your eggs. Then, place your eggs into the cups and pour vinegar over the eggs to completely submerge them. A few minutes after you do this you may notice your eggs have floated up to the surface of the vinegar and a portion of the shell is now exposed. This happens because carbon dioxide gas is released from the reaction of the acetic acid and the calcium carbonate of the egg shell. You can gently tap the egg or roll it over to allow it to sink again. Place your cups in a location where they can't be tipped over and allow to sit overnight.

4. After allowing your eggs to sit overnight in their first vinegar bath, carefully with one hand over the opening of the cup, pour out the vinegar into a sink catching the egg in your hand. The egg should have a wet, powdery sort of feeling at this point. Gently rinse the egg under running water. If the vinegar has completely dissolved the egg, it will look like a yellowish water balloon. Most likely, however, you'll have to place it into a second vinegar bath. Repeat using fresh vinegar and allow to sit overnight. Pour off vinegar and rinse carefully under running water. Gently rubbing the egg will allow all of the shell to be removed. Your goal is to have four rubbery, intact eggs.

5. Next, you'll place each egg into pure water. Begin this step, by labeling each of your glass jars with the identification of each egg. Each egg will have its own jar. Gently place each egg into its jar and fill the jar with pure tap water until its about 3/4 full. Place the jars in a secure place and allow to sit overnight.

6. The next day, find the volumes of eggs like you did earlier in step 2. Prepare your "volume-

measuring" jar as you did earlier by filling it to the line representing the "starting volume." Then, taking your first jar of water with an egg, gently pour off the water, catching the egg in your hand. Do you notice any changes in the egg?

Carefully take the egg and lower it into your volume-measuring container. Mark the level to which the water rises. Record your results in the table below. Repeat this step with your three remaining eggs. Note that there are several more steps to the lab and it's your goal to take at least one egg through all of the steps. Should you accidentally break an egg, just mark in the data table that the egg is no longer present and continue with your remaining eggs. As you work with the eggs, don't allow them to sit out in the air and dry. After measuring the volumes place them back into their labeled jars to await the next step.

7. Your next treatment for your egg will be to place the eggs into corn syrup. Begin by emptying your first jar of water, keeping the egg inside. Then, pour corn syrup over the egg until the egg is submerged. You may find that the egg rises to float in the corn syrup and that's normal. Just stop

Egg Volumes After Pure Water

Egg ID	Total Volume	Starting Volume	Egg Volume

pouring corn syrup when it begins to float. Repeat this step for each of your remaining eggs. Place them in a secure location and allow to sit overnight.

8. The following day, you'll again measure the new volume of each egg. Before doing so, you'll need to first make some observations of the egg in the jar. Look closely through the side of the jar. Can you detect two layers of liquids? Carefully take a blunt spoon and lower it down into the jar alongside the egg while you look from the side. What do you suppose has happened here? Continue by pouring off the liquids from the jar and carefully catching the egg. How does the egg appear?

Could this possibly explain the liquids in the jar? Rinse the jar well and set it aside. One-by-one, measure the new volume of each egg and record your results in the table below. As you measure each egg, return it to its jar. Add tap water to prevent it from drying out. The tap water will be the next treatment.

9. After removing your eggs from the corn syrup treatment, you placed your eggs back into pure tap water. Allow them to sit overnight in a secure location. Before measuring the volumes once again, can you predict how you think they might have changed? Continue by finding the new volumes and record your results in the table below.

Egg Volumes After Corn Syrup Treatment

Egg ID	Total Volume	Starting Volume	Egg Volume

10. After finding the volumes of your eggs following the second water treatment, you'll now place your eggs into granulated table sugar. Quick review: can you recall the scientific name for sugar? _____. Drain the remaining tap water from each jar and place the eggs in the bottom of the jar. Pour dry sugar over each egg until the egg is covered. Observe the eggs

Egg Volumes After Second Water Treatment

Egg ID	Total Volume	Starting Volume	Egg Volume

closely for a few minutes before storing them overnight. Can you see any action taking place? What do you predict will happen?

11. Before removing your eggs from their jars to find the new volumes, observe your eggs. What does the sugar look like? How about the eggs? Carefully pour off the liquid from the jars and find the new volumes of your eggs. Record your results below.

12. The final treatment for your eggs is to place them back into tap water. Carefully rinse the jar and place the eggs and tap water into each. Make a prediction as to what you think will happen. Then consider this question: do you think it would be possible to place the eggs into a solution where they could return to their original volume? How do you think you could determine the make-up of this solution? To get some ideas, read the discussion below.

Egg Volumes After Sugar Treatment

Egg ID	Total Volume	Starting Volume	Egg Volume

Lab Discussion

It should be very obvious from observations and data you collected in this lab that liquids moved in and out of the egg. The concept to be learned here is that water will move readily across a permeable membrane from where it's "wetter" to where it's "drier." Think back about your egg when you first placed it into pure water. Was it "wetter" outside the egg in the water or "wetter" inside the egg? Hopefully, you'll agree that it was "wetter" outside the egg and therefore the water moved inside the egg where it was "drier." This caused the egg to swell. When you placed the egg into the corn syrup, the water inside the egg moved out. The water moved from a "wetter" location to a "drier" location. And finally, when you placed the egg in the "driest" location (which was in the sugar), almost every bit of the water in the egg moved out of the egg.

Let's compare this to substances in general and then we'll take a look at how these concepts apply to living cells. Pretend you are in a room and someone walks in wearing lots of perfume or cologne. At first you likely weren't aware of the new smell, but within a few minutes you could smell it and then likely there would not be a place anywhere in the room where you couldn't sense the smell. The perfume was most concentrated on the person and then, when given the opportunity, it spread throughout the room. Moving from a location of higher concentration (on the person) to an area of lower concentration (the room) is called diffusion. The perfumed diffused through the room.

In the case of your eggs, the water moved from locations where there was lots of water (high concentration) to where there was little water (lower concentration.) This follows the concept above and, in this case, the movement of water from high concentrations to low concentrations across a permeable membrane is called osmosis. No energy is required for this type of movement of a substance as long as the membrane has large enough openings to let the substance pass. There are two other types of movements that cells are capable of accomplishing: facilitated diffusion and active transport. Both of these types of movements require energy as the cell is working against the natural movement of diffusion. Can you recall where this energy comes from? _____ Hopefully, glucose and ATPs come to mind.

In part 2 of this lab, we'll look at plant cells to see how they respond to varying treatments of liquids as you did with the eggs. Finally, we'll discuss the important ramifications of these concepts to human and animal blood cells.

Friendly Biology

Lesson 7 Lab Activity 2: Osmosis and Veggies (Part 2)

PLEASE READ ALL INSTRUCTIONS BEFORE BEGINNING THIS LAB.

Note: this lab activity is a continuation of the previous lab (Part 1) which begins on page _____. If you haven't completed Part 1, please do so before beginning Part 2.

In Part 1 of this lab, you applied various treatments to de-shelled eggs and observed the results. The egg served as a model of a cell for us. In this lab, we'll apply varying treatments to actual cells- plant cells in this case.

Materials you'll need to gather include:

- 4 disposable cups: 8 ounce cups are fine
- Granulated sugar: 1-2 cups
- Corn syrup: 1-2 cups
- 3 4-inch pieces of fresh, raw celery
- 3 pieces of fresh carrots (baby carrots work great)
- 3 pieces of peeled white potatoes (peel the potato and then cut into French fry-sized pieces)

Procedure:

1. In this lab you will place your vegetable samples into four treatments: air, pure water, corn syrup and granulated sugar. Instead of monitoring changes in volumes with your vegetables like you did with your eggs, in this lab you'll conduct the "bendy" test. The "bendy" test is a relatively non-scientific way of assessing what is happening inside the plant cells which make up your vegetable samples. To perform the "bendy" test, take your vegetable sample following treatment and, while holding it between your thumb and forefinger, see how far you can bend it without breaking it. To record your results, place your vegetable on your data sheet (next page) and draw a line following the inner edge of the vegetable.

2. After you've become proficient at the "bendy" test, you can begin setting up the lab. First, take one of your disposable cups and label it "air." You will eventually place one sample of each of your vegetables in this cup (celery, carrot and potato.) But, don't do that just yet. Set this cup aside.

3. Take the three remaining disposable cups and label one "celery," the second, "carrot," and the third cup, "potato." Set them aside.

4. Next, you will need to perform the pre-treatment "bendy" test on each of your samples. Note that you only have to test one of each kind of vegetable and not all three (unless you want to.) The second sample is to used as a back-up should something go wrong during the treatment process. The third sample will get used in the next step. Record your results of the "bendy" test in the data table provided.

5. Next, take one sample of each of your vegetables and place it in the cup labeled "air." These samples will just be exposed to the air for the next four days. At the end of the lab, you will conduct a final "bendy" test on these samples. Allow them to sit uncovered on your countertop.

6. Take the three other disposable cups you labeled in step 3 and fill each about 3/4 full of pure water. Place the two remaining vegetable samples and place them into their corresponding cup. If possible, make a mark on the side of the cup the record level of water after you put in the sample. Because these samples may tend to spoil if left out at room temperature, cover them with plastic wrap or foil and place them in your refrigerator overnight.

7. The next day, remove the covering and take a look a the level of water in each cup. Did it go up, down or stay the same? _Stayed the same_ What do you think happened? _____. Now, pour off the water from each sample and conduct a "bendy" test on each. Record your results on your data table. You might blot the sample dry with a paper towel to avoid getting water on your data table.

8. The next treatment will be corn syrup. So, empty your cups of any remaining water and place the vegetable samples back into their cups. Pour corn syrup over each sample until they are submerged. If the sample rises to the top, that's okay, just stop pouring. Cover the cups again and place them into your refrigerator.

9. The next day, remove the covers from each of your cups. Recall that when you did this step with your eggs that you found two layers of liquids inside the cups. Do you see something similar here with your vegetables? _Yes_ If so, what do you think has happened? _____ Retrieve each vegetable sample from its cup and conduct a "bendy" test. It's a good idea to rinse each vegetable first

"Bendy" Test Results

Treatment	Celery	Carrot	Potato
Fresh (pre-treatment)			
After Pure Water			
After Corn Syrup			
After Granulated Sugar			
After Exposure to Air			

and blot dry before placing it on your data table to make your measurement.

10. The next treatment for your vegetable samples will be to place them into granulated sugar. Before doing this, make a guess as to what will happen to your vegetable samples. Write it here: _The vegetables will get softer_. Why do you think this will happen? _Because the sugar will make it softer_ Wash each of your cups clean to remove all of the corn syrup and then place your vegetable samples in their corresponding cups. Pour dry granulated sugar over each vegetable sample. Cover each cup and place it into the refrigerator overnight.

11. The next day, remove the cover from each cup. Was your prediction correct? _Yes_ Retrieve each vegetable sample from the cup and gently rinse them under running water. Conduct the "bendy test" and record your results on your data table.

12. Finally, go to each of your samples left exposed to air for the past four days. How do they appear? _moldy_ Conduct a "bendy" test on each sample. Are the results what you expected? _Yes_ Why or why not? _rotten vegatables_

Discussion of Lab Results

Look back at your "bendy" test results now for each of your samples. As you moved from the "fresh, untreated" condition to the "water treatment" condition, was it more difficult to bend the vegetable sample? _easier_ The best explanation for this is that water moved into the plant cells causing them to swell. The water moved from the wetter location (in the water) to the drier location (inside the cell) just like it did with your egg in the earlier lab. This swelling is known a turgor.

In the next two treatments, your vegetable was moved into drier and then even drier conditions. The water inside the plant cells moved out of the cell and into the "dry" corn syrup and then even more so in the granulated sugar. Again, we see water movement (osmosis) from wetter to drier. And then, with the vegetable samples which were allowed to sit in the open air you may have noticed varying results depending upon how dry the air is in your home. The samples may have become very dehydrated or maybe not as much as they did in the sugar treatment.

Think now about foods that are dehydrated such as raisins, figs, prunes, apples or even meats used

to make beef jerky. Why do you suppose dehydration is used? _____.
If you said to preserve them, you are correct. As water is removed from these foods, the environment which is conducive for growth of decaying molds and bacteria is minimized. In other words, these organisms that would like to use the food as a food source find it difficult themselves to carry on life processes in such a dry environment. As long as the food is kept dehydrated, it can be preserved.

Before we end this lab activity, this is a good time to apply what you have just learned about osmosis to the procedures which must be followed very strictly when fluids are administered to persons or animals which have become dehydrated or unable to take in enough liquids on their own. You likely know these procedures as giving fluids IV where IV is the abbreviation for intravenously or within a vein. Because IV fluids are intended to go straight into one's circulatory system in the same location as blood cells, the fluids that are given must match the fluids already found in the blood. Otherwise, if the concentration of "stuff" in the IV fluids does not match that of the fluids of the blood, water will begin to move into or out of the blood cells. Either of these actions are detrimental to the life of the blood cells and can result in major problems for the individual being treated.

To understand this process more clearly, let's first look at what we normally find in the fluid part of the blood. Among many small components dissolved into the fluid or watery portion of the blood (plasma) we find the sodium chloride level to be of great importance. In other words, it is critical that the level of sodium chloride, commonly known as table salt, be maintained in order for blood cells to maintain their correct size and shape (not swollen nor shrunken due to osmosis.) This specific amount is 0.9% and this quantity of salt mixed into pure water creates the IV solution referred to as a normal saline solution. Normal in this terms means that it is the appropriate level of concentration and saline refers to the presence of salt or sodium chloride. When this preparation of IV fluid is prepared, it is said that it is isotonic with blood fluids. The prefix iso– means the same and the root word -tonic refers to concentration. Therefore, the IV fluids match the fluids of the body.

Let's consider two situations now: suppose an IV solution was prepared which was not isotonic. In this case, let's pretend that instead of the solution being mixed to 0.9%, it was mixed at 2.7% which would be three times the normal amount. If this fluid were to be given to a patient, we'd be putting fluids in that are too concentrated. This is like placing your egg or vegetable samples into a solution like the corn syrup where the water content was "drier" than the level of water inside the cells. Because there would be less water outside the cell, by osmosis, water would move out of the blood cells.

Recall the "rule" of osmosis where we said water will move to places where it's drier. In this case, it's "drier" outside the cells. Because water is moving out of the cells of the blood, the cells shrink and become misshapen. Misshapen red blood cells are said to be crenated. Because they are misshapen they tend to be removed from circulation by the spleen. It's the job of the spleen to make sure all red blood cells within circulation are of correct size and shape. Any red blood cells not meeting the correct specifications are removed from circulation. Therefore, if one gives IV fluids that are too highly concentrated in sodium chloride, red blood cells will be removed from circulation and the condition of anemia results. Because it's the job of red blood cells to deliver oxygen to all cells of the body to enable the cell to make energy from glucose (cellular respiration) having anemia results in cells being low on energy. The whole organism is then sluggish and pale and unable to accomplish any of its daily tasks efficiently. Eventually, if the anemia worsens, major problems occur in organs of the body which are being starved for energy and things "go downhill" quickly for the individual.

Fluids which are prepared at too high a concentration are known as being hypertonic. The prefix hyper– means high or greater. Again, the root word—tonic refers to concentration. A hypertonic solution has high concentration of the substance that is dissolved into it and, consequently, less water. You might think of these solutions to be too "strong." An everyday example of a hypertonic solution is something like frozen concentrated juice or canned soup. With these foods, you must add water to get them to the appropriate concentration which would be isotonic with what you normally expect them to be.

Finally, let's look at just the opposite situation. Suppose instead of mixing the IV fluids at 0.9%, you mixed them at 0.3% or one-third the correct amount. The solution would be considered to be at a low concentration and we would refer to it as being hypotonic. The prefix hypo– means low. You may be familiar with the term hypothermia meaning too low of heat or being too cold. Hypotonic solutions are low concentration or a common descriptor is dilute or weak.

Now, how would this affect the cells of one's blood? Think back in this lab where you placed your egg or vegetable samples into very low concentration solutions, a solution that was very weak. This solution was actually the pure water you used! It had a concentration of sugar at the zero level. What happened to your egg? Did it swell up? What about the plant cells? Did they swell also making the vegetable more difficult to bend? The same thing will happen to your blood cells if you give IV fluids which are hypotonic. Water will move into the cells, and again we are mainly interested in the many red blood cells found in the blood, causing them to grow like little balloons within the circulation.

Can you guess which organ might recognize them as being the wrong size? Yes, the spleen will say, "Hey, you red blood cells! You're too big! I'm gonna take you out of circulation." Again, if red blood cells get taken out of circulation in large numbers, anemia will result and, again, we would see symptoms of an individual being pale and unable to accomplish any tasks without a great effort.

So, we can readily see here on the cellular level why it is extremely important to give IV fluids which provide the same level of sodium chloride and other components normally found in the blood. The fluids given should be isotonic with the blood. Hypertonic solutions will cause cells to shrink or crenate. Hypotonic solutions will cause cells to swell. Isotonic solutions will allow cells to remain their normal size and enable them to remain in circulation.

> Hypotonic = low concentrated solution; weak or dilute
> Isotonic solution = same level of concentration as naturally present
> Hypertonic solution = highly concentrated solution; strong.

Diagram of an Animal Cell

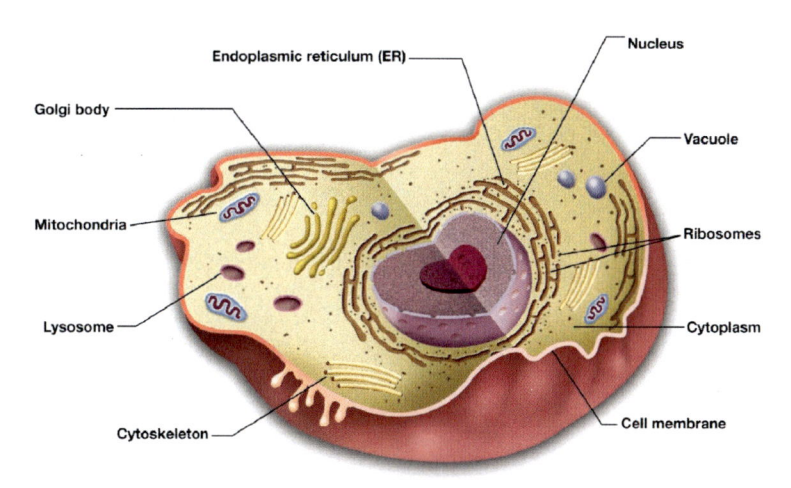

Lesson 8: The Cell (Part 2)

    In the last lesson, we introduced the study of cells, cytology. We learned about three very important cellular organelles: the cell membrane, nucleus and mitochondrion. We learned that the cell membrane functions like a skin for the cell, allowing the passage of certain substnces but not others. We learned that the nucleus functions like the brain of the cell, controlling all of its activities. We likened it to being a cookbook or owner's manual for the cell. We learned that the mitochondrion functions to convert glucose into energy-containing molecules (ATPs) for the cell. We likened it to being a power plant for the cell. In this lesson, we will continue our exploration of organelles by examining the golgi body, vacuole and endoplasmic reticulum.

    The golgi body is very important to cells which create some sort of product. It is the golgi body which packages up the products made by the cell and then moves these products out of the cell. You might think of the golgi body as the post office or UPS service. When viewed beneath a microscope, the golgi body resembles a stack of pancakes.

The job of the golgi body is to package products made by the cell. The golgi body resembles a stack of pancakes.

As the product to be shipped out of the cell gathers within the golgi body, portions at the edge of one of the layers break off to form what is known as a vesicle. The vesicle is like the package ready to leave the post office. It migrates down to one of the "gates" in the cell membrane to be exported from the cell.

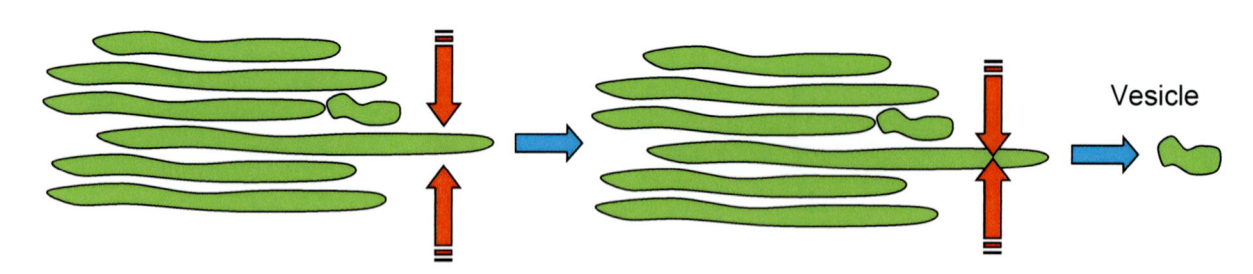

When the cell is ready to package and send out some product, a portion of one layer of the golgi body where the product is located, begins to pinch inward. creating a vesicle.

The next organelle we will learn about is called the vacuole. The job of the vacuole is to store things for the cell. You might think of the vacuole as being a closet or a storage container. Earlier in this lesson when we were discussing the shape of the nucleus of cells, we mentioned how in adipose cells (fat tissue cells) the nucleus is usually found smashed up against the side of the cell membrane. This happens because of the vacuole in the cell. The vacuole is storing so many molecules of lipids that it is bulging at the seams and, in the process, causes all of the other cell organelles to be pressed against the cell membrane. Vacuoles are locations where cells store things. In the case of fat tissue cells, the vacuole is storing fat.

The vacuole functions like a storage container.

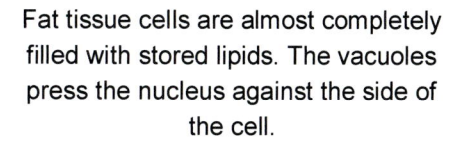

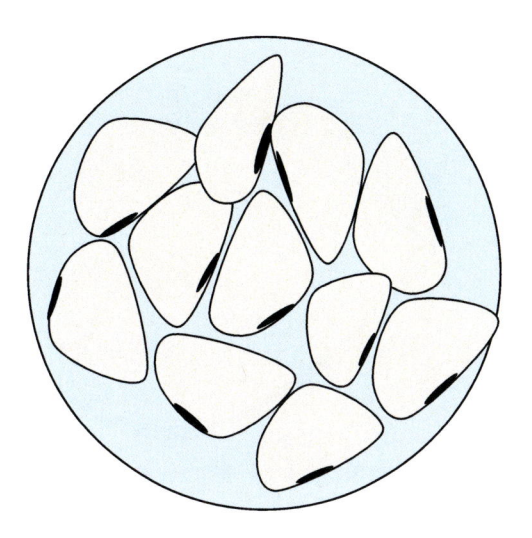

Fat tissue cells are almost completely filled with stored lipids. The vacuoles press the nucleus against the side of the cell.

Another job that the vacuole may have in certain cells is that the vacuole can serve as a means of locomotion. Certain unicellular microorganisms (prefix uni– meaning one, and micro– meaning very small, so these are very small, one-celled creatures) that live in pond or sea water can draw water from their surroundings into their cell vacuole. Once it gets full, the vacuole squeezes, forcing the water out of a small hole causing the creature to scoot forward. It works like an inflated balloon that's released into the air: as the air rushes out of the balloon, the balloon gets propelled forward! The tiny microorganism repeatedly pulls water in and pushes it out which allows it to move about its environment. The vacuole can store substances for the cell or in some cases allow the organism to move about.

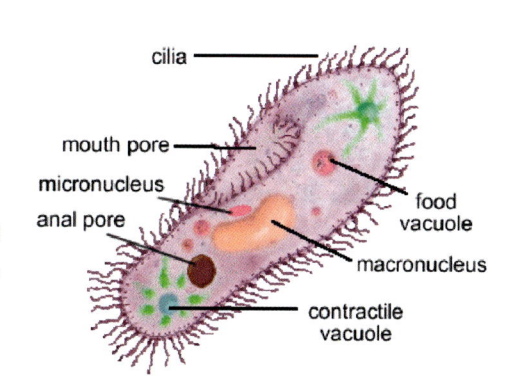

The contractile vacuole of the unicellular Paramecium allows it to move about its watery environment.

The lysosome is another organelle found in cells. Note in its name is the prefix lyse-. To lyse means to cut, as we've learned earlier in our discussion about enzymes. The root word –some (liked we discussed earlier in regard to chromosome) means body. Therefore, the name of this organelle literally is the cutting body. The job of the lysosome is to store enzymes used within the cell. Recall that we learned that enzymes function like scissors to cut things or like matchmakers to speed up reactions between substances in the living creature.

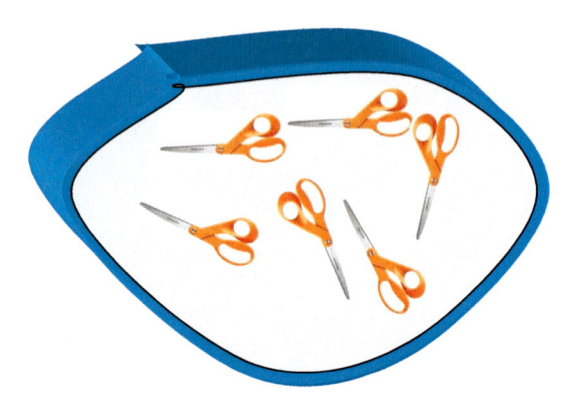

The lysosome is a storage location for enzymes.

Let's pause and review what we've discussed so far about cellular organelles. In the last lesson we learned that:

- The nucleus functions as the control center of the cell;

- Chromosomes carry all of the information regarding the job of the cell and are divided into subunits called genes;

- If an organism has a nuclear membrane, it is classified as a eukaryote and if not, its is classified as a prokaryote;

- The mitochondrion is the organelle responsible for converting glucose into energy-containing molecules known as ATP's.

In this lesson, so far we've learned that:

- The golgi body is the organelle responsible for packaging and shipping of cell products;

- Cells store "stuff" in their vacuoles and some very small creatures use their vacuole for locomotion;

- Enzymes, which can lyse or cut substances, are stored in a cell's lysosome.

Diagram of an Animal Cell

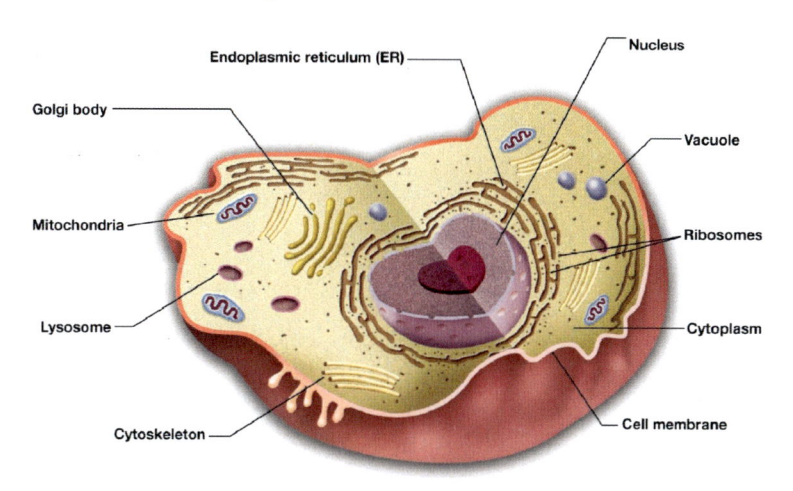

Let's move on now to an organelle which functions like a highway or transportation system throughout the cell. This organelle is known as the endoplasmic reticulum or ER for short. The prefix endo- means within and the root word -plasmic refers to the cytoplasm or jelly-like substance which fills the spaces between the various organelles. So, the first word in the name of this organelle means "within the cytoplasm."

The second word, reticulum, comes from the term reticulated which means lacy-like. So, together, endoplasmic reticulum refers to a lacy network within the cytoplasm of the cell. The endoplasmic reticulum is a web-like system (lacy) of membranes and tubes which function to move various substances around in the cell. You might think of the endoplasmic reticulum as a system of roads, bridges, canals, and tunnels which allow substances to be moved from one place to another within the cell. Secondary jobs of the endoplasmic reticulum include the production of lipids and the processing of carbohydrates.

The endoplasmic reticulum functions like a highway or transportation system throughout the cell.

There are two main types of endoplasmic reticulum: smooth endoplasmic reticulum (SER) and rough endoplasmic reticulum (RER). The difference between the two types has to do with the presence of yet another cell organelle known as the ribosome. Smooth endoplasmic reticulum does not have ribosomes. Rough endoplasmic reticulum has ribosomes.

The endoplasmic reticulum can be smooth endoplasmic reticulum (SER) or rough endoplasmic reticulum (RER)

Ribosomes are organelles where proteins are built within cells. Recall from Lesson 5 where we discussed how proteins (polypeptides) are built from building blocks known as amino acids. You can think of ribosomes as being the little factories where amino acids get assembled into polypeptides.

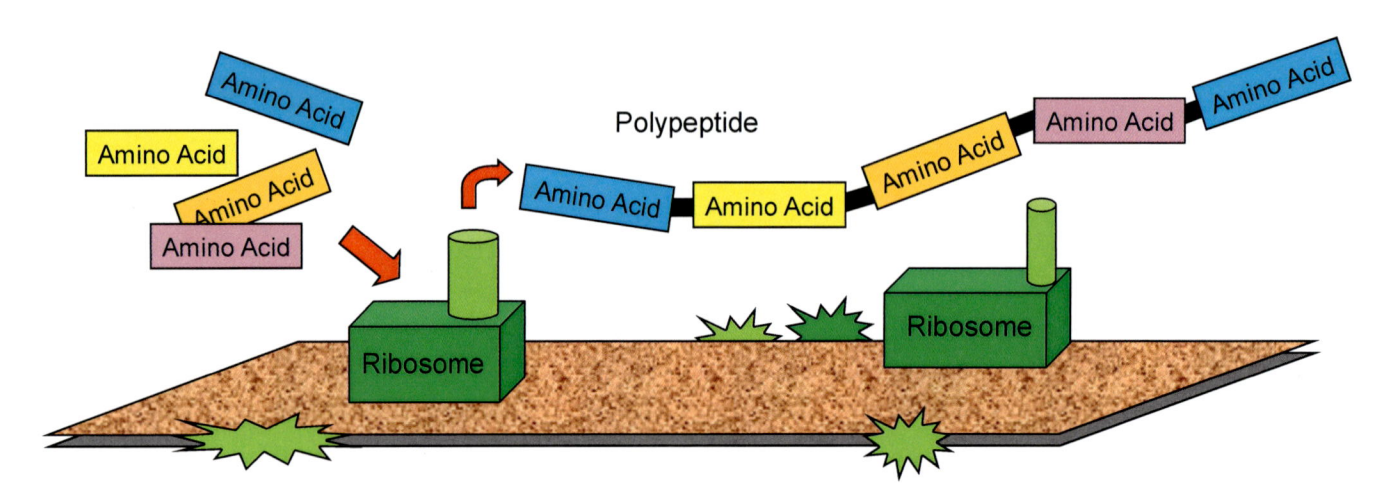

Ribosomes on rough endoplasmic reticulum function like little factories where proteins (polypeptides) are built from individual amino acids.

The endoplasmic reticulum is also thought to function in the folding of proteins. Recall, again, from Lesson 5 where we discussed the folding of proteins. It is here within the endoplasmic reticulum that this correct folding takes place. Folding of proteins is important in that the function of the protein is directly related to its overall shape. In other words, for a protein to function correctly, it must be correctly folded. It is thought that there are four levels of folding organization: the first, second, third and fourth levels. Biologist refer to these levels as primary, secondary, tertiary and quarternary.

To understand these levels of organization, let's do an activity. You'll need a roll of toilet paper and a marker. First, take the roll of toilet paper and unroll approximately six feet of it. Repeat this step to create two six-foot pieces. These strips of toilet paper represent two polypeptide chains.

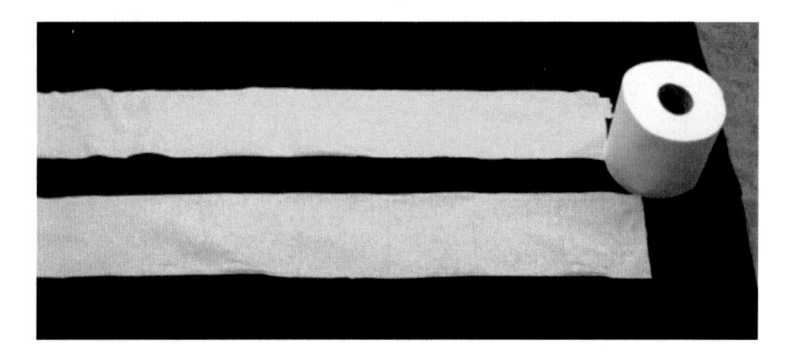

What do you suppose each little square of tissue represents in your polypeptide chains? If you said, amino acids, you're correct! To indicate that each square is an amino acid, using your marker, write the first letter of an amino acid on each square. For example, let's say the first square is the amino acid glutamine. Write the letter "G" on this square. The second might be alanine so write an "A." Choose another amino acid for the third square and write its first letter. Repeat this process for the length of each strip of toilet paper (polypeptide.) Note that if you're using a permanent marker, the ink may bleed through onto the surface below and it may need to be protected.

The two polypeptides you now have represent the first or primary level of protein folding. The primary level of organization "just" consists of the amino acids in long chains.

The secondary level of organization can be of two types. For the first type, take one of your polypeptides and twist it on itself. This shape is known as the alpha helix shape for a polypeptide.

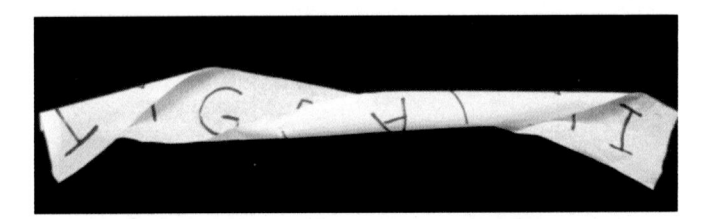

The second type of secondary level of organization is known as the beta-pleated shape. To represent this shape, take your second polypeptide and fold it into pleats as you would fold a piece a paper to create a handheld fan.

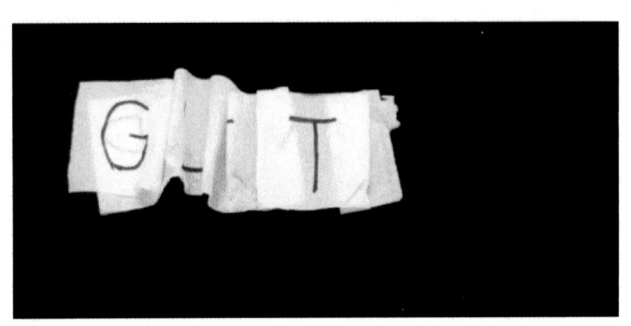

Let's move on to the third or tertiary level of protein folding. At this level of folding each polypeptide chain simply folds upon itself to create a unique overall shape. So, carefully, take your alpha-helix polypeptide and fold it over and over onto itself in a random sort of way. While we are doing this randomly, on the molecular level, this folding is quite organized and specific in order for the protein to function correctly. Do the same for your beta-pleated polypeptide chain.

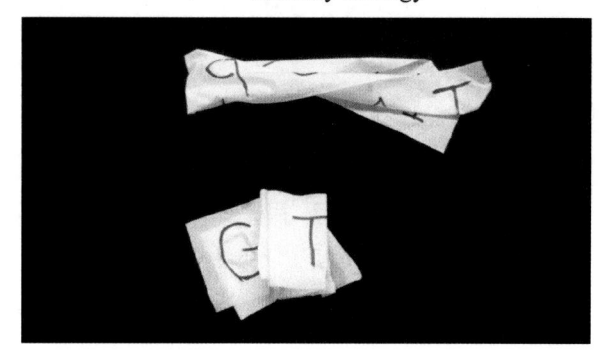

The fourth or quartenary (think quarts or quarters indicating four) degree of folding is where polypeptide chains fold themselves into each other. So, to represent the quartenary level, take your folded polypeptide chains and fold each into the other. Obviously, some of your earlier folding efforts will now be spoiled, but you should have the idea of these four levels of protein folding organization.

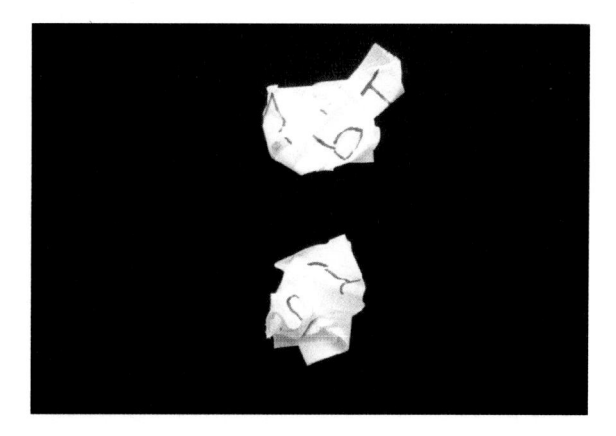

Let's pause and review what we've discussed so far regarding the endoplasmic reticulum. We've learned that:

- The endoplasmic reticulum is a lacy network of membranes and tubes which provides a means of transportation within the cell;

- There are two main types of endoplasmic reticulum: smooth and rough where rough endoplasmic reticulum has ribosomes and smooth ER does not;

- Ribosomes are cell organelles which manufacture polypeptides from amino acids;

- Proteins get folded correctly in the endoplasmic reticulum;

- There are four levels of protein folding: primary, secondary, tertiary and quartenary.

Before we leave our discussion of reticulum in cells, we need to look at one other type which is known as sarcoplasmic reticulum. Sarcoplasmic reticulum is smooth ER which is found primarily in muscle cells. Its name sarco- refers to flesh. Muscles, which make up meat, are referred to as flesh, hence the name sarcoplasmic. The job of sarcoplasmic reticulum is to keep levels of the element calcium at proper levels within the muscle cells which is very important for muscle contraction to take place. If levels of calcium move outside acceptable levels, muscles may become rigid and, in the case of heart muscle, this can result in death. Obviously, the sarcoplasmic reticulum has a very important role in proper muscle function.

The next two cellular organelles that we will discuss are related in that one is made from the other! The first organelle is the microtubule. Microtubules make up the second organelle known as centrioles. Let's begin first with microtubules.

As you would guess by its name, microtubules are very tiny tubes that cells utilize like a skeleton. Microtubules provide strength and stability to a cell and are referred to as the cytoskeleton of the cell. In addition to providing strength, microtubules have the ability to stretch and contract much like our skeletal muscles. This allows for movement within the cell. The microtubule's ability to cause movement is very important when the cell gets the signal to divide. In this case, microtubules are used to make what are known as spindle fibers.

As we mentioned earlier, centrioles are made of microtubules. Centrioles are structures which are also important at times of cell division. We will discuss the rolls of the microtubules and centrioles in greater detail in Lesson 9.

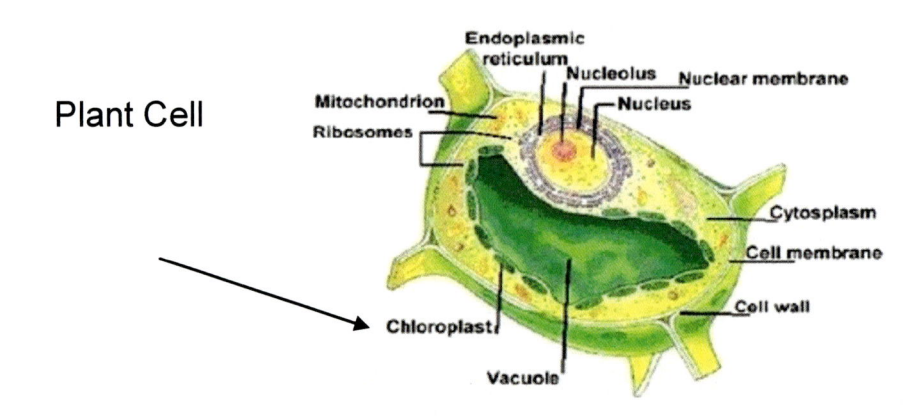

The final cellular organelle we will discuss in this lesson is the chloroplast. This organelle is unique in that for the most part, it is only found in plants. Its name comes from the green-colored pigment found within it which is called chlorophyll. The prefix chloro– refers to the color green. Chlorophyll is the pigment that gives plants their green color. The job of the chloroplast is extremely important in that this is the location where photosynthesis takes place.

The term photosynthesis comes from two words: photo and synthesis. Photo refers to light and synthesis refers to making, so we have a process whereby something is made from light. In this case, plants utilize the energy from sunlight to make glucose molecules.

Recall the formula for glucose: $C_6H_{12}O_6$. In photosynthesis, plants are able to take carbon dioxide (CO_2) and water (H_2O) and create glucose. The glucose is then utilized as the fuel for the plant. This allows plants to be self-sufficient when it comes to acquiring their supply of food. A very "handy" by-product of this reaction is the release of oxygen. Recall that organisms, like you and I, utilize oxygen for converting glucose to ATPs in aerobic respiration. Plants, therefore, are vital for our survival!

But wait. Think back to our discussion of respiration. Can you recall the very important by-product or waste material of respiration? Yes, it was carbon dioxide. So, we as we "burn" our glucose to produce ATPs, we give off the very ingredient plants need to produce their own glucose. Saying that this is an amazing relationship would be a major understatement. Together, we provide what each of us needs! Amazing!

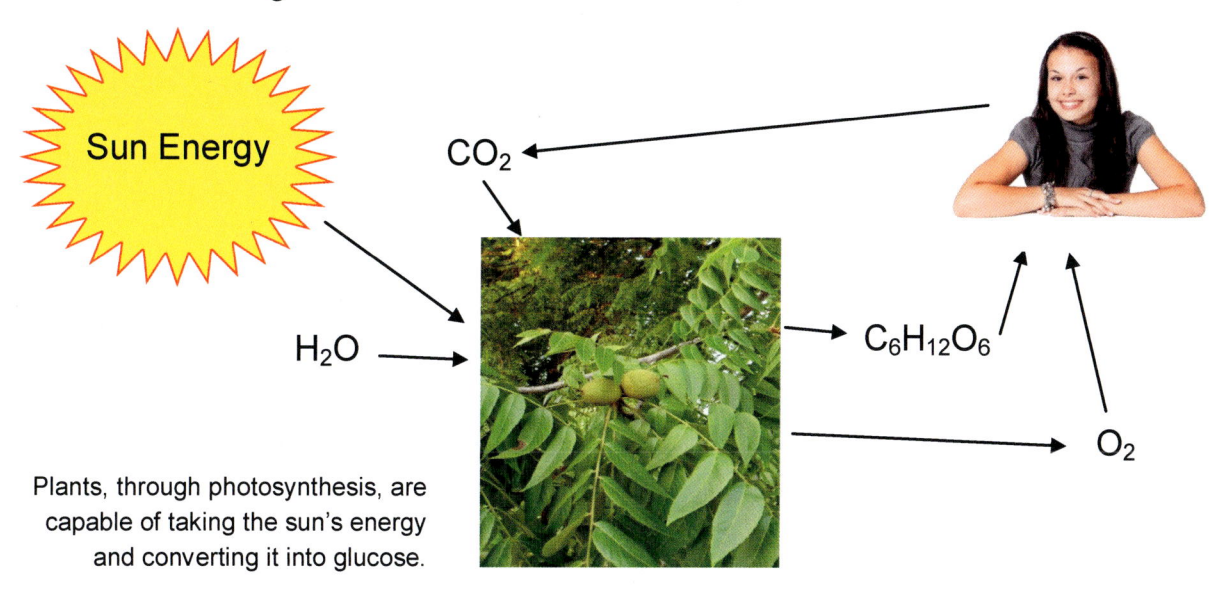

Plants, through photosynthesis, are capable of taking the sun's energy and converting it into glucose.

Let's pause now and review what we've learned about these last three cell organelles. We learned that:

- The sarcoplasmic reticulum is important especially in muscle cells where it regulates calcium levels;

- Microtubules are like tiny tubes and provide strength and integrity to a cell, allowing cells to contract and move and, along with centrioles, playing a vital role in cell division;

- Chloroplasts are specialized organelles found in green plants which carry out photosynthesis;

- Photosynthesis is the process where carbon dioxide, water and sunlight are converted into glucose. The glucose can then be utilized by the plant itself or used as fuel for other living creatures, like you and I. A waste product of photosynthesis is oxygen which is vital for our survival. Our waste product of carbon dioxide from cellular respiration is used by plants in this process. Consequently, we are dependent upon each other.

Friendly Biology

Lesson 8 Activity: Making a Cell Model

PLEASE READ ALL INSTRUCTIONS BEFORE BEGINNING THIS ACTIVITY

The purpose of this activity is to give you practice at recalling and applying the concepts you've learned in this lesson as well as Lesson 7 regarding the organelles found in cells. In this activity you will construct a model of a cell demonstrating all of the parts you have learned. A list can be found later in these instructions. This model may be unlike a cell model you have built or seen before. Instead of the components illustrating the shape, size or dimension of the organelle, the components of this model must demonstrate the function or job of that cellular organelle.

For example, in Lesson 7 you learned about the nucleus of cells. In our discussion we made the analogy that the nucleus of the cell functioned like a brain or control center of the cell. We also said the nucleus could be compared to being a cookbook which contained all of the recipes of the cell. So, for your cell model you might choose something like a television remote or a gaming control or possibly a clay model you make to resemble a brain.

Let's look at another example. Consider the mitochondrion. What was the main function of the mitochondrion in a cell? _____. Think about what you might use to represent this job. Could something like a battery represent a mitochondrion?

Now, in addition to using all sorts of items to represent the function of each organelle, your cell model needs to be three-dimensional. In other words, your cell model needs to have depth and not just be cut-out items pasted to a poster board. Consider starting with a cardboard box which itself could function as the cell wall. Then on the interior of the box, you will place an object which could represent the cell membrane. Because the cell membrane works like a gate or door to allow certain things to pass and others not, consider using a toy fence or doorway made from building blocks. Continue by adding items to your box that represent all of the cell organelles presented in these last two lessons. You can attach these objects to the wall of the box or suspend them from the top using string or yarn.

To assist you in creating your model, look at the chart provided below. The first column lists all of the cell components you should include in your model. The second column asks you to identify the primary function or job of that cell component. In the third column, you are given the opportunity to brainstorm ideas of items you might use to represent that cell organelle in your model.

Cell Organelle	Function of Job	Item Which Might be Used for this Organelle
Cell Wall		
Cell Membrane		
Nucleus		
Chromosomes		
Mitochondrion		
Lysosome		
Vacuole		
Smooth ER		
Rough ER		
Ribosomes		
Golgi Body		
Microtubules		
Centriole		
Chloroplast		

Once you've completed your model, present it to your parent/teacher and your family or class. Make sure to identify each component and tell why you chose the objects you did to represent each cell organelle.

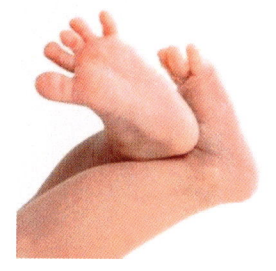

Lesson 9: Cell Division

In Lesson 1 we explored five features that all living things share. We said that living things move, living things reproduce, living things require a source for energy, living things grow and develop and living things respond to their environment. In Lesson 2, you were introduced to some basic chemistry which laid the foundation for Lessons 3-5 where you learned about carbohydrates, lipids and proteins. In Lesson 6, the importance of understanding pH was introduced and then in Lessons 7 and 8 you learned about the names and functions of organelles within cells. In this lesson, we'll return to one of the first five features of living things that we examined in Lesson 1: living things grow and develop. We'll begin by looking at how living things grow at the cellular level.

Let's think now what it means when we say a living creature grows. Think about a baby. How do you know when growth has taken place in a baby? Obviously, the baby is larger and it needs larger-sized diapers and clothing. Because we now know that living creatures are made up of cells, it makes sense to say that living things increase in size by increasing the number of cells from which they are made. In other words, living creatures grow because the number of cells from which they are made increases. In this lesson we'll explore how cells increase in number, a process known as mitosis.

First, let's consider the life cycle of the cell. Cells in living creatures have specialized jobs that they perform on a daily basis. For example, the job of your skin cells is to protect the parts of your body found within. The job of red blood cells in your blood is to carry oxygen from your lungs to the cells who need it to create energy. So, for most of a cell's life, it is going about this assigned job. This phase of a cell's life is called interphase. The prefix inter– refers to in between and the root word –phase refers to time. Interphase is the time between the times when the cell is dividing to make new cells. It is the time when the cell is doing its own, assigned job.

> Inter = between
>
> Phase = period of time
>
> Interphase = time between when cell is doing its "normal" job.

At certain times during this period of interphase, a cell may receive a "notice" that the time has come for it to divide. This division will result in two "new" cells which then return back to interphase and continue completing the job of the original cell. The original cell is called the parent cell. The two cells which result from the division are called daughter cells. In the case of animals and humans, the "notice" that arrives to stimulate the cell to divide is in the form of a hormone known appropriately as growth hormone. Stimulation for cell division in plants is also controlled by various hormones.

One parent cell undergoes mitosis to result in two daughter cells.

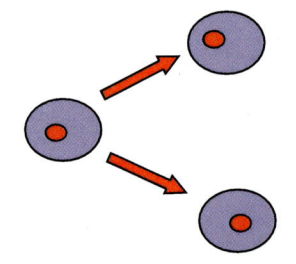

Doing "normal" job.

Interphase → **Mitosis**

Cell division taking place.

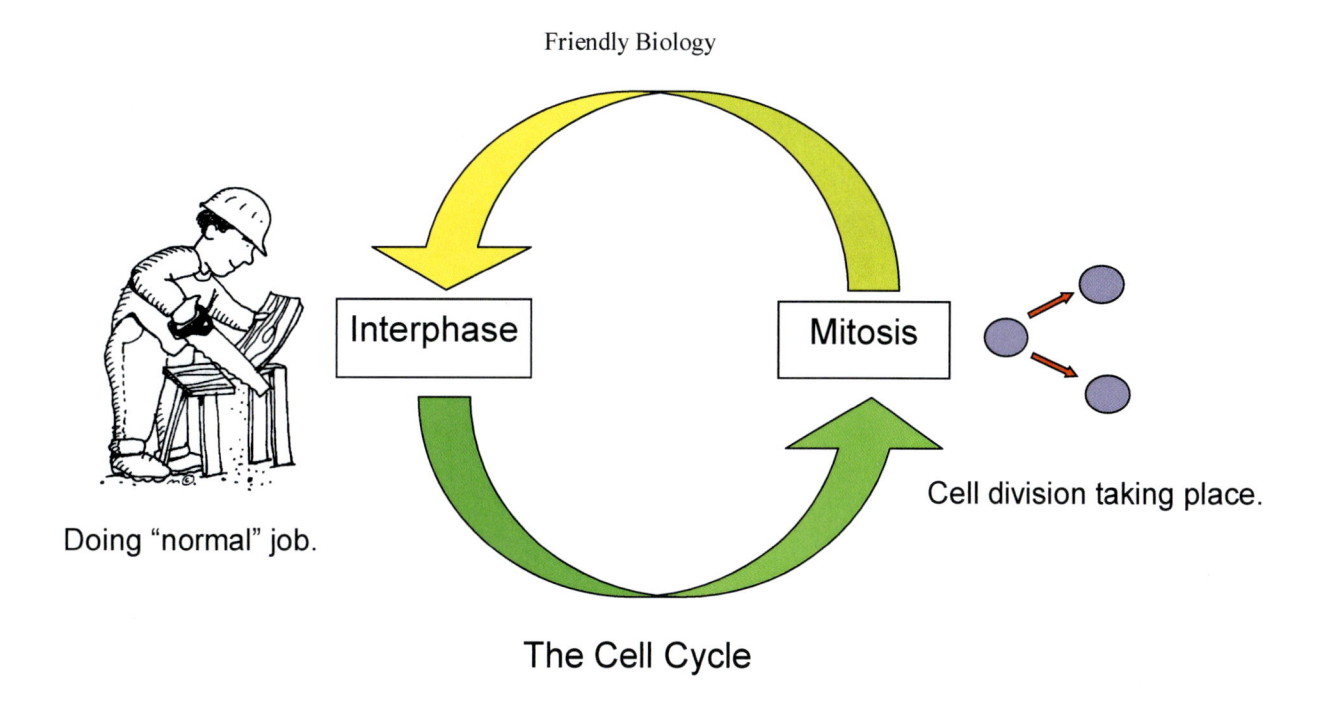

The Cell Cycle

    Hormones are chemicals created in one location of the body, usually by a gland, which have their effect in some other location of the body. In the case of humans and most animals, growth hormone is made by a small little gland which sits beneath the brain called the pituitary gland. Growth hormone moves out of the cells of the pituitary gland and into the blood stream where it is carried all about the body. Cells which respond to growth hormone respond by dividing to make more of their kind. So, our bones lengthen and broaden. Our muscles get bulkier and stronger. Our internal organs grow accordingly. Growth hormone is the signal for cells to move from interphase to the phase of cell division, as we stated above, known as mitosis.

The pituitary gland produces growth hormone which stimulates cells to divide.

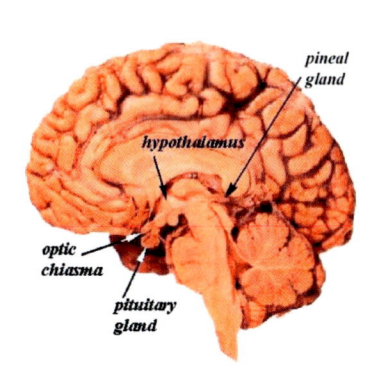

pineal gland
hypothalamus
optic chiasma
pituitary gland

Let's look more closely now at mitosis. If you think for a moment about one cell dividing to make two, it would make sense that both daughter cells would need a copy of the cell's "owner's manual" or nucleus, right? Recall from Lesson 7 where we made the analogy that the chromosomes and genes of cells were like the chapters and recipes in a cookbook. In order for each cell to perform its assigned duty during interphase, each daughter cell definitely needs to know how to go about performing that task. Consequently, creating a copy of the parent cell's nuclear contents (recipes) is one of the first steps in mitosis. You might liken this process as when all of grandma's recipes in the cookbook are passed down from the present to the next generation of cooks. A copy of each recipe has to be made first! But before these copies can be made, there is some preparation of the recipes that has to take place.

While the cell is in interphase, doing its regular job, the contents of the nucleus, primarily the chromosomes, are all spread out or uncoiled in order for cell activities to take place very efficiently. You might think of the chromosomes spreading out as like taking all the pages of recipes from the cookbook chapters and laying them all out to be readily seen by everyone interested in knowing what to do. The material which makes up chromosomes is known as chromatin. So, to be more precise, we can say that the chromosomes uncoil into long strands of chromatin. When the hormone "notice" arrives at the cell to begin dividing, the chromatin begins to re-coil back into the shape of the chromosomes. The pages of recipes of the cookbook are being reassembled back into their binder. While the chromatin is uncoiled it is not readily visible with light microscopes. However, it's when it gets all re-coiled that it becomes visible as chromosomes.

This early stage of cell division (mitosis) is known as prophase. The prefix pro– means first or beginning. Again, the root word -phase indicates time so we have the event or time happening first. The "re-appearing" of the chromosomes is a visible sign that mitosis is about to begin. The nuclear membrane which surrounds the nucleus begins to get dismantled at this stage, also.

> Pro = early, initial, beginning or before
> Phase = period of time
> Prophase = beginning phase of mitosis

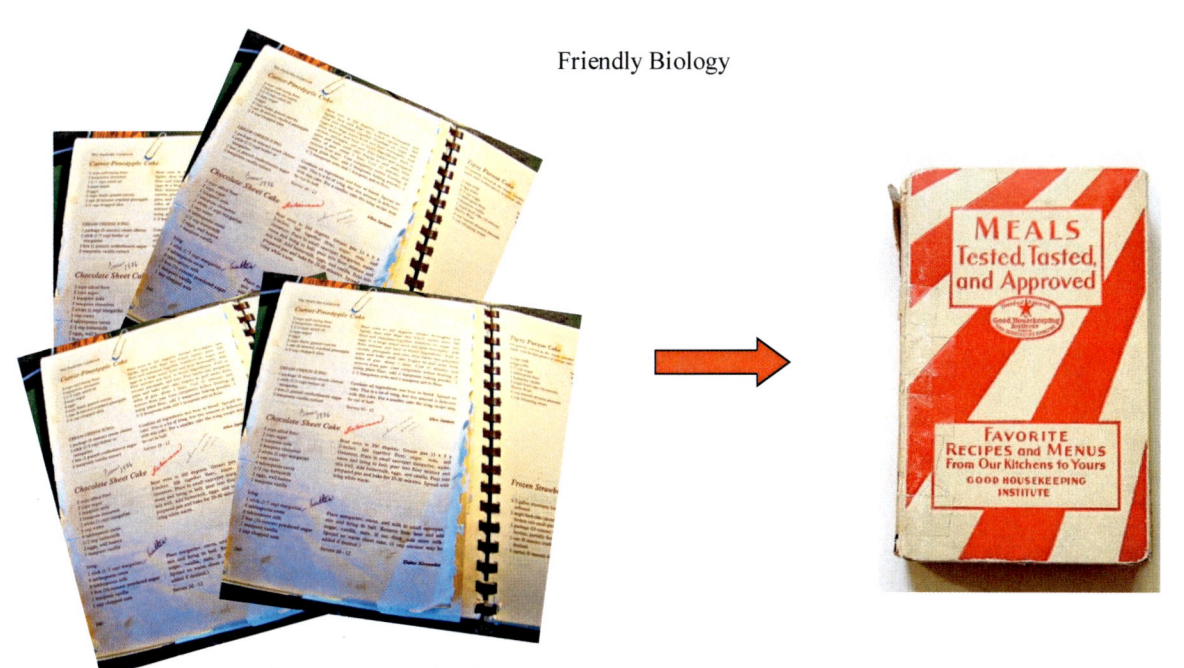

In the first stage of mitosis known as prophase, chromatin condenses back into visible chromosomes. The pages of recipes are placed back into the cookbook.

Let's pause and review what we've discussed so far regarding the process of growth in living organisms. We've said that living things, for the most part, grow by increasing the number cells present. Unicellular organisms grow also, but only increase in size, rather than number. For most of a cell's life cycle, cells are going about their assigned tasks which is known as interphase. A "notice" or signal (hormone) arrives at the cell which prompts it to begin mitosis, the process of cell division. The first stage of mitosis is called prophase and the re-appearance of the chromosomes is a visible sign that mitosis is beginning. Let's continue.

A second important event that also takes place during prophase is that copies of the chromosomes are created. You might think of this event as when each chromosome gets taken to a copy machine and a duplicate set is created. With our cookbook analogy, this is the step where a copy of each chapter (chromosome) holding the many recipes is made. Each tiny gene (recipe) must get copied exactly correctly for the new daughter cells to function properly.

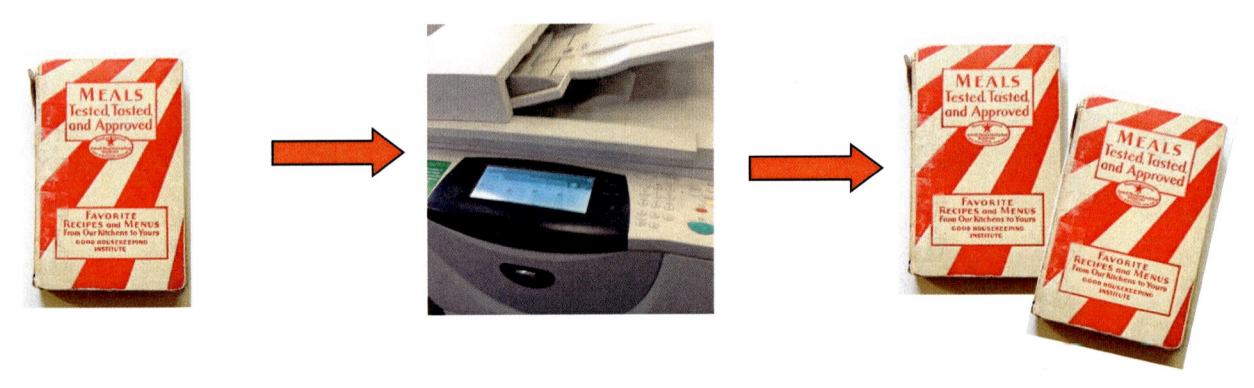

A second important event in prophase is the fact that the chromosomes make copies of themselves. One set is now available for each daughter cell.

Sometimes, however, the copying process doesn't work quite correctly and the new cells have sets of instructions which aren't quite right. This change in information is called a cell or gene mutation. Sometimes cell mutations aren't a big problem for the living creature, however, sometimes it is these mutations which result in cells doing things that they aren't suppose to be doing. For example, sometimes cell mutations cause cells to grow too large or divide too rapidly. These cells may be identified as being cancer cells. These cells usually are unable to perform their jobs correctly or may just outnumber normal cells around them which results in serious problems for the "owner" of the cookbook.

For now, let's go back to our discussion of normal cell division. Let's recap what we've learned so far about prophase. We've said that in prophase, the chromosomes become visible due to the coiling-up of chromatin and they create copies of themselves.

The next phase of mitosis is known as metaphase. The prefix meta– refers to adjacent. During metaphase tiny little threads of microtubules form what are known as the spindle fibers. The spindle fibers connect the now visible chromosomes to opposite ends or poles of the cell. The connecting point on the chromosome is called the centromere. In animals and the simpler plants, the connecting point at the poles of the cell is the centriole. It is during the connecting process that the chromosomes become aligned side-by-side (adjacent to one another) along an imaginary plate in the central area of the cell. You might think of this imaginary plate as being like the equator around the earth. The chromosomes line up along this plate and are readily identifiable during metaphase.

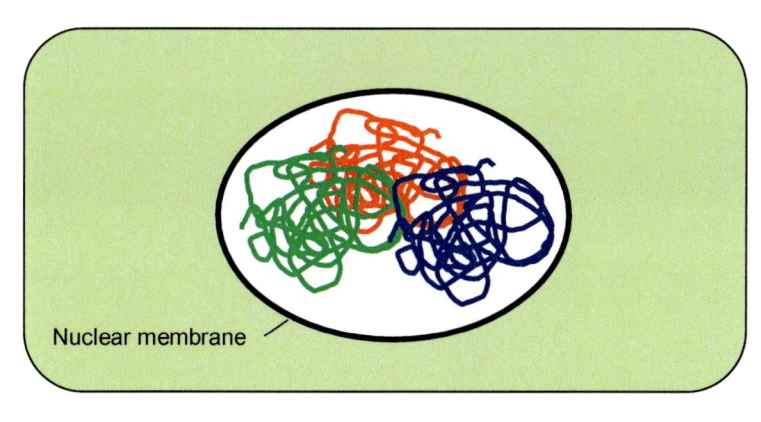

During **interphase,** the chromosomes are uncoiled strands of chromatin being used by the cell for instructions. The nuclear membrane is intact.

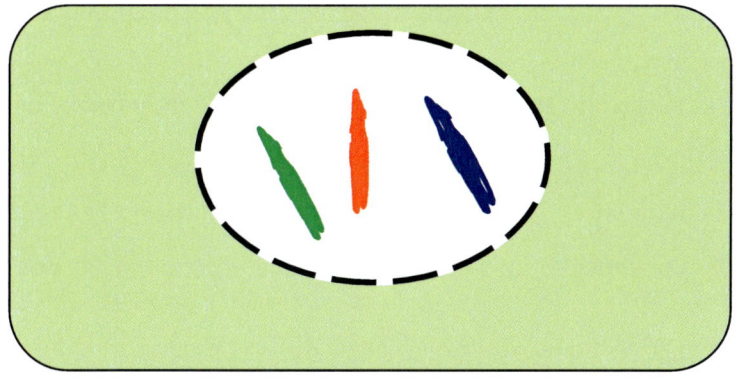

Nuclear membrane

In **early prophase**, the strands of chromatin coil up to become visible chromosomes. The nuclear membrane begins to disintegrate.

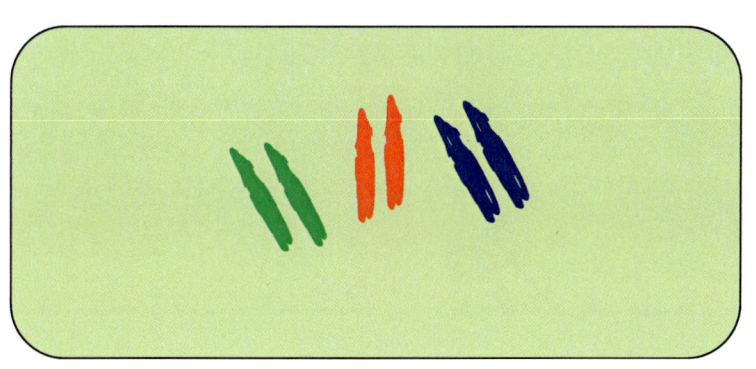

Later, in prophase, the chromosomes make copies of themselves. The nuclear membrane completely disintegrates.

Meta = adjacent to or beside one another

Phase = period of time

Metaphase = time of mitosis when chromosomes align themselves beside each other.

During **metaphase**, the chromosomes become aligned along a central plate within the cell. Spindle fibers attach one member of each pair of chromosomes to centrioles arranged at opposite poles of the cell.

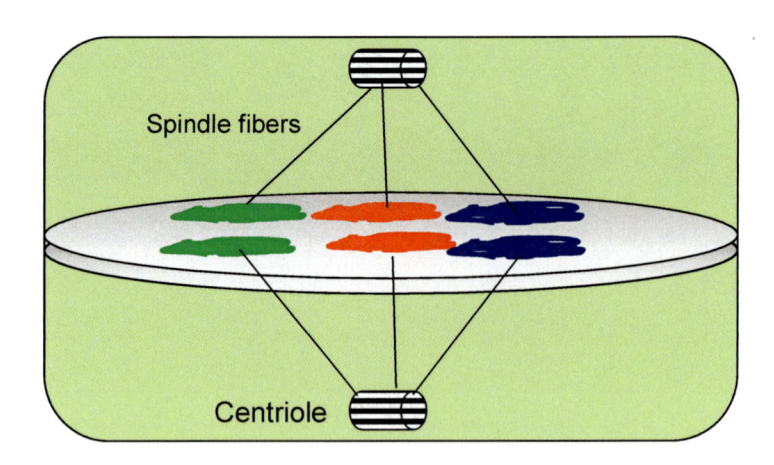

As mitosis continues, the spindle fibers begin to contract which causes one member of each copied chromosome to move towards opposite poles of the cell. This movement towards the poles signifies that the third stage of mitosis is underway. This phase is known as anaphase. The prefix ana– refers to opposite or against which describes the movement of chromosomes toward opposite poles of the cell.

> Ana = opposite, opposing or against
>
> Phase = period of time
>
> Anaphase = time of mitosis when chromosomes move to opposite poles of the cell.

In **anaphase**, the spindle fibers contract pulling the pairs of chromosomes towards opposite poles of the cell.

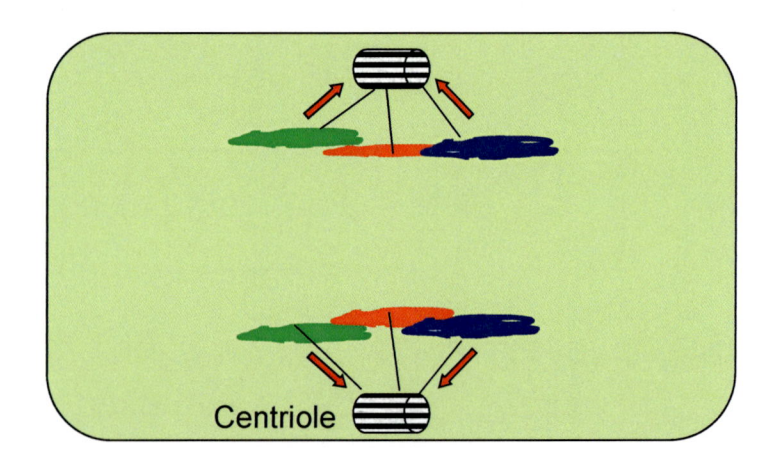

Eventually, each copy of each chromosome makes its way to the opposite pole of the cell. At this point, the chromosomes begin to uncoil once again and a nuclear membrane begins to reappear around each set. Each "end" of the cell now has a complete set of chromosomes ready to begin functioning in the daughter cell. This phase is known a telophase. The prefix telo– refers to final or ending so this phase marks the ending phase of mitosis.

In addition to the chromosomes uncoiling and the nuclear membrane reappearing, the cell membrane begins to "pinch" inward along the equator of the cell eventually allowing for the parent cell to split into two daughter cells. The process of pinching inward is known as cytokinesis. The prefix cyto– refers to cells and the root word kinesis refers to movement or action. After cytokinesis is complete, the two resulting daughter cells return to interphase and begin performing their regular jobs within the cells. In most cells, the organelles found within the parent cell get separated into two sets. However, we'll find out in a later lesson about one situation where this is not the case.

> Telo = ending or final
>
> Phase = period of time
>
> Telophase = final phase of mitosis where parent cell splits into two daughter cells.

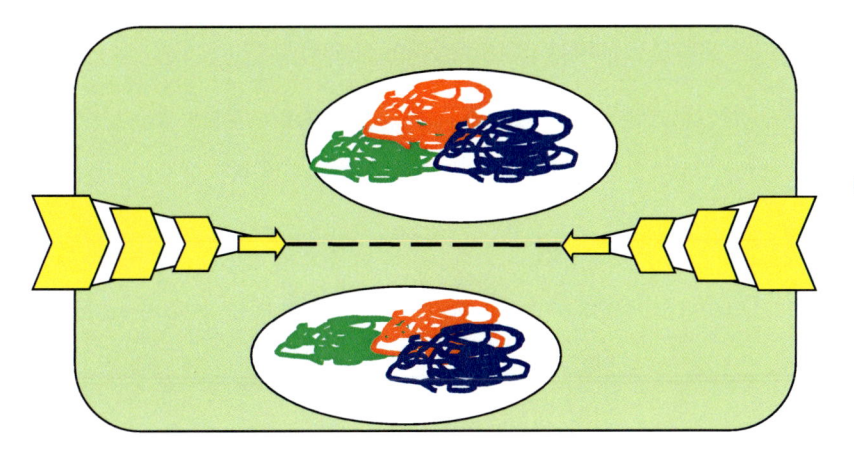

In **telophase,** the chromosomes uncoil themselves once again. A new nuclear membrane forms around each collection of chromatin. The cell membrane begins to pinch inward.

Eventually, the cell membrane pinches all the way across the cell and two daughter cells result. This process is known as **cytokinesis**. Mitosis is complete and the cell returns to interphase.

Cyto = cell

Kinesis = movement or action

Cytokinesis = movement of cell membrane inward.

Let's stop and review now what we've learned about how living things grow by increasing the number of cells from which they are made. We learned that:

- Interphase is the period of time when the cell is going about its regular job when a signal, usually in the form of a hormone, initiates mitosis to begin;

- Mitosis can be divided into four main stages based upon what we can observe happening within the cell. The four stages are phophase, metaphase, anaphase and telophase;

- Prophase is the first stage of mitosis when the chromatin condenses to form visible chromosomes, the chromosomes make copies of themselves and the nuclear membrane disappears;

- Metaphase is the second step in mitosis. During metaphase, the pairs of chromosomes align themselves along the equator of the cell allowing them to readily be identified and studied;

- Anaphase follows metaphase when the chromosomes get pulled towards opposite poles of the cell;

- Telophase is the final stage of mitosis. During telophase the chromosomes begin to uncoil and the nuclear membrane reappears. The cell membrane begins to pinch inward in a process known as cytokinesis.

Friendly Biology

Lesson 9 Lab Activity: Creating a Mitosis Model Display

PLEASE READ ALL INSTRUCTIONS BEFORE CONDUCTING THIS ACTIVITY.

The purpose of this activity is to create a series of models of a cell undergoing mitosis. To create these models, you will use a batch of homemade playdough. The recipe for the playdough is found below. If you have another modeling clay or compound such as Sculpey^R, you are welcome to use your choice. Following is the recipe for the playdough.

Ingredients you'll need:

- 2 cups of white flour
- 1/2 teaspoon salt
- 4 Tablespoons of Cream of Tartar
- 2 Tablespoons of oil—any vegetable oil will work
- 2 cups of water
- Food coloring or model paints
- Medium saucepan and wooden spoon.

Procedure to prepare playdough:

1. Measure all ingredients (except food coloring or paints) into the medium saucepan.

2. Heat over medium heat stirring constantly. The mixture will be very lumpy at first, but as it cooks (3-5 minutes) it will form a ball in the pan.

3. When the ball forms, remove the pan from the heat and allow the dough to cool for 5-10 minutes.

4. Carefully remove the dough from the pan and, using extra flour, knead it to form a soft, pliable dough.

Once your playdough has been prepared, you can begin creating your models. The idea here is to create a model of a cell in each of the stages of a cell's life cycle: interphase, prophase, metaphase, anaphase and telophase. You may choose to follow the diagrams in the text or use another resource book or online website to view cells in various stages. Create a plan to show these stages of mitosis. You might choose to flatten a ball of dough into a disk shape to first create the cell shape. A narrow ribbon or "snake" of dough can be used to create the cell membrane. A small, flattened ball can represent the nucleus of the cell during interphase. You can then repeat these steps as you prepare the next part of the display which will be prophase.

If you would like to color the playdough, you can do so by taking portions of the dough and adding food coloring. Form a depression in the dough and add some drops of the coloring. Knead the dough to evenly distribute the color. You may also choose not to use food coloring, but instead use model paints to color the models later in the process. Note that food coloring can stain your hands and clothing. Consider working over newspaper to make cleanup easier.

When you have models created for all stages of mitosis, allow them to air dry. If you've chosen to use another modeling compound like Sculpey^R, bake it according to package directions. Once your models are dry or cooled, create labels to apply to your model to show the important structures. Finally, share your model display with your family or class members.

Lesson 10: Chromosome Duplication

In Lesson 9 we explored the process of cell division known as mitosis. In that discussion, we introduced the idea that in order for each of the two daughter cells to have their own complete set of chromosomes, duplication or copy-making of the chromosomes had to take place. Recall that this step took place during prophase. Let's look more closely now at how this copy-making takes place. In order to do so, we must first look more closely at how chromosomes are thought to be made.

If we return to our analogy of the nucleus of a cell as being like a cookbook, recall that we said the chromosomes were analogous to chapters in the cookbook. We then said that genes could be considered analogous to the recipes within the chapters of the cookbook. In other words, genes provide the vital information for completion of a particular aspect or feature of the living creature. For example, we said that specific genes exist for eye or hair color, skin color, nose shape, ear size and on and on. Information for ALL parts of the organism (how these parts appear and how they function) are found in the genes of that organism! Let's go one step further now and explore what makes up genes.

Nucleus	Chromosomes	Genes
Cookbook	Chapters	Recipes within each Chapter

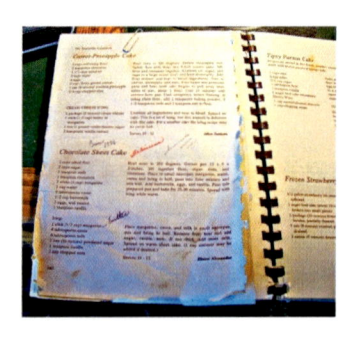

Think about our cookbook analogy once again. If we said that genes are like individual recipes within chromosome "chapters," what, exactly, *is* a recipe? Recipes consist of lists of ingredients and sequences of sentences and phrases which, when completed in the correct sequence, make sense to the user. As long as the sentences and phrases flow in a comprehensible order, the recipe is useful to the person using it. Likewise, genes are composed of sequences of "sentences" which are chains of molecules known as DNA.

Recipe ⟷ DNA molecule.

The letters DNA are the abbreviation for the term deoxyribonucleic acid (dee-ox-ee-rye-bow-new-clay-ic acid). The name deoxyribonucleic acid refers to the chemical components from which it is formed. The ribo- portion refers to the carbohydrate or sugar compound known as ribose found in DNA. This carbohydrate has lost some of its oxygen atoms which is indicated by the prefix deoxy-, meaning without oxygen. So, the first part of the name deoxyribo– refers to the carbohydrate portion of the DNA molecule which is lacking in some oxygen atoms.

The -nucleic acid part of the name refers to another set of molecular compounds known as the bases. Before we look further at the bases, let's take a moment and first examine the structure of DNA. In 1953, biochemists named James Watson and Francis Crick discovered DNA to be built very similar in shape to a ladder. A ladder has two vertical components known as the rails and then a series of horizontal steps called the rungs. In the DNA molecule, the "rail" portions are made up of alternating ribose molecules and phosphate groups. One finds a ribose molecule attached to a phosphate group attached to a ribose attached to a phosphate, etc., going "up and down" the sides of the DNA molecule.

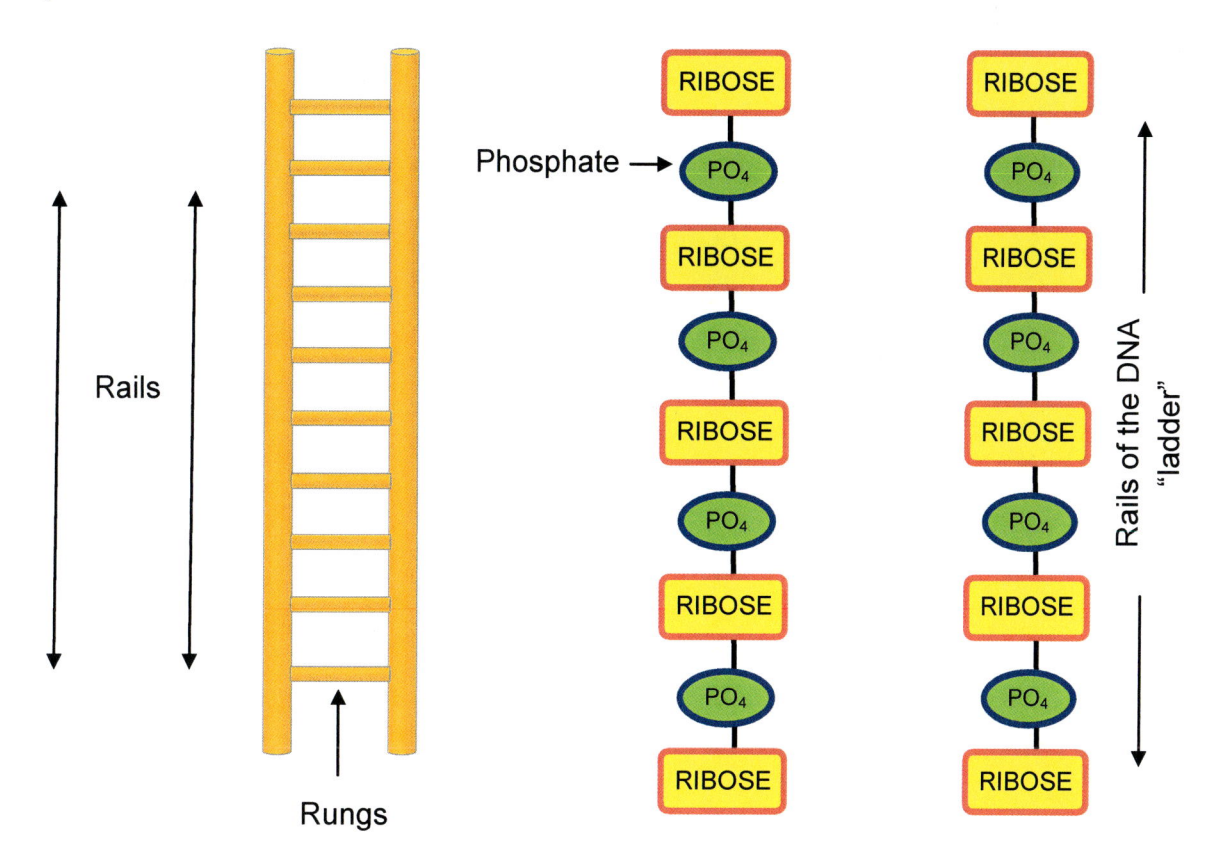

Friendly Biology

The "rung" portions of the DNA "ladder" are attached to the ribose sugar molecules and function to connect the two rails.. These rungs are made up of the bases. There are 4 of these bases and they are divided into two groups: the purines and the pyrimidines. There are two purines and two pyrimidines. The bases which are purines are named adenine and guanine. The bases which are pyrimidines are named thymine and cytosine.

Four bases make up the rungs of the DNA "ladder."

Purines: Adenine and Guanine
Pyrimidines: Thymine and Cytosine

A pair of bases create one rung of the ladder: one purine and one pyrimidine. You might think of each rung being one-half purine and one-half pyrimidine. Recall from above we said that there are two purine bases and two pyrimidine bases. When creating the "rungs" of the DNA "ladder," the bases pair up in a specific manner. They always pair like this: adenine makes a rung with thymine while cytosine forms a rung with guanine. So, we have rungs made up of either an adenine/thymine pair or a cytosine/guanine pair.

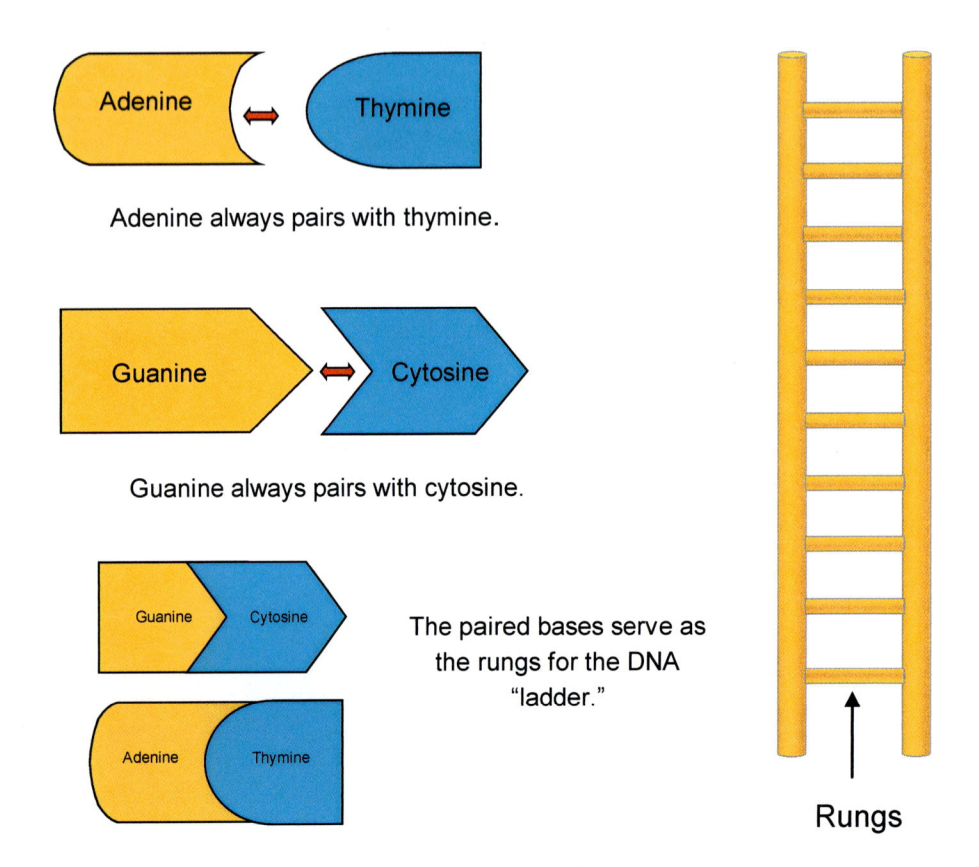

Adenine ⇔ Thymine

Adenine always pairs with thymine.

Guanine ⇔ Cytosine

Guanine always pairs with cytosine.

Guanine Cytosine

Adenine Thymine

The paired bases serve as the rungs for the DNA "ladder."

Rungs

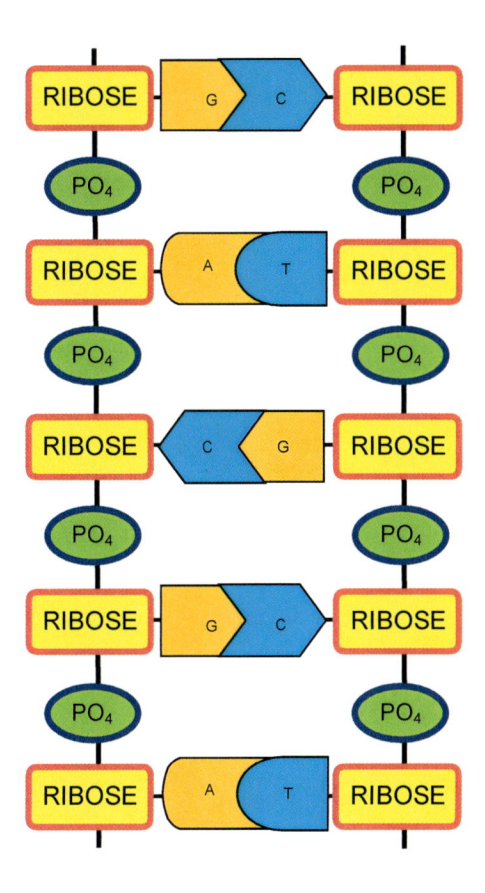

In addition to realizing that DNA was very similar to a ladder, Watson and Crick also discovered that DNA was twisted. This twist is called a double-helix.

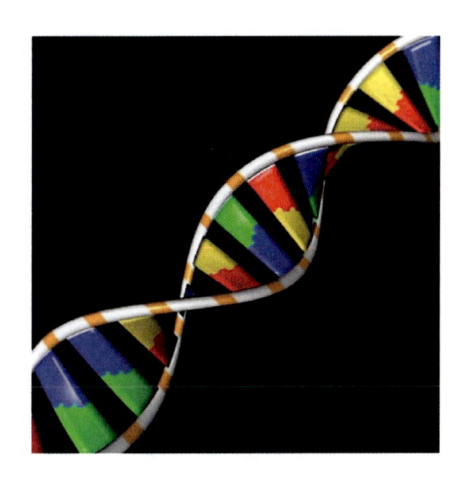

Double-helix or twisted ladder shape of the DNA molecule.

The trio made by one ribose sugar molecule, one phosphate group and one base make up what is known as one nucleotide. Like the sentences and phrases which make up the recipe in the cookbook, it is the sequence of these nucleotides which creates the gene's code.

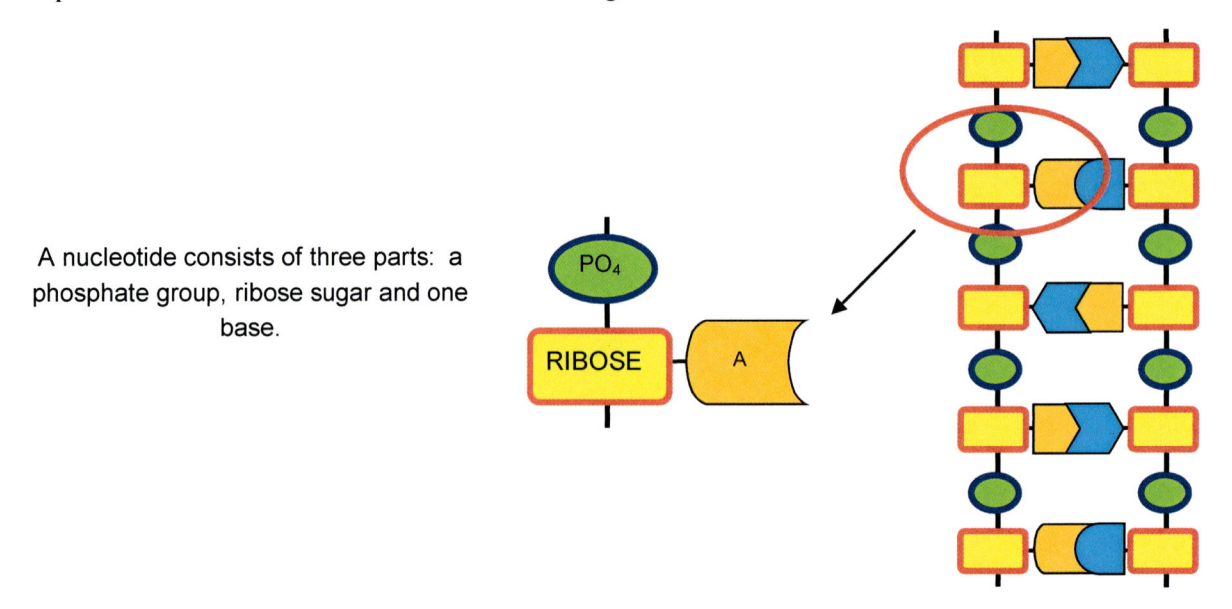

A nucleotide consists of three parts: a phosphate group, ribose sugar and one base.

To better understand what this means, let's return to our earlier discussion of how chromosomes create copies of themselves during prophase. Recall that the chromosomes are made up of very long strands of chromatin which is actually DNA tightly coiled together. If one examines the bonds which hold the molecules of DNA, themselves, together, we find that the bonds between the sugars and phosphate groups of the "rails" are very strong. These bonds are covalent bonds where the electrons forming the bonds are shared back and forth between adjacent molecules (Lesson 2). However, the bonds between the bases which make up the rungs of the DNA ladder are much weaker bonds. These bonds are hydrogen bonds and are readily split apart.

When the time comes for the chromosomes to duplicate themselves, enzymes move in and begin breaking the weak hydrogen bonds on the rungs of the DNA ladder. Recall that enzymes work to cut or lyse molecules, in this case a strand of DNA. You might imagine this as like "unzipping" the DNA strand into a right and left hand string of nucleotides. As soon as the unzipping takes place, other enzymes work to bring in floating nucleotides present in the area which follow the base pairing rule we discussed above. Recall that purine bases always bond with pyrimidine bases: adenine with thymine and cytosine with guanine. Eventually, enough floating nucleotides fill up all available "open" spaces and two "new" strands of DNA are formed. Because of the specific manner in which

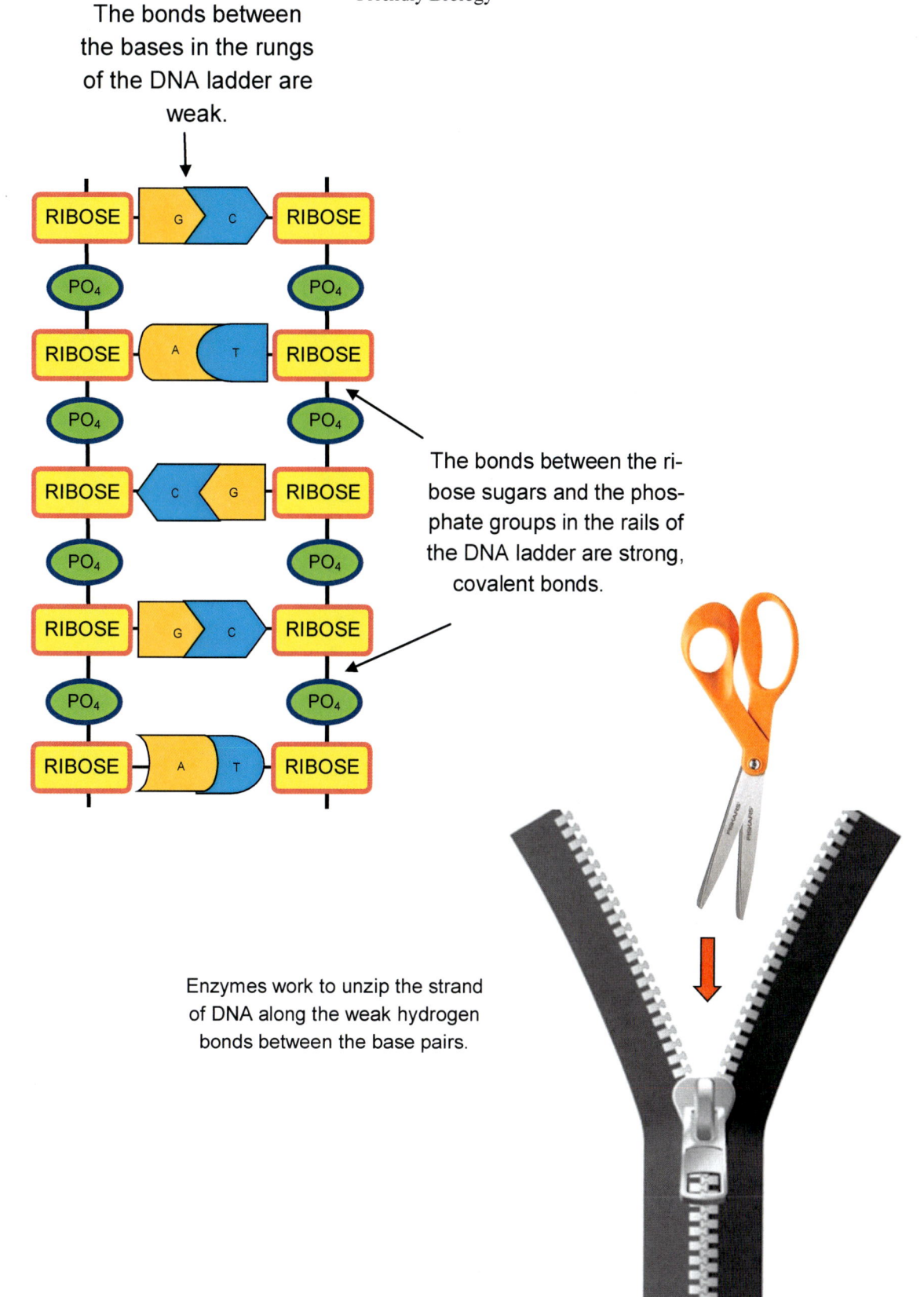

The bonds between the bases in the rungs of the DNA ladder are weak.

The bonds between the ribose sugars and the phosphate groups in the rails of the DNA ladder are strong, covalent bonds.

Enzymes work to unzip the strand of DNA along the weak hydrogen bonds between the base pairs.

the bases link together, the two "new" strands are identical to the original strand of the parent cell. Look at the diagrams here to see how two identical strands of DNA can form from one parent strand.

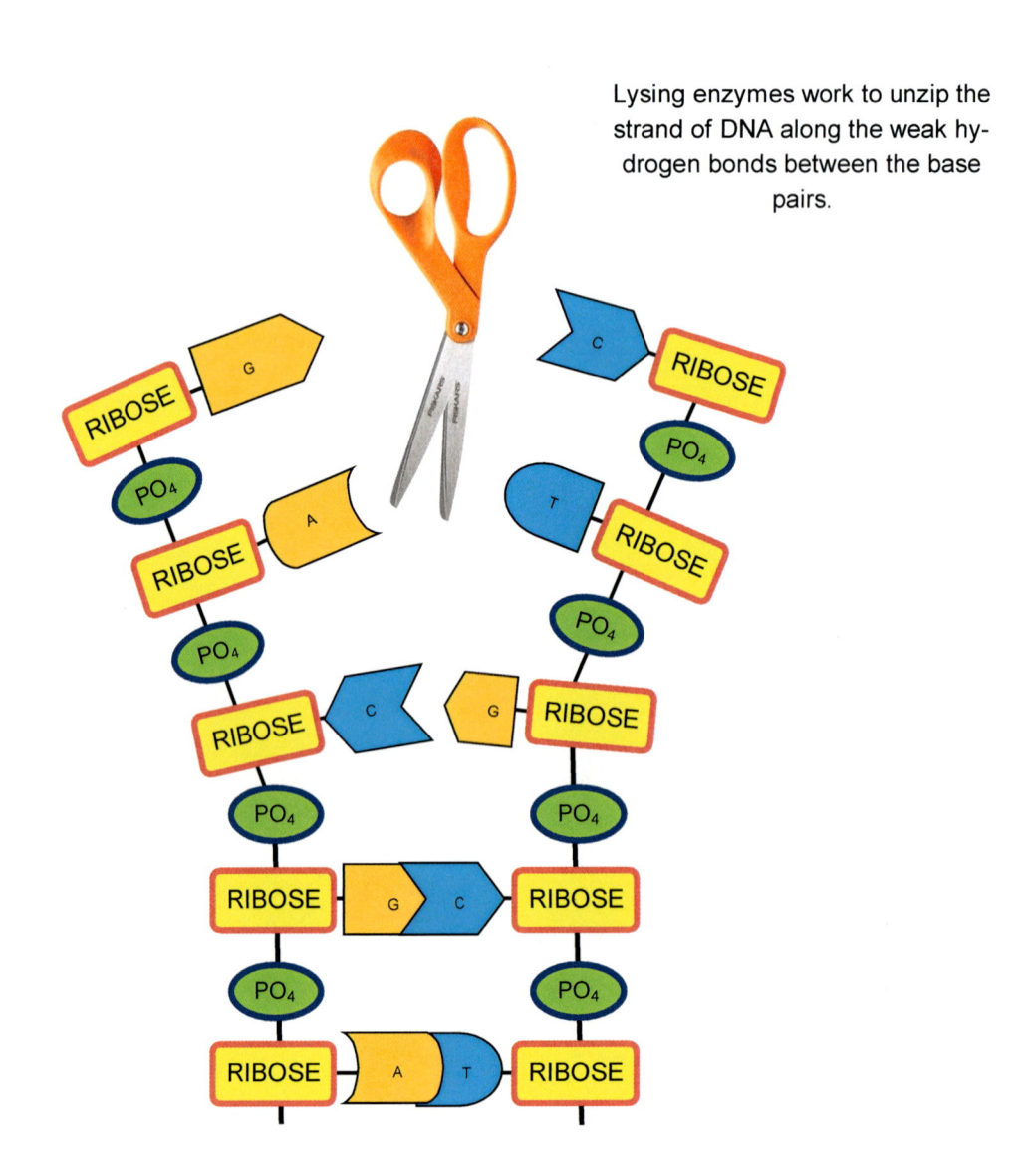

Lysing enzymes work to unzip the strand of DNA along the weak hydrogen bonds between the base pairs.

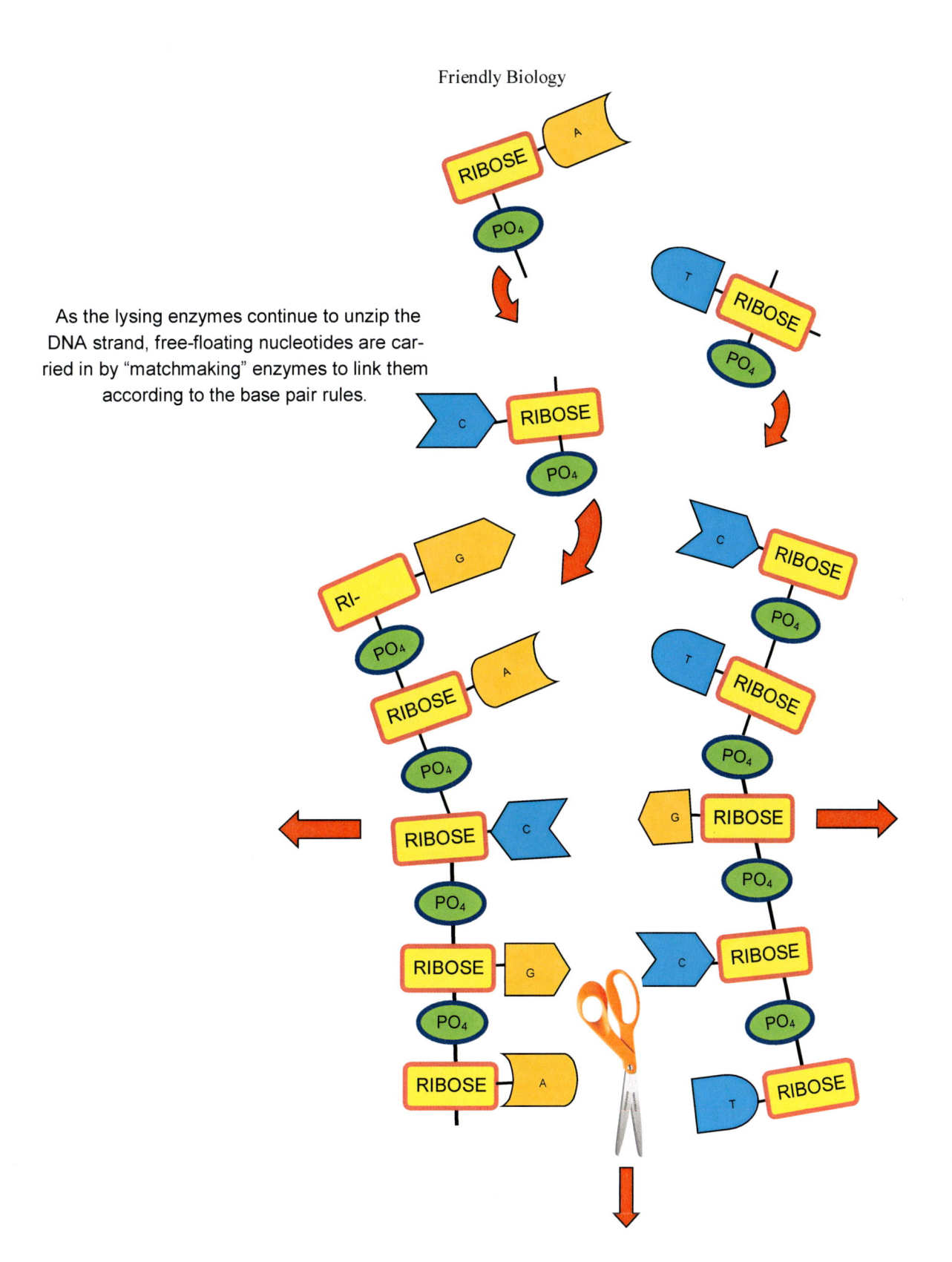

As the lysing enzymes continue to unzip the DNA strand, free-floating nucleotides are carried in by "matchmaking" enzymes to link them according to the base pair rules.

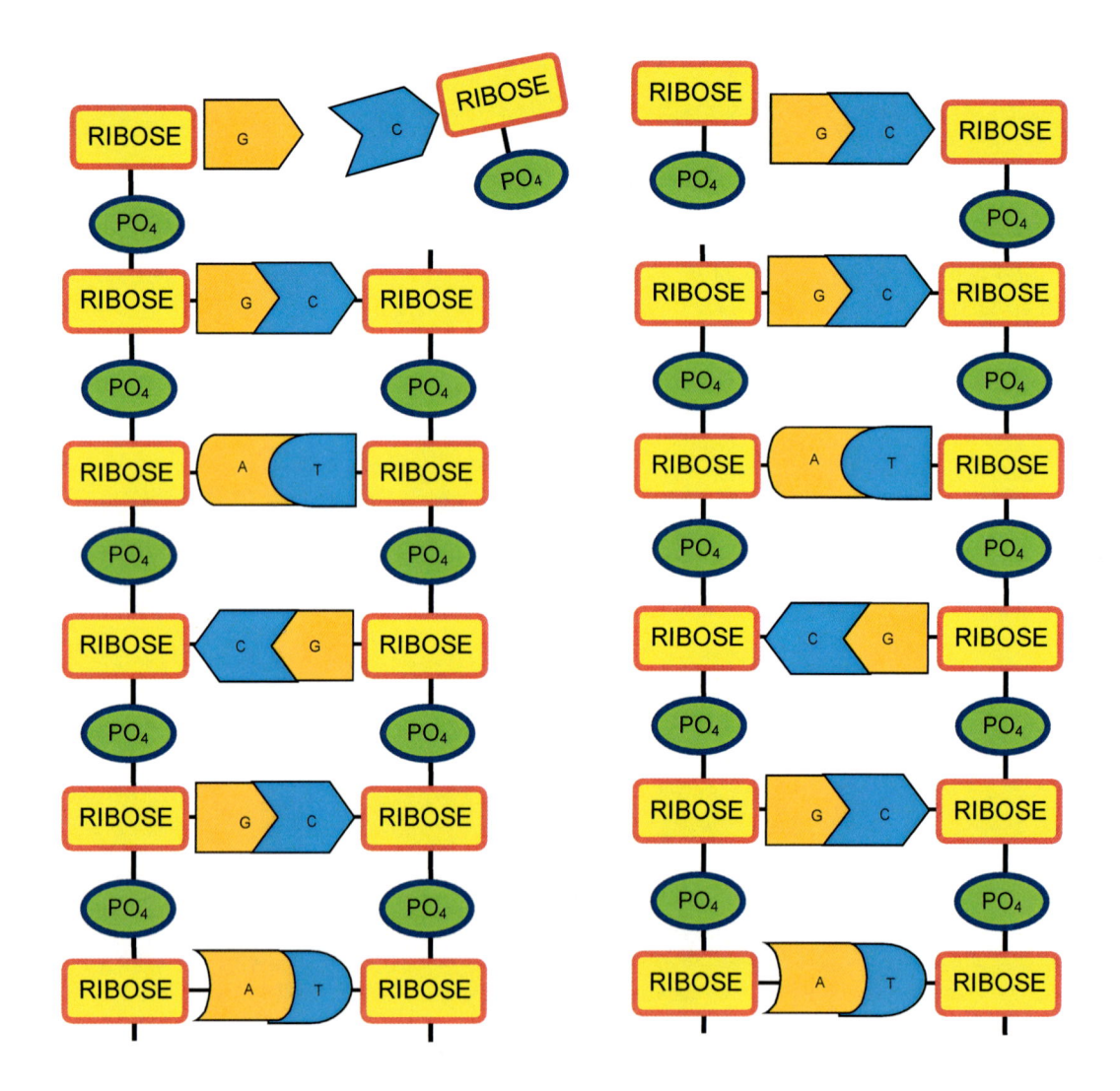

Eventually, two identical strands of DNA are produced. Each daughter cell can now have its own copy of each recipe within each chapter of the cookbook.

Let's pause and review now what we've learned about how chromosomes duplicate themselves. We said that:

- Chromosomes, when uncoiled, consist of long strands of DNA;

- DNA is the abbreviation for deoxyribonucleic acid;

- The structure of DNA is very similar to the shape of a twisted ladder where the vertical component of DNA is made up of alternating ribose sugar units and phosphate groups and the horizontal component of DNA is made up of a pair of bases, one of the pair being a purine and the other being a pyrimidine;

- Adenine and guanine are the purines while thymine and cytosine are the pyrimidines. Adenine always links with thymine. Guanine always links with cytosine;

- When the need arises for the cell to create a copy of its DNA, enzymes "unzip" the DNA strand by splitting it along the hydrogen bonds between the base pairs of the "rungs" of the DNA "ladder." Free-floating nucleotides with corresponding bases link to their desired partners through the work of "matchmaking" enzymes and two identical copies of DNA eventually results. One copy will now be available for each daughter cell.

Lesson 10 Activity: Building a DNA Model

PLEASE READ ALL INSTRUCTIONS BEFORE BEGINNING THIS ACTIVITY.

The purpose of this activity is to build a model of DNA showing the steps of replication. The way you choose to construct your models will be your choice and we will make some suggestions in these notes. Your model should show four basic steps:

- A section of DNA which has been designated for replication.//
- The section of DNA being "unzipped" by an enzyme.
- Free-floating nucleotides moving in to the unzipped DNA segment.
- Two "new" strands of identical DNA.

Your DNA model should include all components of DNA which are the phosphate and ribose groups plus the bases. Ideally, your model should show how DNA is a ladder-like structure which is twisted. Showing how the bases have a special matching arrangement would also be very nice. Your models can be two or three-dimensional. Components which could be used include pipe cleaners, florist wire, electrical wiring like Romex^R used in home wiring. The nucleotide units can be cut from colored paper or cardboard.

Once your models are constructed, label as many components as you can. When complete, present your model to your family or fellow classmates.

Lesson 11: Protein Synthesis

Recall from Lesson 5 that proteins serve as the structural and functional components of living creatures. Proteins are the "bricks, boards and mortar" as well as the "bricklayer, carpenter and plumber" that allow a living thing to live. The "secret" to making these proteins is found in one's genes or "recipes". Let's explore in greater detail how the genes in living things get "read" which results in the synthesis of proteins.

Recall from Lesson 10 that genes of living things are made up of very long strands of DNA. Recall, too, that the structure of DNA is similar to that of a ladder where the vertical portions consist of alternating sugar and phosphate molecules with the rung portions being bonded pairs of nitrogen base molecules of adenine, thymine, cytosine and guanine. We learned earlier that the adenine base always bonds with the thymine base while the cytosine base always bonds with the guanine base. We noted that it was the sequence of these bases which resulted in the production of specific proteins. Let's explore now, in greater detail, how this is thought to take place.

When the need for the production of a particular protein arises, the portion of DNA responsible for creating that protein is located along the DNA strand. This is similar to locating the specific recipe in our cookbook. This portion of the DNA strand gets "unzipped" by enzymes to expose the nitrogen base pairs. This unzipping process is very similar to the process which happens when DNA makes copies of itself during mitosis. The exposed set of nitrogen bases now becomes a template on which another type of nucleic acid, known as RNA, can form.

RNA, which is the abbreviation for ribonucleic acid, is much like DNA, however, there are three big differences. First, unlike DNA, RNA is not a double strand; it is only a single strand of linked nitrogen bases. Second, instead of having the sugar deoxyribose present, it has the sugar ribose present. And, third, RNA while being composed of nitrogen bases like DNA, RNA consists of adenine, cytosine, guanine and the "new" nitrogen base, uracil. RNA does not have the base thymine present. Looking at the first stage of "reading" the DNA "recipe" now will help you see how these three differences make sense in the overall process of building new proteins from DNA.

So let's suppose that we have located the section of DNA or "recipe" that we'll be using as a template on which we will build the needed protein. Let's say that the needed protein is fibrin that we use in our blood to form clots when we suffer an injury. We've located the fibrin "recipe" on the gene and will now unzip that section of DNA to expose the nitrogen bases. Complementary base pairs which make RNA will now align themselves appropriately with the exposed DNA bases. Cytosine and gua-

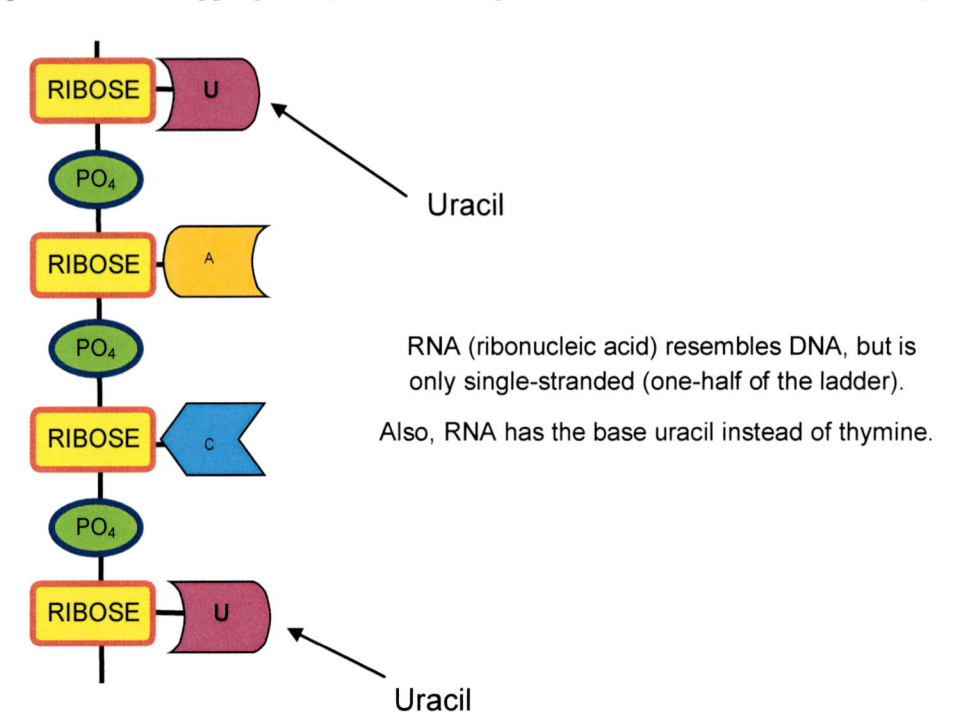

RNA (ribonucleic acid) resembles DNA, but is only single-stranded (one-half of the ladder).

Also, RNA has the base uracil instead of thymine.

nine bases will still align accordingly, however, because there is no thymine base in RNA, uracil will now align itself with any adenine found in the "recipe." Eventually, the complete exposed portion of DNA will have a complementary strand of RNA made from it.

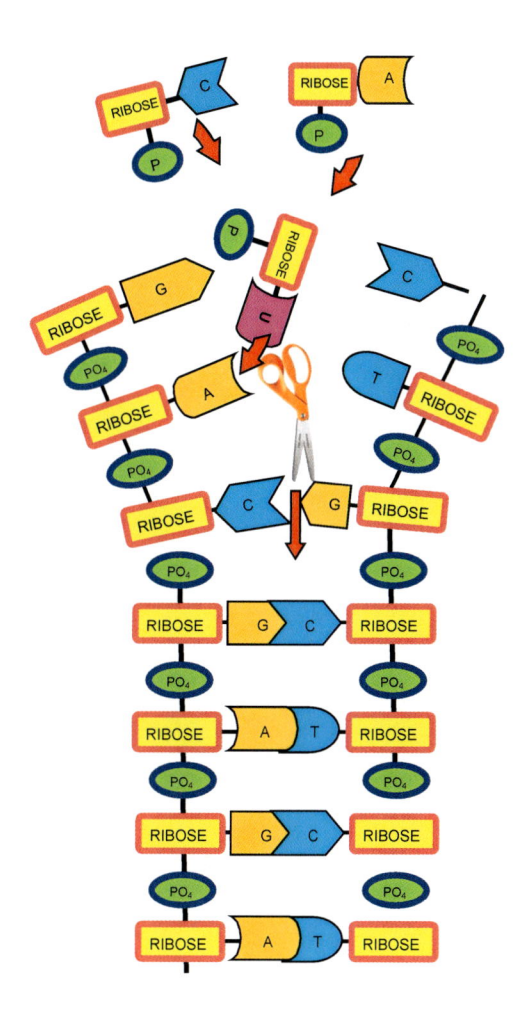

Note how the enzyme (scissors) are unzipping the strand of DNA. This allows the RNA molecules to float in and align themselves with their appropriate base partners. Note that the RNA nucleotides do not bind with the DNA strands.

Note that all of these events we've discussed so far have taken place in the nucleus of the cell. Recall from our earlier discussion of cell parts, that the production of proteins takes place *outside* the nucleus of the cell in the small organelles known as the ribosomes. Recall that it was the ribosomes which make rough endoplasmic rough. So, the strand of RNA that we've made from the exposed DNA within the nucleus of the cell must get transported out of the nucleus and into the cytoplasm to the ribosomes. Because of this need to transport genetic information from one location to another, this RNA strand is given the name *messenger* RNA or mRNA for short.

Before we move to the next step in the creation of a protein, let's review what has taken place so far. The need or request for a specific protein has arrived at the cell. The "recipe" was located along the appropriate section of DNA. That section of DNA was unzipped by enzymes to be used as a template on which a complementary section of messenger RNA could form. This strand of mRNA then moved through the nuclear membrane and out into the cytoplasm to a ribosome where the protein can be built. If we refer back to our cookbook analogy, we can liken this process of taking the copy of the recipe (mRNA) from the main office (nucleus) out to the kitchen (ribosome) from where the preparation of the protein can take place.

At this point, let's look back at the structure of a protein. You learned earlier that proteins are built of smaller subunits called amino acids. You learned that it was the specific *sequence* of amino acids which resulted in specific proteins being made. Each protein required by the living creature has its own specific sequence of amino acids (each recipe is unique). The sequence of these amino acids is found in the DNA. The mRNA that was made "from" the DNA is carrying the information which includes that specific sequence of amino acids! Once it arrives at the ribosome, the ribosome can "read" the mRNA (recipe) to line-up amino acids in the correct sequence to successfully build the correct protein.

Like a recipe can be divided into words to be comprehended, the mRNA can also be divided into segments. Where words in a recipe consist of letters of our alphabet, segments of mRNA consist of the nitrogen bases (A, U, C, G). In addition, where words in a recipe may vary in length, segments of mRNA are only three bases in length. Therefore, a ribosome can "read" the mRNA by noting it consists of "words" three bases in length. Each three-base segment codes for a specific amino acid.

For example, in a cookbook recipe, the letters: b, u, t, t, e and r code for the word butter. In a strand of mRNA, the sequence CUU codes for the amino acid leucine. The sequence GUU codes for the amino acid valine and the sequence AUG codes for methionine. The strand of mRNA arriving at the ribosome can be broken into these three-base segments which in turn tell the sequence of amino acids needed to build the protein. These three-base segments are called codons.

Look at the diagram here. First note how the specific section of DNA needed to prepare the protein has been identified (yellow box) and then unzipped by enzymes.

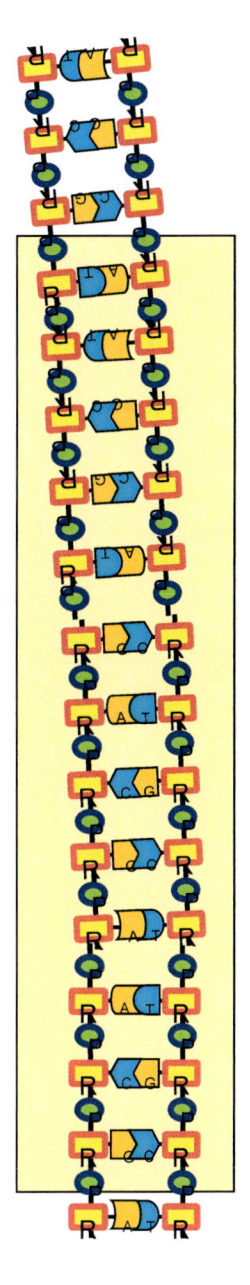

The portion of DNA needed to produce the necessary protein is identified (yellow box on left). Enzymes work to "unzip" that section of DNA exposing the base pairs.

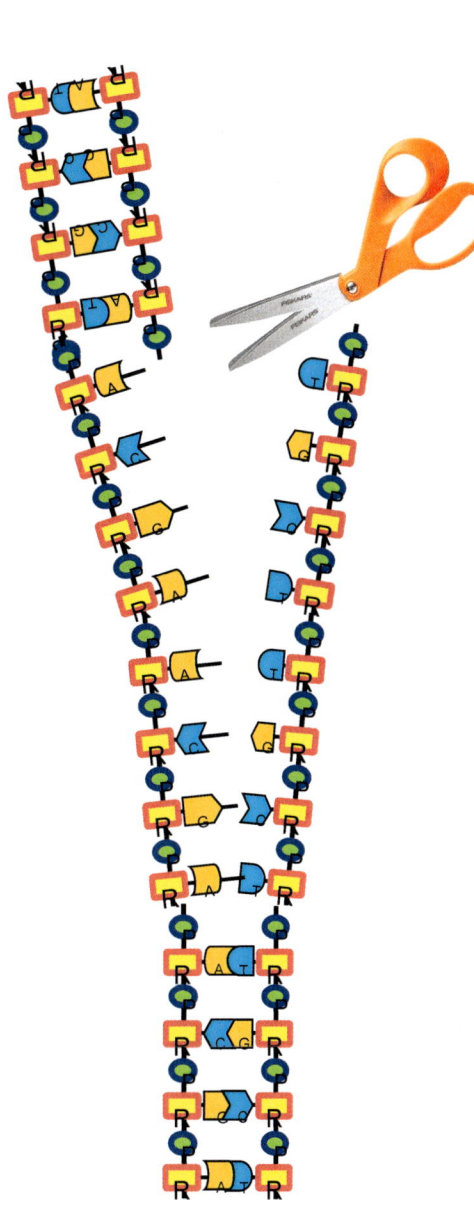

169

Friendly Biology

Look now at how mRNA has begun to form. The bases A, U, C and G have correctly aligned (but not bonded) themselves with the DNA template. These nitrogen bases link together to create a strand of mRNA.

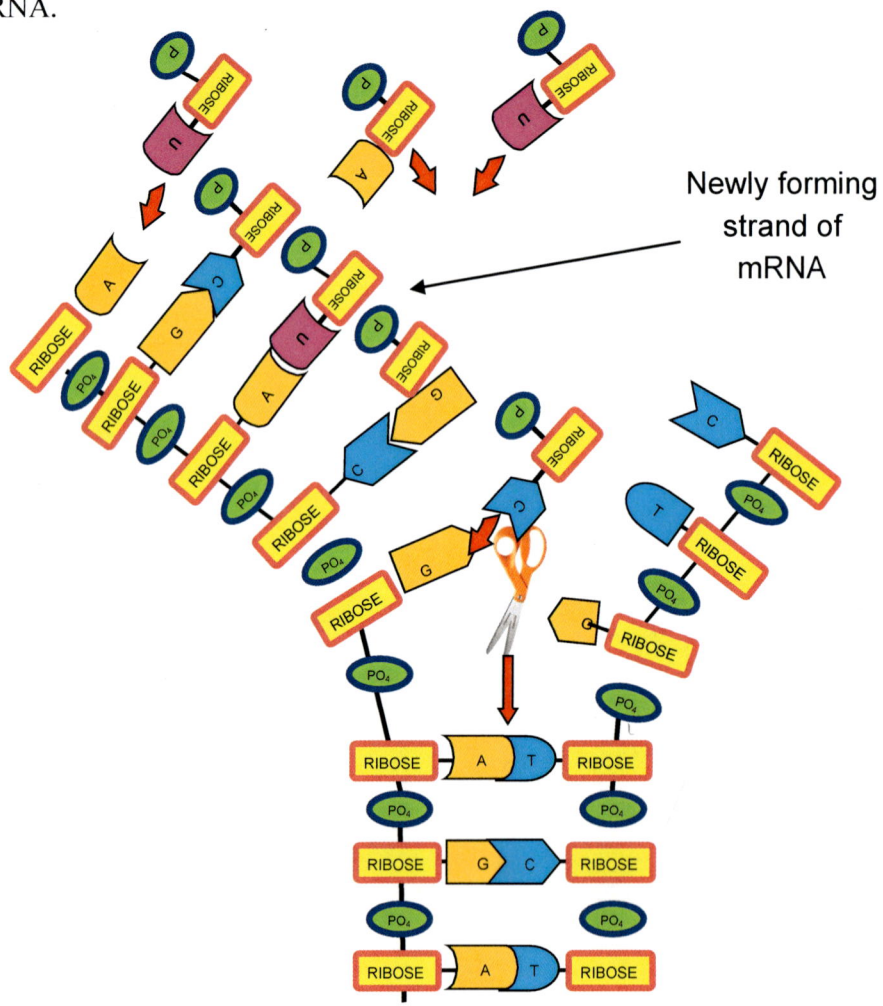

Newly forming strand of mRNA

Now, we can see the strand of mRNA making its way out of the nucleus into the cytoplasm to find a ribosome where the protein can be built.

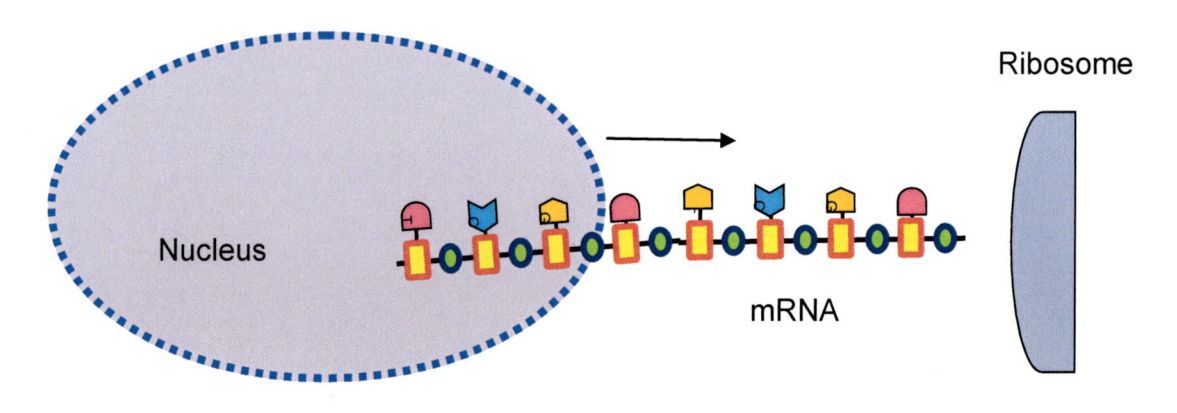

Nucleus

mRNA

Ribosome

Here we can see that the strand of mRNA has arrived at a ribosome and how the strand of mRNA is being divided into three-base segments known as codons. Each codon is specific for a particular amino acid.

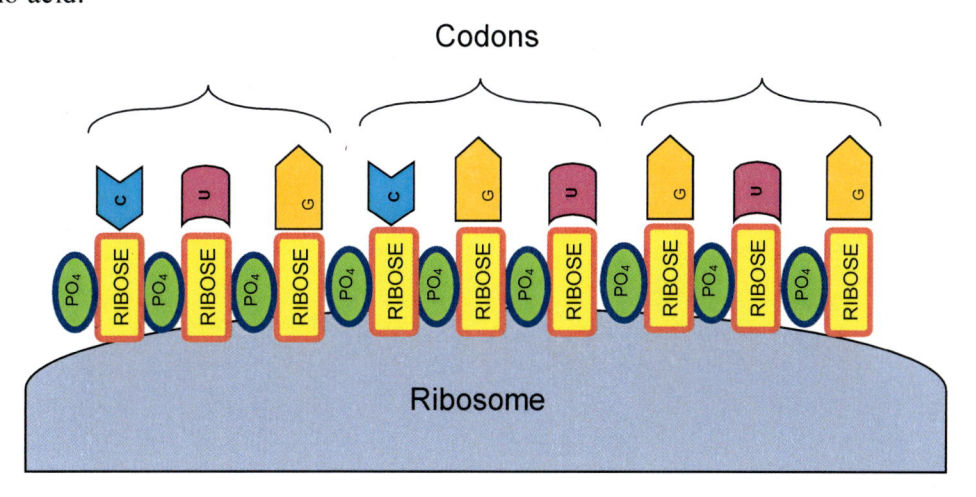

The amino acids that the ribosome will use to create the new protein (fibrin, in our example) can be found out in the cytoplasm of the cell. The origin of these amino acids could have been in the protein portion of a meal we ate earlier in the week. They are now "waiting" to be used to create fibrin for our body's use. Attached to each of these "waiting" amino acids is another form of nucleic acid known as transfer RNA, abbreviated tRNA. Like mRNA, tRNA is single-stranded and contains the nitrogen bases A, C, G and U. A special three-base segment of this tRNA works like a key in that it will only fit into a specific codon in the mRNA strand being used at the ribosome. In other words, each "waiting" amino acid has an attached key which only allows it to be used according to the specific sequence of codons on the mRNA strand.

Look at the diagram here. Note the "waiting" amino acids, "hanging-out" each with its own specific tRNA strand (key). This "key" is known as the anticodon.

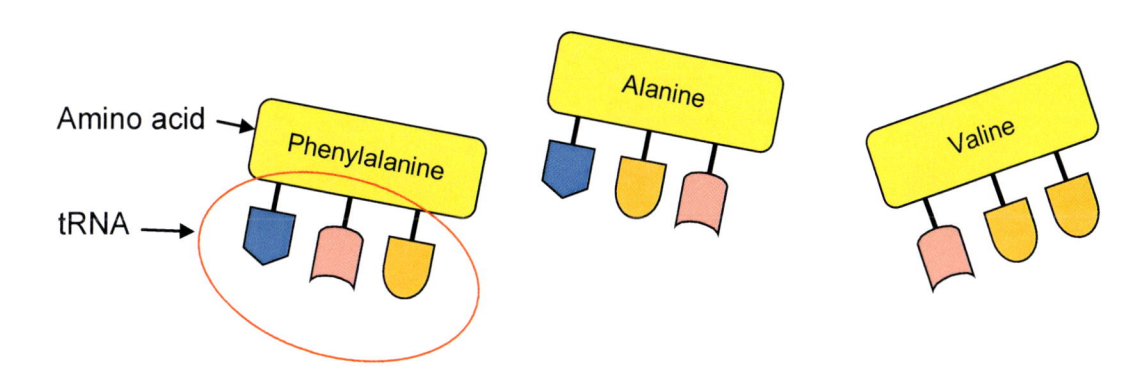

As you can see here, the next amino acid needed to build the protein, has the codon AUG. This sequence of bases codes for the amino acid methionine. Note how methionine has the "correct" tRNA code that will allow it to fit the "lock" and now join the strand of amino acids forming the new polypeptide. The codon of the mRNA fits with the anticodon of the tRNA.

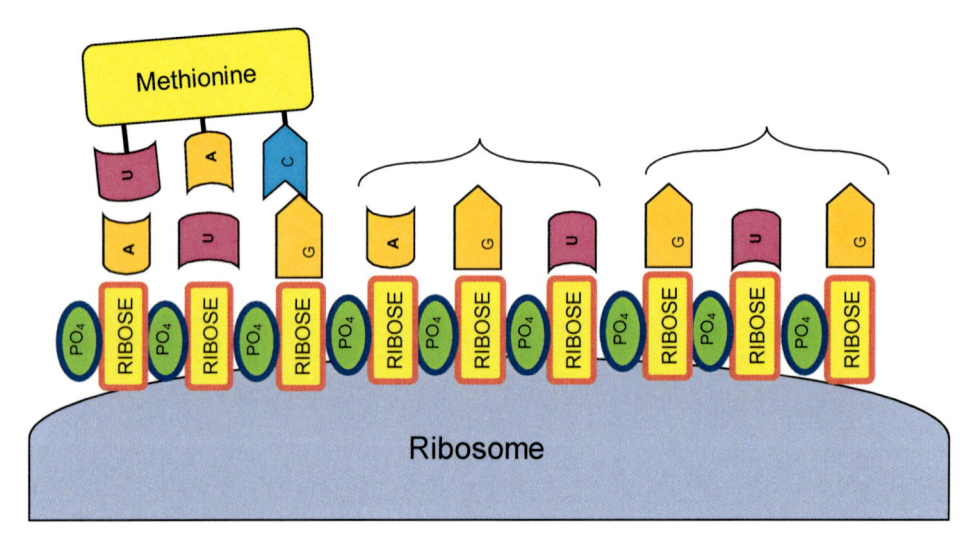

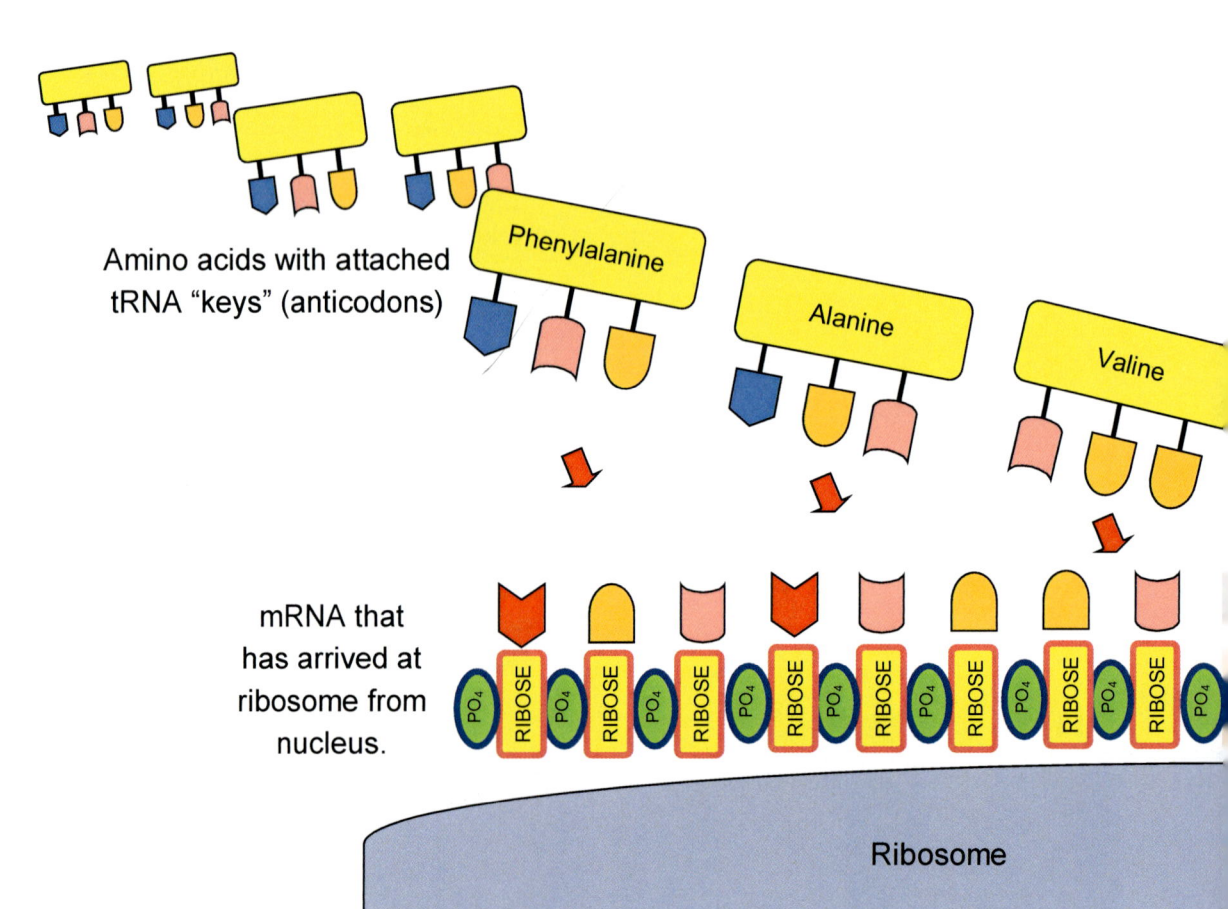

Amino acids with attached tRNA "keys" (anticodons)

mRNA that has arrived at ribosome from nucleus.

This process of transferring amino acids to the ribosome will continue until a specific codon is reached which is called the stop codon. Like its name says, this codon will stop the amino acid linking process and this stage in the preparation of the protein will be complete. Note that a "start" codon is also used by the ribosome to know where to begin the process of building the protein.

Proteins, in general, are extremely large molecules and may consist of thousands of appropriately linked amino acids all based upon the sequence of DNA found in the cell's nucleus. Following this compilation of amino acids, the protein will then be folded and processed into its final form and, ultimately, be delivered to the location where it will be utilized by the body. In our example, the protein fibrin will be delivered to the blood.

If we look at the process of creating proteins from DNA as a whole, we can break it into two big events. The first event is the process where information found in the DNA gets communicated to the mRNA. This event is given the name transcription.

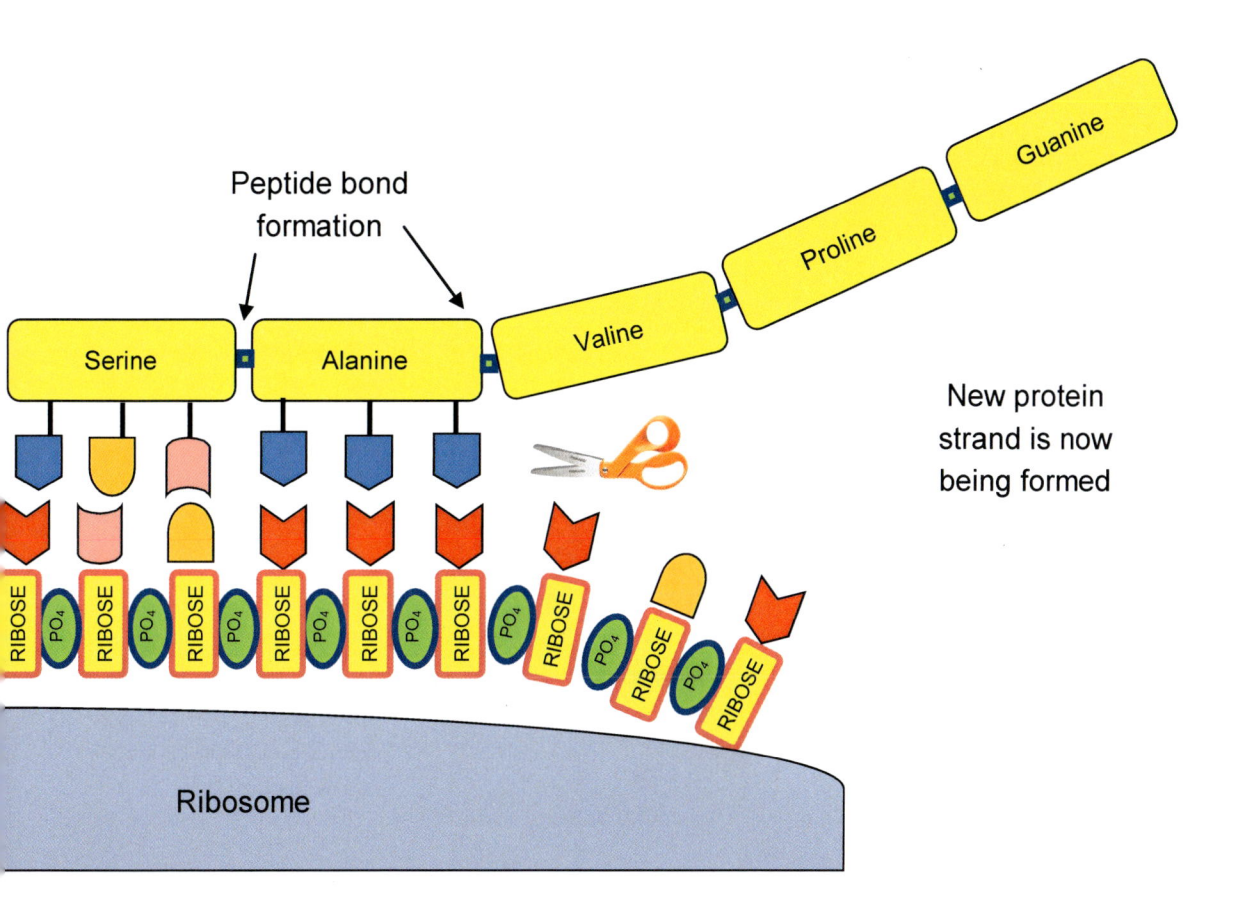

The second big event in this process is when the information found on the mRNA gets communicated to the ribosomes which in turn link together the appropriate sequence of amino acids to create the desired protein. This event is given the name translation.

To summarize, transcription is "going" from DNA to mRNA. Translation is "going" from mRNA to correct amino acid sequencing. tRNA functions like a key to assure that appropriate amino acids get linked in the correct sequence.

This entire process is truly amazing. It is almost unfathomable that this process goes on continually in living creatures all day, everyday, to keep all body systems functioning as they should. Thousands upon thousands of base pairs and amino acids all getting where they should be at the right moment is truly an amazing feat. Before we end this lesson, let's briefly explore what happens, however, if things don't go exactly according to plan.

When cells undergo cell division, whether it be mitosis or meiosis, there are times when the duplication or separation of the chromosomes does not happen correctly. In the case of mitosis, if we refer back to our analogy of the chromosomes as being like recipes in a cookbook, the new copies which result may have experienced some unexpected changes. Instead of the "recipe" saying to add 4 cups of flour, it may now say add 6 cups of flour. Obviously, the outcome would not be what would be expected. Where the original DNA code may have stated the protein needs four methionine amino acids in the sequence, the "new" recipe may now state six methionine amino acids. The resulting protein from this alteration may or may not function as expected!

In the case of meiosis, if the separation of the pairs of chromosomes in the primordial cells did not happen correctly, the genes which go to make the sex cells (gametes) would not have the correct information for the potential new offspring. Again, we would see unexpected results.

This Angus steer shows evidence of a genetic mutation. An extra limb has grown from the neck region of his vertebra.

This alteration in chromosomes, and, more specifically, in the genes of the chromosomes, is known as a genetic mutation. As with the recipe analogy, the resulting phenotype from this "new" arrangement of genes, may (or may not) have detrimental consequences for the resulting daughter cells or offspring. In humans and animals, these unexpected results are known as birth defects or, more scientifically, a teratogenic event. Depending upon the degree of alteration of the gene, the results of the change varies in severity.

Some alterations may be very mild and may go unnoticed while some genetic mutations can be extremely severe and are identified as being lethal mutations. The results of these mutations are so dramatic that the new daughter cells (in the case of mitosis) or new offspring (in the case of meiosis) do not survive. The "recipe" is so messed-up that the new cells or new individual cannot carry on life processes and may die before being born.

There are many known causes for genetic mutations and then, sometimes, mutations may just happen and no apparent cause can be found. Identified causes for genetic mutations, known as mutagens or teratogens, include environmental factors such as overexposure to sunlight or radiation. Excessive exposure to the sun causes changes in the DNA of skin cells which results in abnormal growth and multiplication of the cells and is commonly known as skin cancer or melanoma. If you've ever

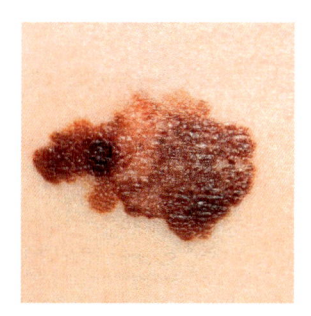

Excessive exposure to sunlight may result in the formation of skin cancer. The DNA of skin cells is modified by the sun's radiation. This form of cancer is known as known as a melanoma.

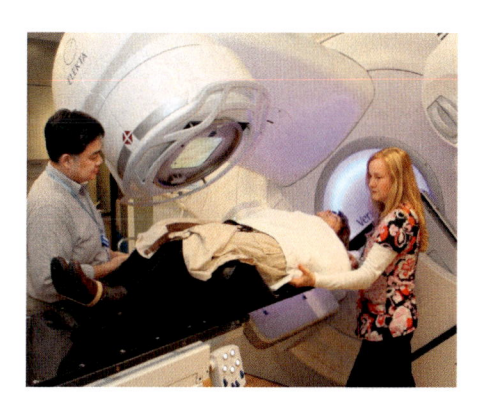

Cancer cells have DNA which has malfunctioned for various different reasons. By bombarding those abnormal cells with radiation, oncologists hope to further damage the DNA of the cancerous cells causing them to reduce their activity or die.

had x-rays made of yourself you may recall how the persons taking the x-rays always step behind a shield or wall before the machine is triggered to make the films of your body part. They may also shield other parts of your body from the radiation produced by the camera by having you wear a lead-filled apron or vest. The radiation which exposes the x-ray film can cause unwanted changes in the DNA of cells in your body.

Exposure to various medications and chemicals may also result in mutations in cells. In the late 1950's a drug called thalidomide was commonly dispensed to women experiencing morning sickness (nausea) in the early weeks of pregnancy. It was discovered later that this medication caused genetic changes in the developing baby which resulted in malformed arms and legs. Some infants were born missing one or both arms. Toxins (poisons) found in cigarette smoke are also known to cause genetic changes in developing babies.

On the other hand, use of mutagens to cause cells to malfunction or die can be useful. Once a person is diagnosed with certain forms of cancer, controlled amounts of radiation may be used directly upon cancer cells to cause them to stop their uncontrolled growth and hopefully die. Radiation treatments can be relatively successful in the treatment of certain forms of cancer.

This ends our discussion of protein synthesis. In this lesson we learned that:

- The sequence of amino acids is determined by the DNA found in one's genes;

- A "copy" of the DNA is made through the process of transcription. This copy is in the form of messenger RNA (mRNA);

- The mRNA moves from the nucleus of the cell out to the ribosome where the protein assembly takes place. The mRNA bases are in sets of threes called codons. Specific codons code for specific amino acids;

- Amino acids which are available in the cytoplasm have transfer RNA (tRNA) attached to them which function like "keys" to allow them to be attached to the mRNA. These "keys" are known as anticodons. This process is known as translation;

- Inaccuracies in these processes, known as genetic mutations, result in a range of problems ranging minor to severe. Doctors have learned ways to utilize mutations for the good of their patients in certain cancer treatments.

Friendly Biology

Lesson 11 Activity: Protein Synthesis Race Game

In this dice game, you will review and practice the steps involved in protein synthesis. Making careful strategic plans can help you build your protein first and win the game. This game is designed for two or more players.

For the game you will need:

- A copy of the DNA playing chart for each player
- A stack of sticky-notes to be shared among players
- One die which can be modified for the game
- A pen or pencil for each player.

Before play begins, you'll first need to prepare the die. By placing the following letters on the die. Using a small sticker over the existing dots works well. No particular order is required:

- A representing adenine
- U representing uracil
- C representing cytosine
- G representing guanine.
- E representing enzyme, and
- R representing ribosome

Rules for play:

Level 1 of the Game: Transcription

1. Each player should receive a copy of the DNA Segments playing chart. Note on the chart there are segments of DNA which code for a specific polypeptide (protein.) Of all the segments of DNA available, each player chooses one segment for the first round of play.

2. Because the DNA segment is double-stranded, only one half will be utilized to build the protein. Each player has the choice as to which side of the DNA strand they will utilize to build their protein (either the top or bottom half). In order to make this choice, each player has to first

"unzip" their chosen DNA strand by rolling an "E" on the die. Recall that the "E" represents an enzyme which can "unzip" the DNA segment. Play begins by one player rolling the die. If the player does indeed roll an "E" the player can "unzip" the DNA and indicate so by placing the sticky edge of their sticky-note over the side of the DNA the player is NOT going to use. See diagram below. If the player rolls something other than an "E," the die his handed to the next player who repeats the process. Play moves around the group in a like manner. If a player rolls an "R," he or she rolls again and then can roll an extra time.

3. Once a player has successfully "unzipped" the DNA, the player continues by building a strand of mRNA alongside the exposed DNA strand. To build this mRNA strand, they roll the die. Each letter they roll on the die represents a possible base they can add to their mRNA strand. As letters are rolled, the player writes them below (or above) the exposed DNA bases according to the appropriate pairing rules learned in the lesson. Each player has one opportunity to collect one base for each turn. If a letter is rolled isn't needed, the die is handed-off to the next player in line. The goal of

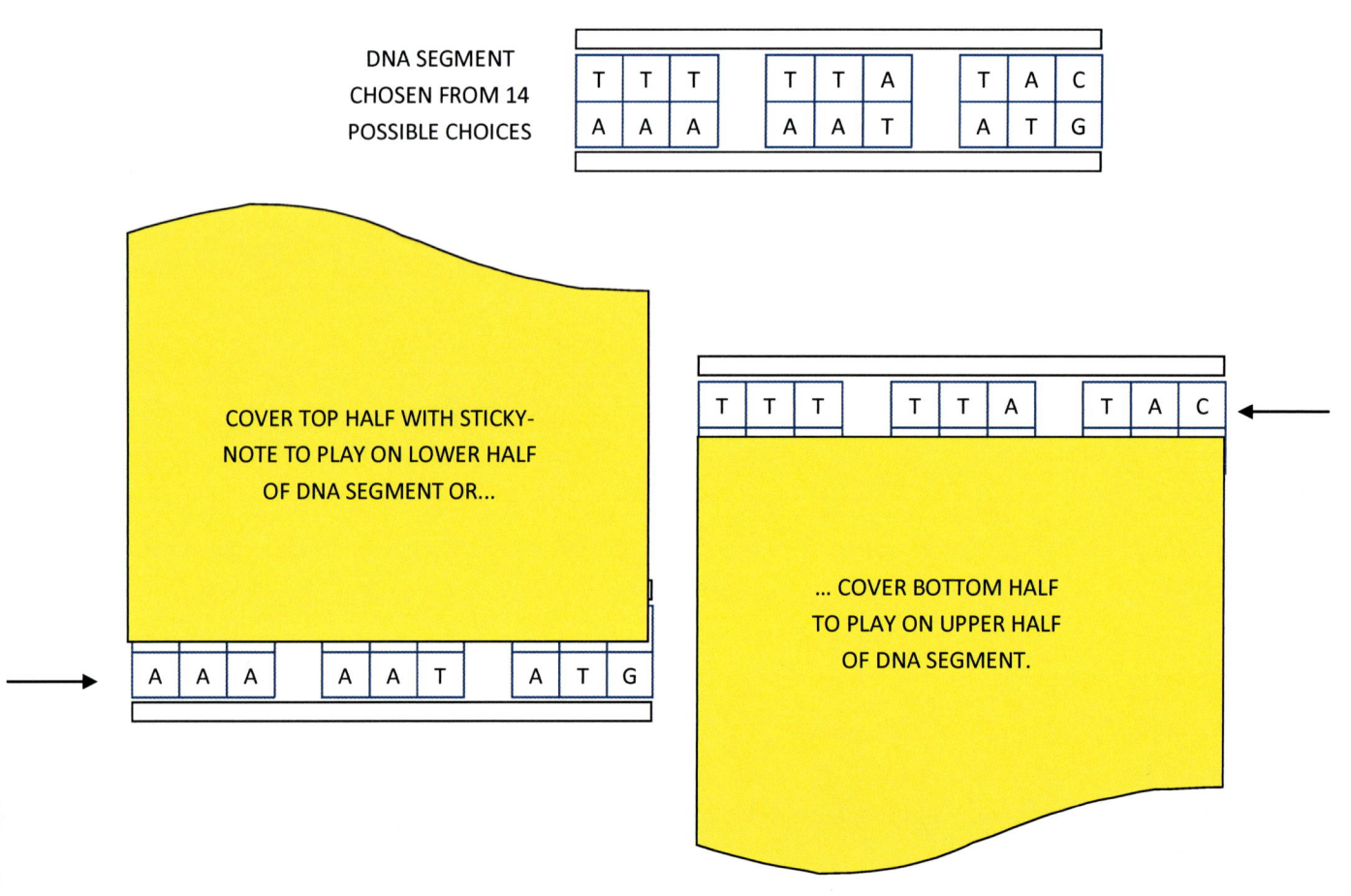

DNA SEGMENT CHOSEN FROM 14 POSSIBLE CHOICES

T	T	T		T	T	A		T	A	C
A	A	A		A	A	T		A	T	G

COVER TOP HALF WITH STICKY-NOTE TO PLAY ON LOWER HALF OF DNA SEGMENT OR…

| A | A | A | | A | A | T | | A | T | G |

| T | T | T | | T | T | A | | T | A | C |

… COVER BOTTOM HALF TO PLAY ON UPPER HALF OF DNA SEGMENT.

this part of the game is to completely build the mRNA strand. If the letter "E" is rolled on the die, the player simply rolls the die again. If the letter "R" is rolled, the player rolls the die again and then earns an extra roll of the die. Once a player completely builds his or her mRNA strand, he or she is ready to advance to Level 2 of the game.

Level 2 of the game: Translation

1. In this part of the game, the action now takes place out at the ribosome on the endoplasmic reticulum of the cell. In order to "get" to the ribosome, players who have become eligible for Level 2 must roll an "R" on the die.

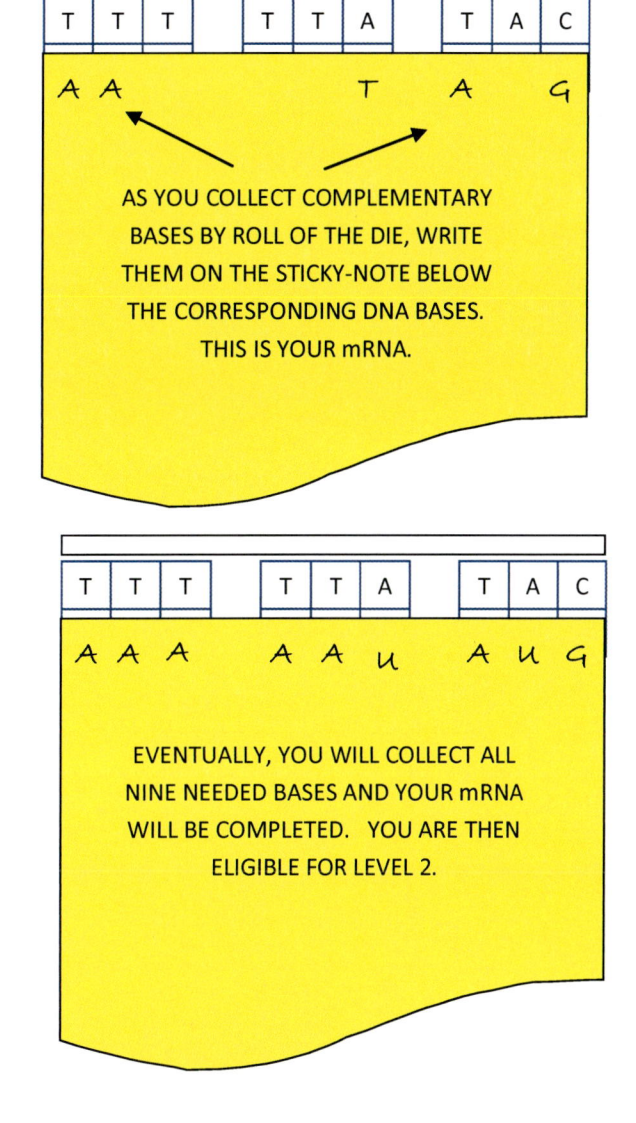

T	T	T		T	T	A		T	A	C
A	A					T		A		G

AS YOU COLLECT COMPLEMENTARY BASES BY ROLL OF THE DIE, WRITE THEM ON THE STICKY-NOTE BELOW THE CORRESPONDING DNA BASES. THIS IS YOUR mRNA.

T	T	T		T	T	A		T	A	C
A	A	A		A	A	U		A	U	G

EVENTUALLY, YOU WILL COLLECT ALL NINE NEEDED BASES AND YOUR mRNA WILL BE COMPLETED. YOU ARE THEN ELIGIBLE FOR LEVEL 2.

2. After a player successfully rolls an "R," the player can begin collecting bases which will make up the anticodons found on tRNA which are attached to amino acids. By examining his or her mRNA strands, each player should be able to list on the sticky-note the needed tRNA bases just beneath the mRNA bases. See diagram below.

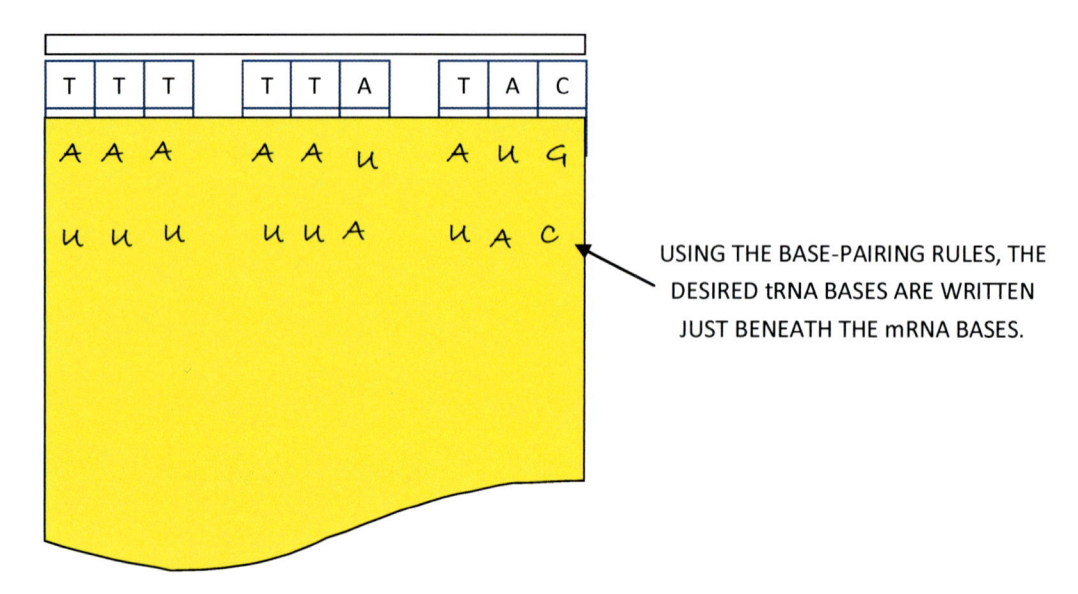

T	T	T		T	T	A		T	A	C

A A A   A A U   A U G
U U U   U U A   U A C

USING THE BASE-PAIRING RULES, THE DESIRED tRNA BASES ARE WRITTEN JUST BENEATH THE mRNA BASES.

3. On their turns, players roll the die and collect the bases they need to complete their three sets of anticodons. As a base is collected, its letter is crossed off the list of needed bases. One base can be collected per turn. See diagram. If the letter "E" is rolled, the player rolls again. If the letter "R" is rolled, the player rolls again and then earns a second turn.

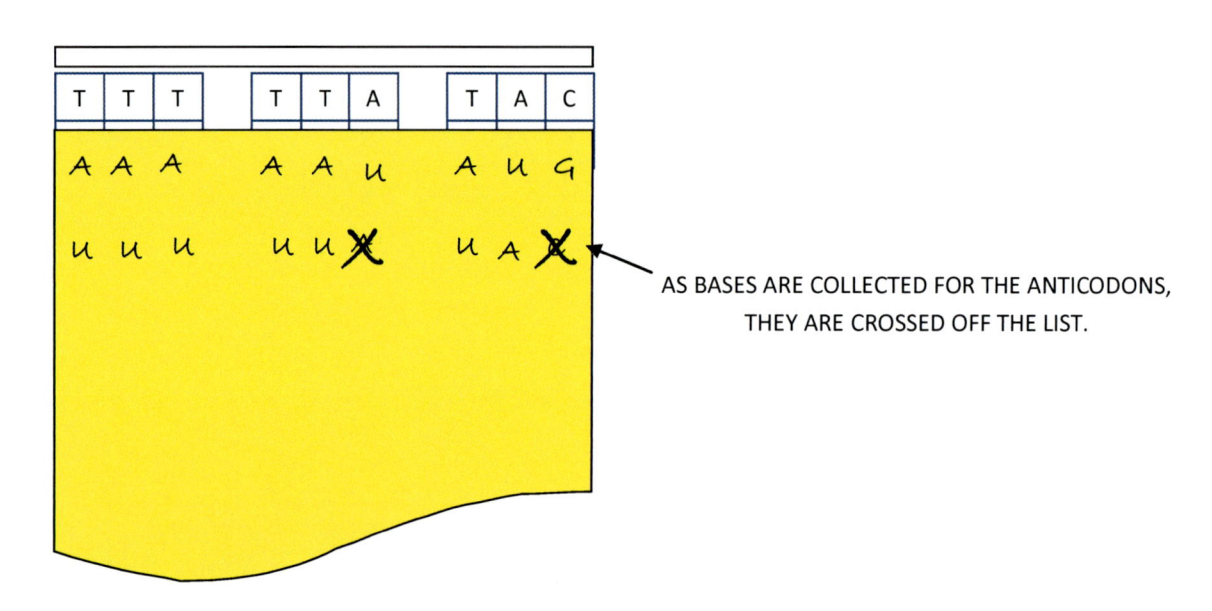

| T | T | T | | T | T | A | | T | A | C |

A A A   A A U   A U G
U U U   U U X̷   U A X̷

AS BASES ARE COLLECTED FOR THE ANTICODONS, THEY ARE CROSSED OFF THE LIST.

4. Play continues until one player has collected all nine of his or her anticodon bases. At this point, he or she consults the Amino Acid Codon/Anticodon Chart to find the amino acids linked in his or her protein. These amino acids, listed in order from left to right, are then written as the final entry on the sticky-note.

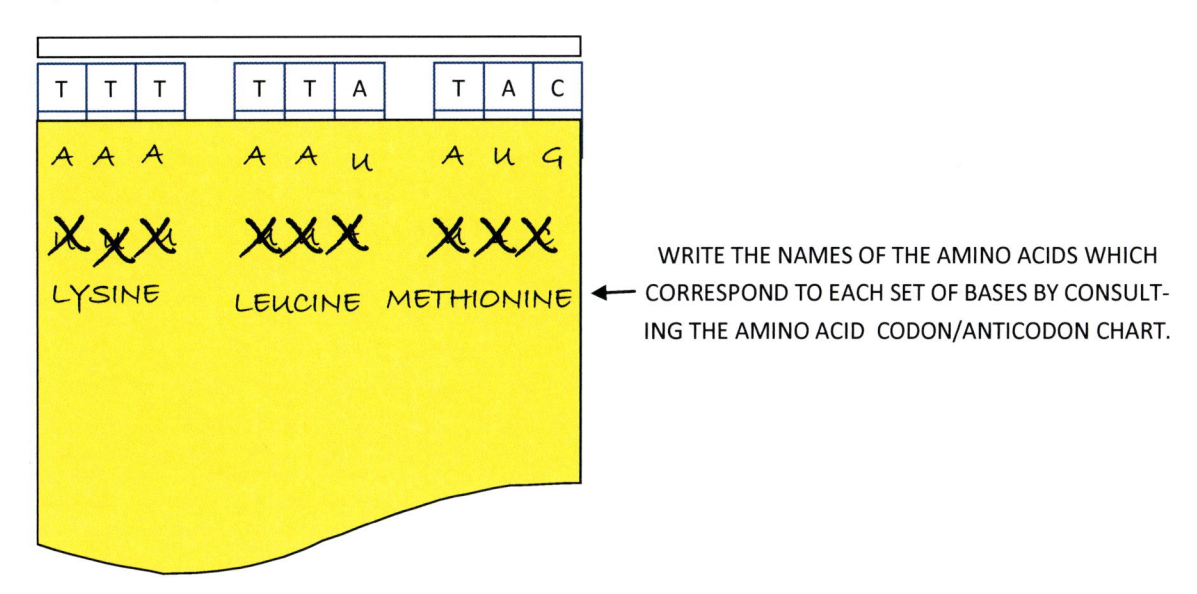

WRITE THE NAMES OF THE AMINO ACIDS WHICH CORRESPOND TO EACH SET OF BASES BY CONSULTING THE AMINO ACID CODON/ANTICODON CHART.

5. At this point, the player who has written down the three amino acids he or she has linked together based upon the original DNA sequence, checks his or her work by consulting the Answer Key. If the amino acids listed are correctly listed (and in the correct order), he or she wins the game. Other players continue to play for second, third, etc. place positions.

DNA SEGMENTS

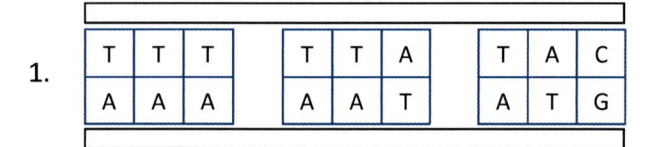

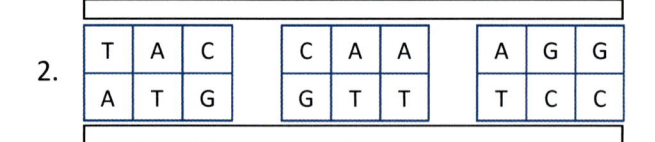

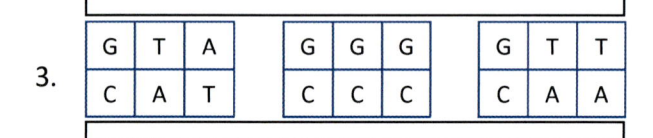

1.
T	T	T		T	T	A		T	A	C
A	A	A		A	A	T		A	T	G

2.
T	A	C		C	A	A		A	G	G
A	T	G		G	T	T		T	C	C

3.
G	T	A		G	G	G		G	T	T
C	A	T		C	C	C		C	A	A

4.
C	G	A		G	T	A		G	T	T
G	C	T		C	A	T		C	A	A

5.
T	T	T		T	C	C		C	C	T
A	A	A		A	G	G		G	G	A

6.
A	G	G		T	C	C		C	G	A
T	C	C		A	G	G		G	C	T

7.
G	T	T		T	A	C		T	T	T
C	A	A		A	T	G		A	A	A

8.
T	T	T		C	A	A		G	T	T
A	A	A		G	T	T		C	A	A

9.
T	T	A		C	A	A		G	G	G
A	A	T		G	T	T		C	C	C

10.
T	T	T		T	A	C		G	T	A
A	A	A		A	T	G		C	A	T

11.
G	T	A		G	T	T		C	C	T
C	A	T		C	A	A		G	G	A

12.
T	C	C		C	A	A		T	A	C
A	G	G		G	T	T		A	T	G

13.
T	T	T		T	A	C		G	G	G
A	A	A		A	T	G		C	C	C

14.
T	C	C		G	T	T		C	G	A
A	G	G		C	A	A		G	C	T

CODON (mRNA), ANTICODON (tRNA) LIST and RESULTING AMINO ACID LIST

IF THE DNA strand is...	...the mRNA codon will be...	.. And the tRNA anticodon which attaches will be...	...Which results in this amino acid being placed into the polypeptide chain:
AAA AAG	UUU UUC	AAA AAG	PHENYLALANINE
TTA	AAU	UUA	ASPARAGINE
CCT CCC	GGA GGG	CCU CCC	GLYCINE
TAC	AUG	UAC	METHIONINE
CAA CAT	GUU GUA	CAA CAU	VALINE
AGG	UCC	AGG	SERINE
GGG GGA	CCC CCU	GGG GGA	PROLINE
GCT TCC	CGA AGG	GCU UCC	ARGININE
CGA	GCU	CGA	ALANINE
ATA ATG	UAU UAC	AUA AUG	TYROSINE
GTA	CAU	GUA	HISTIDINE
GTT	CAA	GUU	GLUTAMINE
TTT	AAA	UUU	LYSINE
AAT	UUA	AAU	LEUCINE

PEPTIDE SOLUTIONS FOR EACH DNA SEGMENT

SEGMENT	TOP HALF	BOTTOM HALF
1	Lysine leucine methionine	Phenylalanine asparagine tyrosine
2	Methionine valine serine	Tyrosine glutamine arginine
3	Histidine proline glutamine	Valine glycine valine
4	Alanine histidine glutamine	Arginine valine valine
5	Lysine arginine glycine	Phenylalanine serine proline
6	Serine arginine alanine	Arginine serine arginine
7	Glutamine methionine lysine	Valine tyrosine phenylalanine
8	Lysine valine glutamine	Phenylalanine glutamine valine
9	Leucine valine proline	Asparagine glutamine glycine
10	Lysine methionine histidine	Phenylalanine tyrosine valine
11	Histidine glutamine glycine	Valine valine proline
12	Arginine valine methionine	Serine glutamine tyrosine
13	Lysine methionine proline	Phenylalanine tyrosine glycine
14	Arginine glutamine alanine	Serine valine arginine

Lesson 12: Methods of Reproduction

In Lesson 9, we discussed that living things develop and grow through the process of mitosis where more cells get made. We discussed in Lesson 11 that it was through protein synthesis that more raw materials are produced which can allow a cell to undergo mitosis as well as carry out its assigned role in the living organism. In this lesson, we will go a step further and explore how one complete organism can create another complete organism. This process is known as reproduction.

In some biology books, the discussion of reproduction begins with a presentation of how tiny microorganisms and similar "lower" creatures reproduce. Instead of taking this approach, we'll start our discussion of reproduction with animals, including humans and plants (the "higher" living creatures) and then return to the tinier creatures.

In order for the process of reproduction to make sense, we need first to establish some basic concepts. It is obvious that the babies, also known as offspring, of higher organisms have all of the basic characteristics of their parents. Initially, baby animals including humans, very closely resemble their parents and with time, they grow and develop into organisms which can perform all of the living functions of their parents. In order to do so, they *must* have been provided all of the necessary genetic material from their parents (the complete owner's manual or cookbook)!

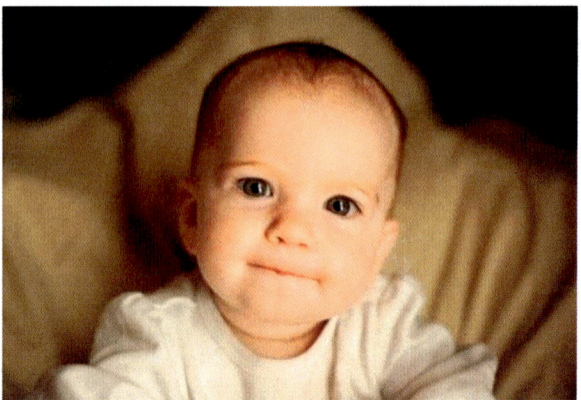

Both flowers and babies inherit all of their characteristics from their parents through their chromosomes.

This genetic material was passed on to them through their chromosomes. In other words, everything about the baby was made possible due to the capabilities and attributes of the chromosomes received from his or her parents. So, in order to understand how the chromosomes of the baby come to be, we must first look a little more closely at the chromosomes of his or her parents. As we mentioned earlier, we'll first look at higher living organisms which are born from two parents. Later we'll look at living organisms which arise from only one parent. So let's begin by looking at creatures which come from two parents.

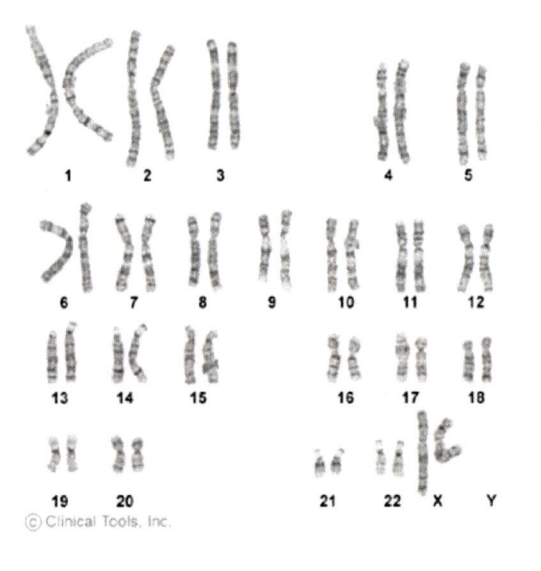

Humans have 23 pairs of chromosomes in almost all of their cells.

If we examine the chromosomes of humans and animals, we find that instead of single chromosomes, there are, in fact, *pairs* of chromosomes. Humans, for example, have 23 pairs of chromosomes for a total of 46 all together. Animals, depending upon their species, have varying numbers of chromosomes, but we find that within a species, each member of that species has the same number of chromosomes. For example, tigers have 19 pairs of chromosomes and all tigers of the same species have 19 pairs of chromosomes. Rabbits have 22 pairs of chromosomes and all rabbits of the same species have 22 pairs of chromosomes and so on.

Mulberry Tree	308	Dog	78	Elephant	56
Carp	104	Bears	74	Cotton	52
Shrimp	92	Fox	72	Pineapple	50
Hedge Hog	90	Deer	70	Potato	48
Pigeon	80	Elk	68	Giant Panda	42
Turkey	80	Horse	64	Oats	42
Chicken	78	Donkey	62	Mango	40
Coyote	78	Cattle	60	Cat	38
Kangaroo	16	Koala	16	Opossum	22
Radishes	18	Ants	2	Wallaby	10
Barley	14	Peas	14	Mosquito	6

Chromosome number of various animal and plant species. Note this is the 2N or total number.

If we examine each pair of chromosomes, we see that the genes present *within* each pair are the same. For example, on chromosome 1 we might find the genes for eye color. Within that pair of chromosomes, we'll find the same set of genes on both members of the pair. In other words, the gene that codes for eye color is present on both chromosomes. Now, the code for a specific color may not be the same, but the code for eye color is present. One chromosome may be coding for blue eye color while the other may be coding for brown eye color. These variations in specific coding for the same attribute are called alleles. In the case of eye color, we have one allele coding for blue eye color and another allele which codes for brown eye color. Or, they could code for the same color! We'll discuss alleles in greater detail in the next lesson. Our point here, is that, in humans we find 23 *pairs* of chromosomes with each member of the pair being the same type as the other.

Alleles for eye color.

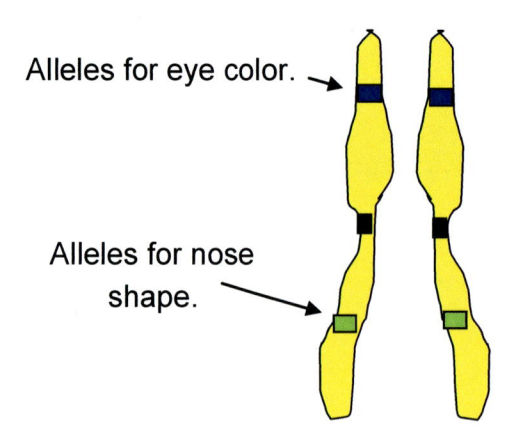

Alleles for nose shape.

On each pair of chromosomes, there are the same genes for specific characteristics. However, the code, itself, may not be identical.

The reason for this pairing of chromosomes is that one member of each pair comes from the father parent and the other member of the pair comes from the mother parent. So, within each of your 23 pairs of chromosomes, one member of each pair came from your mom while the other chromosome came from your dad. Together, this pair of chromosomes provides the information necessary for you to be alive. The father of the baby provides his information in the form of a single cell and the mother does likewise. Together, these two cells join to become one cell which eventually grows and divides into more cells to become the new living creature. Ideally, this new creature has all of the necessary information to carry out all life processes.

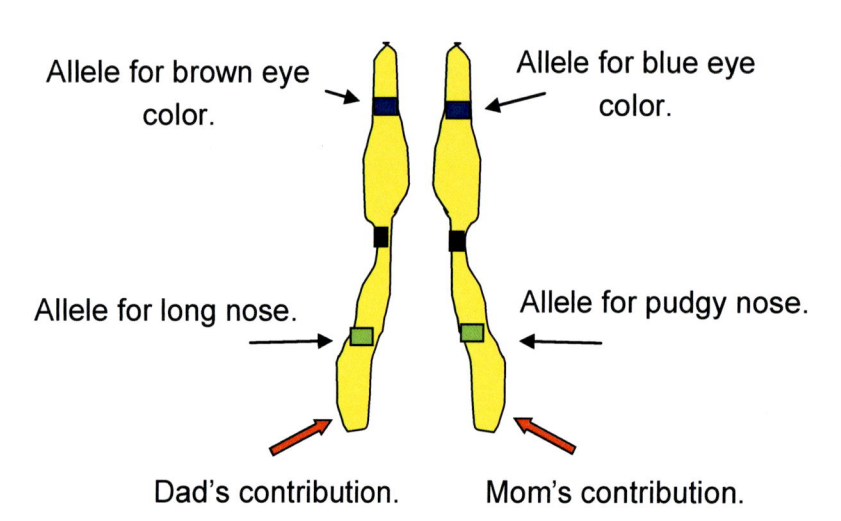

Allele for brown eye color.

Allele for blue eye color.

Allele for long nose.

Allele for pudgy nose.

One member of each pair came from the father while the other member came from the mother. Each contributes genetic information to the offspring.

Dad's contribution.

Mom's contribution.

Living things which have pairs of chromosomes in all their cells are called diploid organisms. The prefix di– refers to two and the root word –ploid refers to chromosomes. We can also describe these creatures as being 2N where the 2 refers to how many members of each chromosome (N) are present. Animals, including humans, and plants are considered to be diploid or 2N.

> Di = two
>
> -Ploid = chromosomes
>
> Diploid = having pairs of chromosomes (2N)

But, now we face a dilemma. We've established the fact that animals, including humans, provide the genetic information for their offspring through the passing on of it through their chromosomes. We've said that the father provides half of this information and the mother provides the other half and that these "halves" come together to make the new "whole." If the father provides a 2N (diploid) cell to the mother's 2N (diploid) cell, we would theoretically get a baby which would be 4N or have *two* sets of each pair of chromosomes!

Obviously, this isn't the case as you and I, which are offspring from our parents, are 2N and not 4N! In order for us to be 2N, at some point 2N cells in our father and mother became 1N cells. And then, later, when these two cells became united in the process known as fertilization, a 2N individual resulted. This reduction in the number of chromosomes (from 2N to 1N) present in these particular cells is the process known as meiosis.

> Meiosis = the process where 2N cells are reduced to 1N cells

Don't confuse the name of this process with the process we discussed earlier: mitosis. Mitosis the process of cell division whereby the daughter cells each end up with the exact same number of chromosomes that the parent cell had. In mitosis, a 2N parent cell divides to create two 2N daughter cells. Now in meiosis, the number of chromosomes which result from the process is cut by half.

> Mitosis = the process where 2N cells divide to produce more 2N cells.

In meiosis, 2N cells divide to result in 1N cells. 1N cells (meaning that only one member of each original pair of chromosomes is present) are known as haploid cells (half the original number of chromosomes). It is when the two haploid cells join that we have 1N + 1N to get a 2N or diploid individual. Two haploids, when brought together, make a diploid!

> Hap = single
>
> -ploid = chromosomes
>
> Haploid = having one chromosome (1N)
>
> or half in the case of 2N individuals.

So, how and where in humans and animals does this process of meiosis take place? If we were to look at samples of cells from every part of our body, we'd find that the cells which have nuclei (mature red blood cells in humans don't have nuclei) they are all diploid (2N) except for the cells produced by the testis of the male and the ovaries of the female. The 2N or diploid cells of our bodies are called somatic cells. Here the term somatic refers to the flesh of the body.

> Somatic cells = all cells of the body which are 2N
>
> Sex cells = cells of the body which are 1N

The cells produced by the testis and ovaries which are 1N or haploid are referred to as the sex cells. Yet another name for these cells is the gametes. And, to even be more specific, the gametes produced by the male are called sperm cells or spermatozoa while the gametes produced by the female are called eggs or ova or, singular, ovum.

> Male sex cells = male gametes = spermatozoa or sperm
> Female sex cells = female gametes = ova or eggs

Let's pause and review. We've said that animals, including humans, consist primarily of diploid cells, meaning their cells consist of pairs of chromosomes (2N.) One member of each pair comes from 1N or haploid cells contributed by each parent. The haploid cell from the dad is known as the sperm cell and the haploid cell from the mom is known as the ovum. The process of creating these haploid cells (2N cells going to 1N cells) is meiosis. In the process of fertilization, these haploid cells join to form a new individual which is 2N or diploid.

Let's take a closer look now at the process of meiosis. As we mentioned earlier, we said that meiosis in animals and humans takes place in the testis of the male and ovary of the female. If we look closely at the development of the testis and ovary in a developing human baby (or animal), we'd find that in the embryo, the cells that make these organs both come from the same location! In the very early stages of development, cells of the testis and ovaries were one in the same. Due to influences of the chromosomes this set of cells either becomes testis or ovaries. In the case of humans, the cells of girls remain in the same location where they begin which is in the upper abdomen and become the ovaries. In males, these cells become the testis and migrate down into the scrotum shortly before birth or very soon thereafter, which is the sack which holds them outside the body. So, it is in the ovaries of females and testis of males that meiosis occurs.

If we examine the cells of the ovaries and testis more closely, we find that there are two main populations of cells present. Certain parts of these organs are made of cells which serve to support the actual cells that become the gametes (ova and sperm cells). They may provide nutrients for the gametes or work to hold or transport the gametes. These supporting cells are 2N or diploid cells. The cells which eventually become the gametes are also originally 2N in chromosome number, but become haploid as they undergo meiosis. The process of meiosis has several steps. We'll examine the main steps to try to avoid confusion.

The cells which eventually become ova or sperm are called primordial sex cells. As we said above, they are originally 2N cells. In the early stages of meiosis, these primordial sex cells actually undergo mitosis, but they don't quite split into two cells. They remain as one cell having 4 sets of chromosomes (4N). As meiosis continues, this cell then divides into 4 cells each having only 1 set of chromosomes (1N). So, theoretically, for each primordial sex cell we begin with, we end up with 4 gametes. Now, this is true in males: one primordial sex cell results in four sperm cells. However, in females, only one of the resulting 1N cells survives. The three other cells are called polar bodies and do not survive. Cells which survive are termed viable while cells which do not survive are termed nonviable. So, in females, for each primordial sex cell we being with, we get one viable ovum and three nonviable polar bodies. The process of creating sperm cells from single primordial sex cells is called spermatogenesis. The root word genesis means to make or create. The process of creating ova is called oogenesis.

One other interesting thing to note is that female humans and animals are born with all of the primordial sex cells that they will ever have for their life. Women do not make more primordial sex cells throughout their lives. Therefore, a woman has a limited number of ova she can produce. Males, on the other hand, do make more primordial sex cells throughout their lives and therefore are not limited in the total number of sperm cells they can produce in their lifetime.

> Spermatogenesis = creation of sperm cells from primordial sex cells.
> Oogenesis = creation of ova from primordial sex cells.

The gametes (sperm and ova) produced from meiosis have the potential for creating a new living organism. Each will contribute its half of the needed genetic information to create a viable living thing (1N + 1N = 2N). The joining of the two gametes, as we mentioned earlier, is known as fertilization.

> **Fertilization = the process in which two 1N (haploid) cells join to result in one 2N (diploid) cell.**

Depending upon the living organism we are considering, fertilization can take place internally (within the female's body meaning that the sperm cells must be delivered to the appropriate location at the appropriate time) or fertilization may take place externally (meaning that both the female and male release their gametes from their bodies in the same location at near or close to the same time, allowing them to mix with the hopes that at least one of each join together to create a new offspring). Once joined, this new 2N cell undergoes innumerable events of mitosis to eventually result in the new living organism.

The fact that this all can happen, whether internally or externally, where two microscopic cells, each carrying an overwhelming amount of information, can join together in the appropriate fashion to result in a new, unique living creature capable of carrying out all life processes, including the possibility of someday reproducing on its own, is truly an amazing miracle!

Let's pause here and review. We've said that:

- The cells of animals, including humans, and higher plants are diploid cells (2N), except for the sex cells, known as gametes, which are haploid (1N);
- Meiosis is the process whereby 2N cells of the testis and ovaries, known as primordial sex cells, become 1N;
- In males, due to meiosis, each primordial sex cell results in four sperm cells;
- In females, due to meiosis, each primordial sex cell result in one ovum and three non-viable cells, known as polar bodies;
- Fertilization is the process whereby a single sperm cell joins with an ovum to result in one cell being 2N which then undergoes mitosis many, many times to eventually produce the new living organism.

There is one more feature of this type of reproduction that is important the realize. When two parents are involved in the creation of a new living organism, the possibility for genetic variation in the resulting offspring exists. The gametes each bring with them variations for characteristics (alleles) to be found in the new offspring. There is a continual mixing of genetic information with the results always being a unique individual. We'll examine this genetic mixing in much greater detail in Lesson 13.

The form of reproduction where two parents, a male and female, are involved is known as sexual reproduction. However, not all living things reproduce in this way. The processes some living creatures utilize where only one parent is involved is known as asexual reproduction, where the prefix a– indicates without. Some living creatures are capable of creating new, fully functional living organisms on their own. What do you think the genetic makeup of these new organisms might look like? You're correct if you say that the genetic makeup of the new organism is exactly like that of the parent organism. Unlike sexual reproduction, asexual reproduction results in, essentially, carbon copies of the parent organism.

> Sexual reproduction involves two parents (a male and female) and results in offspring having a combination of genetic information from each parent.
>
> Asexual reproduction involves only one parent and results in offspring with genetic information identical to the parent.

Let's take a closer look now at various ways living creatures can reproduce asexually. The first method we'll look at is called fission. Fission means to divide or break into two. Organisms which reproduce by fission reproduce by just breaking into two parts. The parent cell breaks into two new cells. You may be thinking that this process sounds a lot like mitosis and you are correct. These organisms—and they generally are the very small, unicellular ones (uni– meaning one or single)—undergo mitosis and then, for the most part, divide themselves into two parts, with each new part having enough chromosomes and other organelles to continue their life processes.

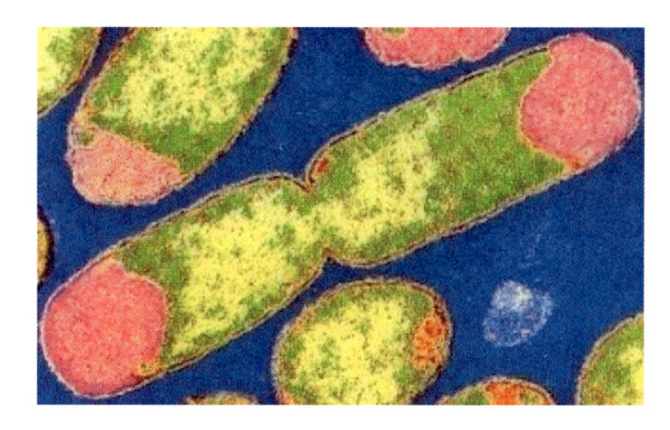

Bacteria divide through binary fission. One cell makes copies of its chromosomes and then splits into two cells.

In some creatures, the split results in two equal parts and this is called binary fission (binary indicating two parts). Bacteria reproduce in this way. One bacterial cell divides into two and after a period of time, those two divide into two and then those two divide into two. You can readily see how bacteria can divide extremely rapidly and what was once a very small population of bacteria can, in a short amount of time, become overwhelmingly large if conditions are right for their division.

Think about leftover food that doesn't get properly returned to the refrigerator. The cold conditions of the refrigerator make it difficult for bacteria to function and ultimately reproduce well. As a result, your food, when stored in the refrigerator does not quickly spoil. However, the warm conditions on the kitchen counter where leftover food might be left sitting are great conditions for bacterial

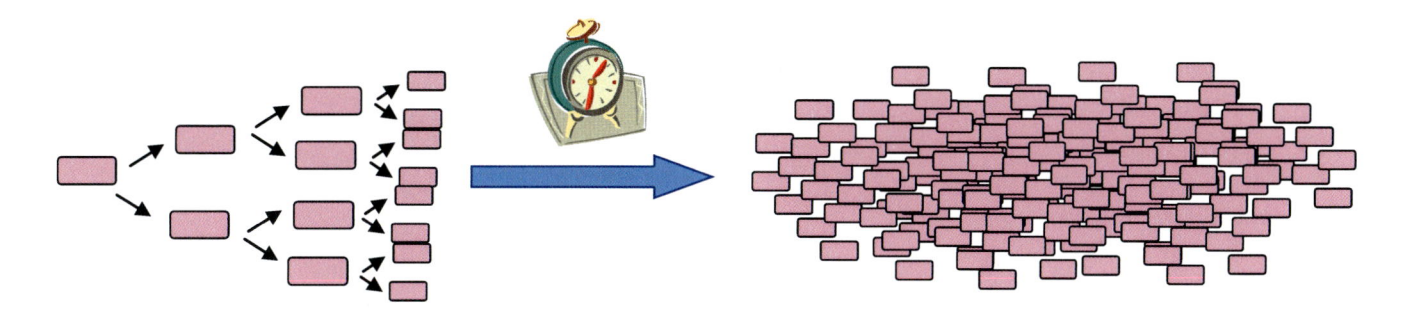

One bacterial cell, through binary fission, can reproduce to produce millions of new bacterial cells in a relatively short period of time when conditions are right.

growth and reproduction. In the hour or so that the food sits out on the counter, the bacteria that got introduced into the food from the serving utensils and serving bowl rapidly begin to undergo binary fission. Several generations may quickly be produced which allows numbers to get so high that the food becomes hazardous for one to safely eat. Some bacteria themselves are detrimental to human consumption (they cause infection of the gastrointestinal system) but many produce toxins or poisons which cause us the very uncomfortable effects of food poisoning. Binary fission, while efficient for bacteria reproduction, can be quite the nemesis for other living creatures like you and me. The "moral" of the story is to make sure you quickly refrigerate foods in adequately working refrigerators to reduce the growth of bacteria.

Another type of asexual reproduction is known as budding. Like fission, budding initially involves mitosis where the original set of chromosomes in the parent cell gets copied. However, instead of the whole parent cell dividing into two halves, in budding only a small portion of the parent cell pinches off and eventually breaks free to live on its own. This process of pinching and breaking off is called budding and, in this case, specifically external budding. The bud which breaks free grows to eventually become the size of the parent cell and then it, too, begins to create buds which break free and the process continues. The yeast used to make bread reproduces by external budding. A water creature known as the hydra, which happens to be multicellular, also reproduces by external budding.

Yeast, which is used to make bread, reproduces by external budding. Chromosomes are duplicated and one copy moves into the bud. The bud eventually breaks free to create a new independent cell.

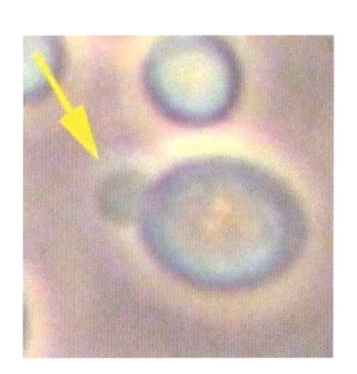

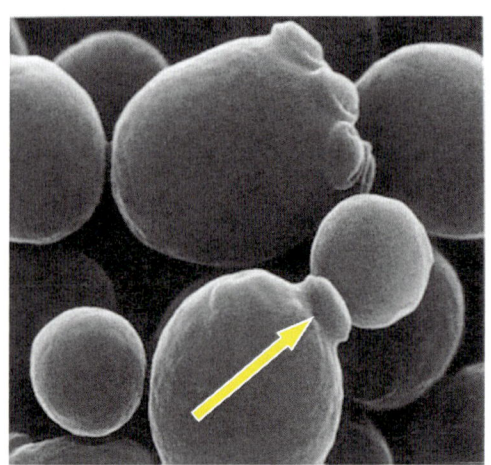

The hydra, as shown in this sequence of diagrams, reproduces by creating an external bud (1). The bud contains identical genetic material to the parent organism. Eventually, the bud breaks free (4) from the parent and begins to live independently.

Internal budding, on the other hand, is a process whereby chromosomal copies are made but, instead of forming a bud on the surface of the parent cell, the new organisms form *within* the parent cell. In some cases these newly formed internal buds eventually eat the parent cell! An example of an organism which utilizes this means of reproduction is an intestinal parasite often carried by cats known as *Toxplasma gondi*. This organism is classified as a protozoan and lives within the cells which line the intestines of cats. While in the cat's intestinal cells, it is capable of utilizing sexual reproductive methods to produce cells known as oocysts. When the infected intestinal cells are shed from the cat's intestine they find themselves getting transported via the cat's feces out into the environment or in situations with indoor cats, the cat's litter box. The person who then comes into contact with the cat's litter may unintentionally get some of these microscopic oocysts on his or her hands and then accidentally ingest them.

Once inside the new host (a human, now), the oocysts begin to grow and develop. However, for some reason not totally understood, the toxoplasma organism only divides by internal budding while in a host *other* than a cat. They reproduce sexually while in cats, but asexually while in other creatures! As we described internal budding above, buds form within the original parent cells (developed oocysts) which eventually consume the parent cell. These "new" cells then grow and they, too, divide by internal budding. Not only can humans be infected by coming into contact with cat feces, any warm-blooded animal may become infected by getting the oocysts into their mouth. In humans, for the most part, infection by the toxoplasma organism doesn't cause any great problems, however, if the person is pregnant at the time of coming into contact with the oocysts, the protozoan can cause problems for the unborn baby. Beyond the lesson learned here about organisms reproducing by internal budding, is the lesson that women who are pregnant or may soon become pregnant should not have the chore of cleaning the kitty's litter box!

While binary fission and budding are two types of asexual production which are primarily utilized by the very tiniest of organisms, yet another type of asexual reproduction, known as vegetative propagation, is used by larger, multicellular organisms. The word vegetative refers to a living portion or part of the parent organism and the term propagate means to share or disperse. In vegetative propagation a portion of the living creature actually moves beyond the parent organism and, over time, begins life on its own.

Strawberry plants are a great example of vegetative propagation. As the strawberry plant grows it sends out specialized stems called stolons. At the far end of the stolon are cells which are developing into tiny, "baby" strawberry plants. These stolons extend beyond the normal perimeter of the plant and the end with the "baby" comes to rest on soil away from the "mother" plant. Contact with the soil stimulates roots to begin growing from the baby plant and the job of the stolon becomes less and less active. Eventually the baby strawberry plant becomes established as an independent organism and the stolon dies.

Strawberry plants reproduce through vegetative propagation. A special stem, the stolon, emerges from the parent plant and touches the nearby soil. Leaves and roots begin to emerge from the tip of the stolon and a new plant develops.

A common houseplant, the spider plant, also reproduces by sending out stolons. If the "baby" spider plant is allowed to touch soil, it will develop roots. Eventually, it can be severed from the "mother" spider plant.

Other plants which reproduce in the way include the iris. If one examines the root portion of the iris you'll readily find an enlarged tube shaped structure which lies horizontal to the surface of the ground or just beneath it. Smaller, more conventional looking roots can be found extending beneath this structure. The enlarged tube-like structure is called a rhizome and functions like a storage cellar for the iris plant, much like potatoes do for a potato plant. Along the rhizome small iris plants emerge and new plants eventually develop. Sweet potatoes and yams are a variation of rhizomes known as tubers.

The bearded iris reproduces through rhizomes. The rhizome functions to store nutrients produced in the leaves of the plant. "Baby" iris plants emerge from the rhizomes having the same genetic make-up as that of the parent iris plant.

199

Many grasses reproduce this way as well. The parent plant sends out lateral "roots" which produce new leafy structures above and roots below. In this manner large colonies of plants with the same or very similar genetic makeup develop. Plant growers and nurserymen can take advantage of this natural process and develop innumerable plants from a few parent plants. Maybe you've taken cuttings from houseplants and rooted them yourself!

Common Brome is a type of grass which reproduces by sending shoots just beneath the surface of the ground. From these shoots new plants develop and the grass quickly spreads across the ground.

So far, we've discussed three types of asexual reproduction: binary fission, external and internal budding and vegetative propagation. A fourth type of asexual reproduction is known as sporogenesis which is the making of spores. Sporogenesis is utilized by many types of fungi and some plants for reproduction. You may be familiar with the puff of spores which arise after disturbing the common mushroom known as the puffball. If you examine the underneath side of a mushroom you've purchased at the grocery store, you'll see several slit-like divisions. It is from between these divisions that mushrooms form their spores. A group of plants which utilize sporogenesis as a portion of their reproductive cycle are the ferns. Let's explore more about the make-up of these spores.

Ferns that you are most likely familiar with that are often used as "fillers" in flower arrangements made by florists are only one phase or portion of the fern's whole life cycle. Ferns have two phases of their life cycle: one phase in which their cells are all haploid and a second phase in which their cells are all diploid. The phase that we see in flower arrangements is the diploid or 2N stage. This stage is known as the sporophyte stage.

The cells of a frond of a fern are 2N and this stage is known as the sporophyte stage. On the underside of the fronds, you can see the sori (small, bead-like nodules) which produce spores. The spores released from the sori are 1N. The spores fall to the ground and develop into the gametophyte stage. The gametophyte stage is much smaller and usually very close to the ground. Eventually, they produce gametes which join together to create another sporophyte which begins the cycle again.

If you examine the underneath surface of the leaflets of the fern, you can see many small black nodules which are capable of producing spores. These cells within these nodules which make spores are known as sporangium. Sporangium cells, which are 2N, undergo meiosis to produce spores. Recall that meiosis results in cells which have half the number of original chromosomes present, therefore spores are haploid or 1N. As the fern plant matures, eventually, these spores are released from the nodules and float or fall to a new location. There they begin the second phase of the fern's life cycle known as the gametophyte stage.

In the gametophyte stage, the spore undergoes many, many cycles of mitosis and becomes an independent plant, however, it remains haploid and also may not resemble the bushy, leaf-like fern plant we are accustomed to recognizing as a fern. Many of these forms are much smaller and lie close to the ground. (Note: this is not true for all plants that utilize sporogenesis. In fact, in some species it's just the opposite and in some, both phases are equally developed). As in the sporophyte stage, parts of the plant, known as gametangium, produce gametes that are 1N. These gametes get released and eventually join with other gametes (as in sexual reproduction) to form, once again, a 2N individual. This 2N individual is the sporophyte stage or "bushy fern" stage we began with in the preceding paragraph. This cycle continues will alternating generations of 2N and 1N individuals.

Another form of asexual reproduction is known as fragmentation. Starfish and lichens are examples of living creatures which possess this very interesting means of reproducing. In the case of a starfish, if a portion of the body is inadvertently broken away, the fragment has the ability to regenerate into a new independent organism.

Starfish can reproduce sexually or asexually. In the asexual form, known as fragmentation, a portion of the starfish may become broken away from the main portion of the body. Over time, the broken fragment regenerates the central disk and other arms and continues life as a complete starfish.

Let's pause now and summarize what we've discussed in this lesson on how living things reproduce. We said:

- Living things either reproduce sexually or asexually. Sexual reproduction involves the participation of two parents that each contribute genetic information for the new baby. Asexual reproduction only involves one parent;

- In organisms that reproduce sexually, there exists a pair of each chromosomes in each of that organism's cells: one chromosome that was contributed by the mother of that individual and one chromosome contributed by the father of that individual. The cells which have both chromosomes present are called somatic cells. Somatic cells are diploid (2N);

- In order for organisms that reproduce sexually to reproduce, they must reduce the number of chromosomes present by half in the cells they contribute to the new offspring. This process of reducing the number by half is known as meiosis;

- Meiosis results in cells which are haploid (1N) and are known as gametes. The female gamete is the egg or ovum. The male gamete is the sperm. In animals and humans, the egg is produced by the ovary and the sperm are produced by the testis;

- The joining of the egg and sperm to produce a "new" 2N individual is known as fertilization;

- Organisms which reproduce asexually provide all the genetic information from a single parent organism;

- The more common methods of asexual reproduction include: binary fission, budding, vegetative propagation, sporogenesis and fragmentation;

- Some organisms employ both sexual and asexual forms of reproduction in their life cycles.

Before we leave the topic of reproduction in living things, we need to re-emphasize the major difference between sexual and asexual reproduction. This major difference is that in sexual reproduction, because of the involvement of *two* parents, there is a "mixing" of the genetic information provided by the parents that is received by the offspring. In simpler terms, as in the case of you and me, some traits you possess came from your mom and some traits came from your dad. The term trait we are using here refers to an observable characteristic in the individual. Examples of a trait could be eye color or hair texture or nose shape or finger length, etc.

On the other hand, in asexual reproduction, where there is only one parent, there is no mixing of genetic information possible. All of the inherited traits come from the one parent. Whatever traits are found in the parent organism will also be found in the offspring.

In our next lesson, we will explore in greater detail how traits get passed from parents to the offspring of creatures that reproduce sexually, a field of study known as genetics.

Lesson 12 Lab Activity: Growing Oyster Mushrooms

THIS LAB ACTIVTY EXTENDS OVER SEVERAL WEEKS.

PLEASE READ THESE INSTRUCTIONS IN THEIR ENTIRETY BEFORE BEGINNING.

The purpose of this great lab activity is to experience first-hand the life cycle of the oyster mushroom. The oyster mushroom is an edible mushroom meaning it is safe to eat. Note that not all mushrooms are edible and some are poisonous causing serious illness. It is important to follow the instructions provided here to insure your experience is successful and not harmful to your health.

Materials you will need for this lab include:

- TP Oyster Mushroom kit. The kit includes mushroom spawn, special growing bags, rubber bands and instructions. This kit is available through Field and Forest Products, N3296 Kozuzek Rd, Pestigo, WI 54157. Ordering online is easy at: http://www.fieldforest.net/Oyster-TeePee-Kit-Small/productinfo/W-TPS/

- 7 rolls of toilet paper

- Source of boiling water

- Cooling rack used for baked goods or colander

- Access to a dark closet or cupboard

- Access to a refrigerator for a short period of time

Procedure:

Before we move right into the procedure, let's briefly examine the life cycle of the oyster mushroom. Mushrooms, in general, have two main segments in their life cycle: the mycelial stage and the fruiting stage. The mycelial stage is that portion of the life cycle where the fungus is growing beneath the ground or within the log and is not usually visible. It consists of mycelium which are chains of tube-like cells that are slowly making their way through the substrate (tree root, log, etc.) that is using for its food supply. You might think of this stage as when tiny little "fingers" of fungi are "invading" the log. A triggering event takes place, which is usually a weather change, that stimulates this non-visible mycelium to move to the fruiting stage. It's in this stage that the visible mushroom develops on the surface of the log or ground.

Friendly Biology

The oyster mushroom is a fungi which belongs to the basidiocarp division meaning that it creates spores from basidia when it is in its fruiting stage. Let's step back a little and talk about the basic parts of a mushroom and then we'll discuss basidia. In the photos below you can see the parts of a simple mushroom. There is the stipe or stalk which holds up the cap of the mushroom. Beneath the cap we find the many slit-like structures known as the gills of the mushroom.

Cap

Stipe or stalk

Stipe or stalk

Gills

Note that it is on the gills that we find tiny, club-like projections called basidia. It is from these basidia that spores are formed. As we discussed earlier in this lesson with ferns, spores in mushrooms are the 1N or haploid reproductive cells. Millions of spores are produced by each mushroom. In the photo below you can see two caps of mushrooms emerging from a planter box sitting on a deck. Notice the brown powdery substance on the surface of the decking. These are the spores which have been released from these mushrooms.

The brown powdery substance beneath the caps of these mushrooms are 1N spores.

As we have mentioned, the spores are 1N or haploid cells. When they are released from the gills they float down onto the ground or surface below in hopes of finding another 1N cell in which to join to become the 2N cell again. In humans, as you know, these cells are the sperm and ova, each distinctly a male or female cell. In mushrooms, instead of having male and female 1N cells, there are what is known as mating types. Some types of mushrooms can have several different mating types in hopes of improving the chances that two compatible spores can meet to produce the 2N phase of the life cycle. In oyster mushrooms, there are four different mating types.

Once two compatible mating types have found each other, the cell membranes join, but the nuclei remain separate. This joining of cells is known as plasmogamy. The cell is technically now 2N but the genetic material has not yet combined. During this phase it divides into more 2N cells which continue to grow and collect nutrients from the substrate on which it is growing. This is the mycelial phase which is usually deep within the substrate and not readily visible.

As we mentioned earlier, a triggering event takes place which stimulates the development of the fruiting body or visible mushroom. Note that the cells of the mushroom are still 2N but the nuclei holding the genetic material have not yet combined. As the gills form, the cells on the surface form the basidia. Finally, the nuclei within the cells join to form the true 2N cells. However, they quickly undergo meiosis to form four haploid cells (which are the spores) and the cycle begins again!

In your TP Mushroom Kit, you will find a bag of spawn. Spawn is the mycelial stage of the mushroom life cycle which has been allowed to grow on a substrate to make it easy to maintain. The substrate your oyster mushroom spawn is growing on is likely a grain. Can you think why a grain might be a great choice for mycelium to grow upon? If you said grain is a concentrated source of carbohydrates (remember amylose from Lesson 3?) you're absolutely correct. The oyster mushroom mycelium utilize the carbohydrates found in the grain to readily grow and be maintained while they are waiting to be deposited in the substrate intended. In the case of this lab, the substrate will be the rolls of toilet paper. Let's go now to the procedure for this lab.

Procedure:

1. If you've ever had the misfortune of getting rolls of toilet paper wet, you may likely know that once they dry out, many times they become covered with mold. This mold originated from spores found in the toilet paper when you purchased it or from spores floating around your house which happened to land on the toilet paper before or after it got wet. Mold is another form of fungus. We'll discuss it in greater detail in a later lesson. Conditions allowed these spores to develop and you ended up with a roll of moldy toilet paper. In this lab, we don't want to have any of these "wild" fungal spores developing and later competing with our intended oyster mushroom spawn. To get rid of them, we'll attempt to kill them by pouring boiling water over the toilet paper rolls before the add the spawn. Here's how to do this:

Bring 2-3 quarts of water to a boil in a large sauce pan. While the water is heating, place your rolls of toilet paper on the cooling rack or in your colander over a sink. You may have to do this step in two batches depending upon the size of your cooling rack or colander and sink. Place the rolls in an upright position (tube going up and down).

Once the water is at a rolling boil, carefully pour the water over the rolls of toilet paper. The idea is to thoroughly drench the rolls with very hot, sterilizing water in hopes to eliminating most of the "wild" mold spores present. Allow the toilet paper rolls to drain into the sink and cool well before going to the next step.

Instead of pouring the hot water over the rolls, you can dip the rolls one at a time into the water. Carefully lower the rolls down into the water using kitchen tongs and allow the water to fully soak the rolls. Then, move them to the colander or cooling rack to drain.

2. Next, find the special bags which came in your kit. These bags have a special vent system to allow the correct air environment to take place inside the bag. Take a cooled roll of toilet paper and place it down inside a bag with the tube facing upward. Roll the top of the bag down to allow easy access to the roll.

3. Now find the bag of spawn which came in your kit. Look at the grains and find the white, lacy-like mycelial cells. Carefully open the bag and pour the spawn down into the tube of the toilet paper. Fill the tube with the spawn; you don't have to pack it tightly. This step is called inoculation. You have inoculated your substrate (TP) with the fungal organisms.

4. Roll the sides of the bag back up and secure the bag closed with a rubber band. Make sure the rubber band is placed above the special ventilation window. Using a marker, write the date you filled the roll on the side of the bag and your name if necessary. Repeat this step with the remainder of your rolls. If you purchased various kinds of mushroom spawn, write the name of each type on the bag, too.

5. Next, place your bags into a dark closet or cupboard that is almost never used. During this time, the mycelium will grow out into the toilet paper and utilize the _____ found in the toilet paper as a food source. Hopefully, the word cellulose popped into your head! Keep the bags in this location for 4-6 weeks. It's okay to peek at them every now and then, but it's important they stay in the dark. This is done to simulate being

beneath the bark of a dead tree where oyster mushrooms like to grow.

5. After 4-6 weeks have passed, you can remove your bags from the cupboard. Look through the side of the bags. You should see a fluffy white mass totally covering each toilet tissue roll. This is the mycelium of the oyster mushroom. If you find what appears to look like black mold or a colored mold-like formation on any of your rolls, your roll may have been contaminated with a wild fungus. These are best thrown out.

6. Recall, that to stimulate the formation of the mushrooms (fruiting bodies,) a triggering event must occur. In this case, you'll simulate a cool weather change for your bags by placing them into the refrigerator for 48 hours. After this time has passed, bring your bags out and open them. The release of carbon dioxide which has built up inside the bag and change in temperature will trigger the fruiting stage of the fungus.

Place the bags in a lighted area and begin watching for signs of mushrooms to appear. At first, you'll see tiny little dark, bead-like nodules forming in clusters on the surface of the mycelium. These will grow into beautiful mushrooms over the next few days. You may find it helpful to mist your rolls daily to keep the fruiting process active. The caps of the mushrooms will grow to about the size of a hen's egg and can be eaten at about any time. To harvest them, gently break them away from the side of the roll. Wash them and use them like you would any fresh mushroom you might buy at the grocery store.

Once it appears that all of the mushrooms have been harvested, if you wish, you can close up the bag once again and place it back into the dark cupboard. Repeat the process all over again until all of the cellulose in the toilet paper has been exhausted. Also, you don't have to stimulate all of your rolls to fruit at the same time. Rolls can be kept up to six months in your refrigerator before taking them out and opening.

Fruiting bodies of the oyster mushroom emerge from the surface of your inoculated roll of toilet paper. Photo courtesy of Field and Forest Supply.

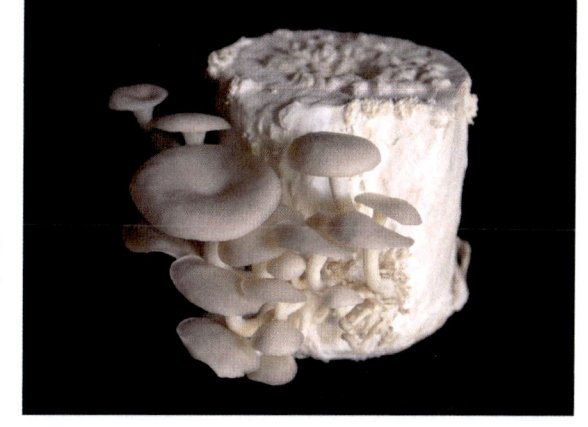

Lesson 13: Genetics

In the last lesson we explored how living things reproduce. We said that some living things reproduce by having two parents contribute genetic information to the new offspring and that this form of reproduction was known as sexual reproduction. Other living things reproduce with only one parent involved and this was known as asexual reproduction. We emphasized the idea that in sexual reproduction, because there were two parents involved, there would be a "mixing" or potential for assortment of the genetic information passed on to the new individual. In this lesson, we'll explore in greater detail how this assortment takes place. The study of this assortment is known as genetics.

In the late 1800's a monk, by the name of Gregor Mendel, was keenly interested in how traits or characteristics in living things were passed on to new generations. He did several studies with plants, pea plants in particular, which revealed many principles of genetics. At the time of his studies, however, many of his contemporaries refuted or simply ignored his results. It was not until years later that work done by other biologists found Mendel's work to be highly credible. From Mendel's work with pea plants the study of Mendelian genetics was born. We'll examine some of the basic principles of Mendel's work in this lesson.

Gregor Mendel, 1822-1884, proved the existence of paired units of heredity, today known as genes.

First, we need to introduce two terms: genotype and pheonotype. Genotype refers to the record or list of genes held by an organism. You might think of the genotype as being analogous to the recipes in the "cookbook" analogy we used in Lesson 7. This set of recipes makes up the genotype of the individual.

Phenotype, on the other hand, is the results or what gets revealed due to the genes being present. If we use our cookbook analogy, we can say that phenotype would be the collection of prepared dishes and foods made from the recipes (genes). Traits such as eye color, hair texture, nose shape, etc. are all examples of phenotype. The phenotype also includes behaviors observed due to the genes present in the individual. Genotype is the record of genes while phenotype is the result of that collection of genes.

The blue, purple and green coloration of the feathers of this homing pigeon are examples of the phenotype of this bird.

Friendly Biology

> Genotype = the record of genes present in an individual.
> Phenotype = the results of the genes present in an individual.

In Lesson 12 we introduced the term allele, which is important in the study of genetics. Allele refers to a possible variation in a gene. For example, we all have a gene which codes for eye color. An allele of the eye color gene would be one of the many possible colors for eye color. Alleles for eye color would include brown, blue, green, hazel, etc. In our cookbook analogy, we can think of an allele as being like a possible flavor for cookies. The recipe (gene) is for cookies and the possible flavors (alleles) might include chocolate chip, peanut butter or oatmeal!

Cookbook = full set of chromosomes

Dessert chapter = a particular chromosome

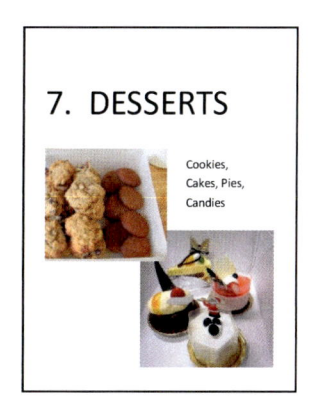

Cookie section = a particular gene on that chromosome

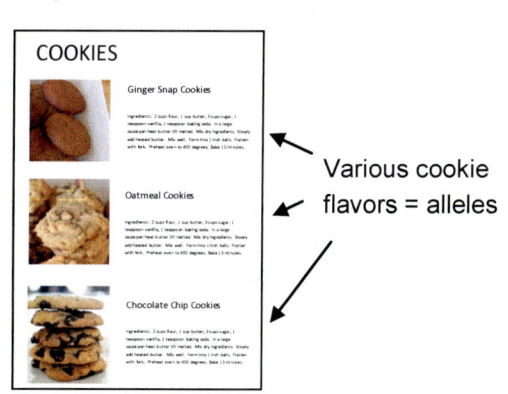

Various cookie flavors = alleles

In his experiments, Mendel observed several characteristics of his pea plants. One trait he observed in particular was whether the peas (the seeds themselves) were smooth or wrinkled on their surface (the seed coat). In this case, the texture of the seed coat would be the trait of interest and either being smooth or wrinkled would be alleles of that trait.

In growing many, many pea plants Mendel discovered some fascinating things about how various alleles appear in new generations of the plants. The first "big" idea he learned is known as the Law of Segregation. The Law of Segregation basically says that in diploid organisms (2N) during the formation of the gametes (1N), one member of each pair of chromosomes goes to each gamete. Recall that the primordial cells which undergo meiosis to form the gametes, like the other somatic cells of the body are 2N, meaning they have two chromosomes (one from the mother and one from the father). At meiosis, the pairs of chromosomes split and the resulting gametes receive one member of each pair.

Recall also from our earlier discussions of meiosis, that each primordial cell first undergoes an intermediate stage of mitosis to form two total sets of chromosomes (4N) and then each of these sets undergoes meiosis to result in four gametes (each being 1N). We learned that in males this resulted in four sperm cells. However, in females one viable egg cell results with three non-viable polar bodies. Within each of these gametes, we will find one member of each pair of chromosomes.

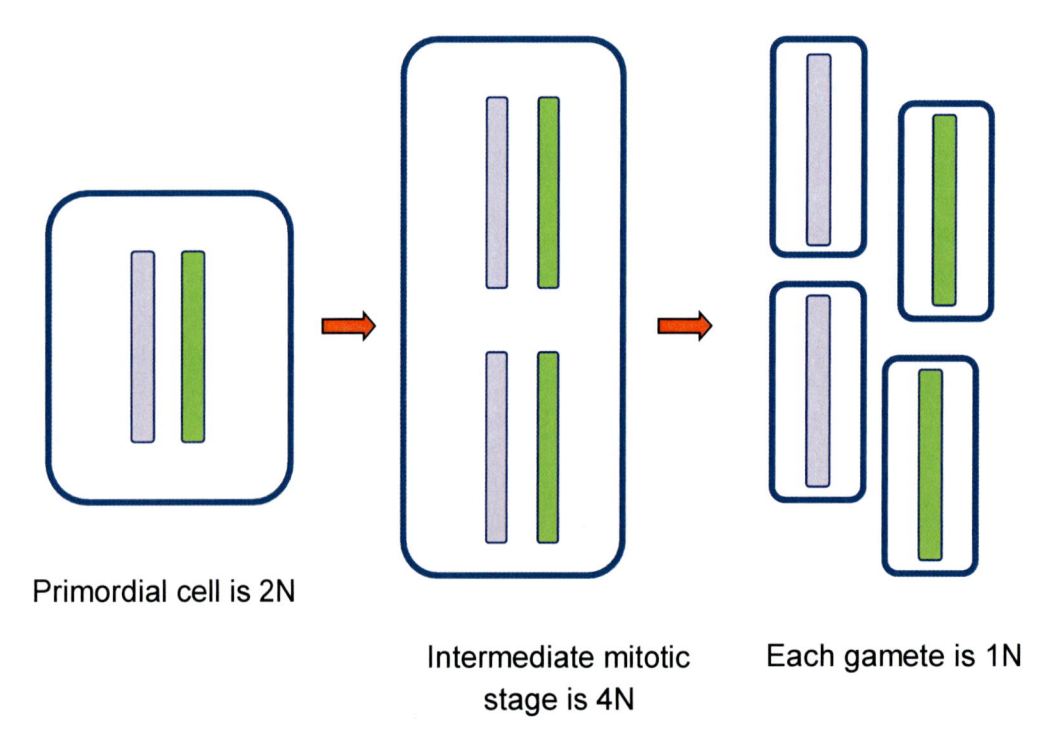

Primordial cell is 2N

Intermediate mitotic stage is 4N

Each gamete is 1N

Let's look at a theoretical example of the Law of Segregation. Suppose a married couple, John and Mary, desires to begin their family. John's cells are all 2N or diploid meaning he has two sets of all genes (one set came from his mom and one set from his dad). Note that this is John's genotype. In his primordial cells in his testis (which eventually make his sperm cells) during meiosis, because the 2N cells get reduced to 1N gametes, some sperm cells will receive the genes which came from his dad and some sperm cells will receive genes from his mom. The sperm cells get either one or the other; the genes are separated or segregated.

Likewise, with Mary. She has genes from her parents, too. Her primordial egg cells are 2N and when her ova form during meiosis, each ovum will only have 1 set of chromosomes, 1N, and therefore one allele for each trait. The ova she produces will either have genes from her dad or her mom. Again, the genes are separated or segregated.

At fertilization, as we learned in Lesson 12, the new baby will be formed from the combination of one sperm cell and one egg cell. In the example of John and Mary, the genotype of their baby depends upon which exact sperm cell from John joins with which exact ovum from Mary. This new combination of genes determines the genotype of little Junior.

Let's pause and review what we've learned so far. We said that:

- The earliest studies of genetics was done by Gregor Mendel who studied pea plants;

- Mendel developed the Law of Segregation which stated that each trait or characteristic of an individual is the result of two genes: one from the male parent and one from the female parent. During meiosis when the sperm and egg cells are made, these genes are segregated or separated. Each sperm or egg gets either a gene from that individual's mom or dad;

- At fertilization, the new individual's genotype is the result of the joining of each "half" of the code.

Let's look now at the second law of genetics developed by Mendel known as the Law of Independent Assortment. Earlier in our discussion we stated that each sperm or ovum that gets produced receives only one allele either from the mom or dad of that individual. The Law of Independent Assortment says that each trait is independent of other traits when it comes to determining which genes go to which new ova and sperm cells. For example, if we look at two traits, say eye color and hair color, the Law of Independent Assortment says that one kind of allele for eye color doesn't always go with one kind of allele for hair color. They get passed on independently from each other. However, we should note that this law was later found not to be totally correct as some traits are found to be linked to one another. Nonetheless, Mendel was the first to realize that traits "went" to gametes in an independent manner.

Let's look now at how the genes one inherits from one's parents get expressed. We said earlier in our discussion that this expression or revelation of the genes was called the phenotype of the individual. We need to realize first that while each of us has genes from both our mom and dad, for most of our genes we only see the results of one OR the other and not necessarily a blending of the genes. In other words, in the case of eye color, we either have one color of eye (which came from either our mom OR dad) and not a mix of the two colors. This expression of either one OR the other (but not a mix of the two) led to the discovery of the idea of dominance and recessiveness of genes. While both genes are present, only one gets expressed. The one form (allele) that *gets* expressed is known as the dominant allele. The form of the gene, while still as present as the other, that does *not* get expressed is known as the recessive allele.

> Dominant allele = variation of the gene which, when present, always gets expressed. In other words, if it's there, you observe the results.

> Recessive allele = variation of the gene which, even though it's present, is not expressed. In other words, it's there, but you don't observe the results.

When studying these concepts of dominance and recessiveness, Mendel, developed a system of lettering genes to note whether a gene is considered a dominant allele or recessive allele. A trait is given a particular letter. The upper case or capital form of that letter indicates the gene is a dominant allele. The lower case or "little" form of that letter indicates that form of the gene is recessive.

> Dominant alleles are noted using upper case letters:
> H, R or W for example.

> Recessive alleles are noted using lower case letters:
> h, r or w for example.

Let's look at an example. Suppose we are looking at the gene for hair color and we'll use the letter H. The form of the gene which is dominant will be given the symbol of upper case, H. The recessive form of the gene will be given the lower case symbol, h. Dominant alleles are symbolized with upper case letters while recessive alleles are symbolized with lower case letters.

Let's go back to our couple, John and Mary. If we were able to examine John's genotype we might find that for hair color he has the following genotype: HH. This means that John inherited the dominant allele (H) from his dad and the dominant allele (H) from his mom. If his genotype were Hh, we would say he inherited the dominant allele H from his dad and the recessive allele h from his mom. If his genotype were hh, we would say that John inherited the recessive allele h from both his mom and dad. A fourth possibility would be hH which means that John inherited the recessive allele from his dad and the dominant allele from his mother. The first letter in the genotype is the allele inherited from the dad and the second letter is the allele inherited from the mom. If we looked at Mary's genotype for hair color we would find the same possible combinations of alleles. She could have HH, Hh, hh or hH genotypes.

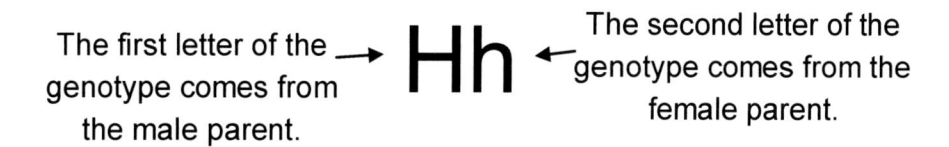

The first letter of the genotype comes from the male parent. → **Hh** ← The second letter of the genotype comes from the female parent.

Let's go one step further now and look at what happens when John and Mary each produce their respective gametes (sperm and ova). Let's look at John first. Let's pretend that John's genotype is Hh (he inherited the dominant allele from his dad and the recessive allele from his mom.) As we stated earlier, during meiosis each primordial cell makes a duplicate of itself and then each duplicate splits into two gametes. This results in four sperm cells for each primordial cell. Each sperm cell will have either the H genotype or the h genotype. See the diagram on the next page.

With Mary, let's pretend she has the same genotype as John, Hh. The one ovum that results from each primordial cell (remember, she makes four from each primordial cell, but only one becomes a viable ovum) will either have an H genotype or an h genotype. So, if we were to examine the gametes of both John and Mary, we would find some to be H and some to be h. Some gametes would carry the dominant allele and some would carry the recessive allele.

John's primordial cell has the genotype Hh

Intermediate mitotic stage results in two copies of each chromosome

The 4 resulting gametes have these genotypes

Mary's primordial cell has the genotype Hh

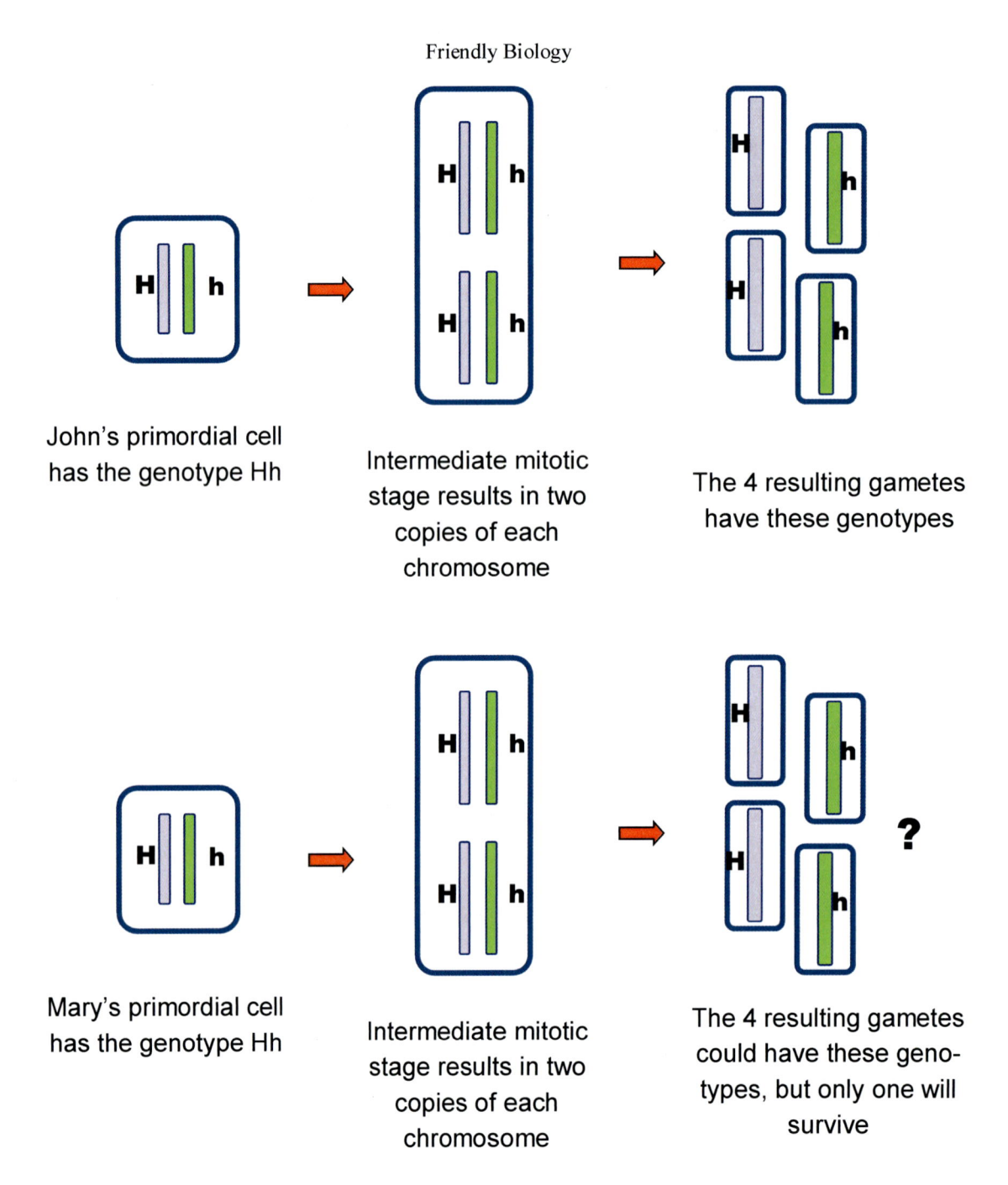

Intermediate mitotic stage results in two copies of each chromosome

The 4 resulting gametes could have these genotypes, but only one will survive

To predict what the possible outcomes might be for John and Mary's baby, Junior, we can use a special type of chart known as the Punnet square, named for the geneticist Reginald C. Punnet. Across the top of the Punnet square we will list the possible genotypes for John's sperm cells. Down the left side of the chart, we will list the possible genotypes for Mary's egg cells. Then, by combining these possibilities, we can make predictions as to what hair color Junior might have!

First, let's insert the possible genotypes for John's gametes across the top. Recall from above that we said his sperm cells could possibly be H and h. Now, list the possible genotypes for Mary's gametes down the left side. Again, like John, she could produce ova with H and h genotype. Now, let's do some combining of these gametes to see the potential outcomes.

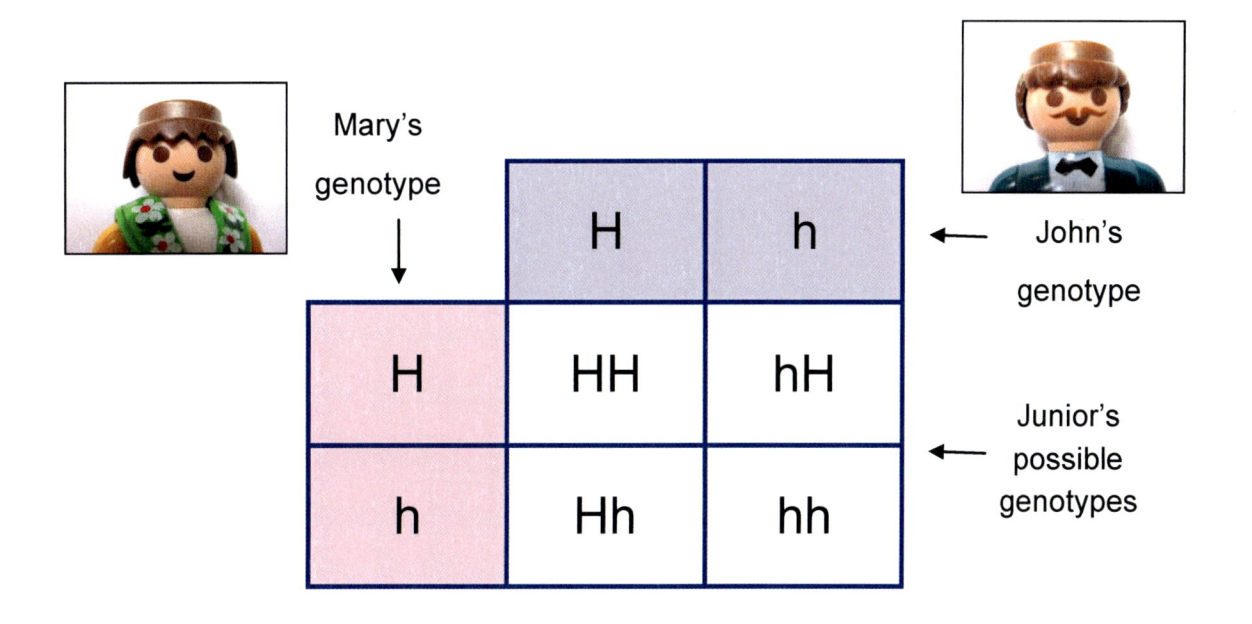

Mary's genotype ↓

	H	h
H	HH	hH
h	Hh	hh

← John's genotype

← Junior's possible genotypes

- Note that we could have an H from John join with an H from Mary: HH

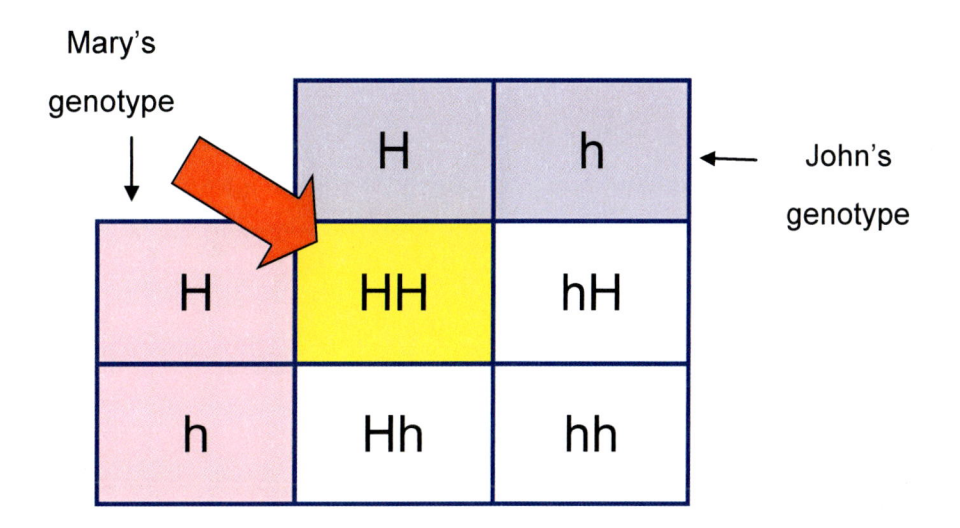

Mary's genotype ↓

	H	h
H	**HH**	hH
h	Hh	hh

← John's genotype

- We could have an h from John join with an H from Mary: hH

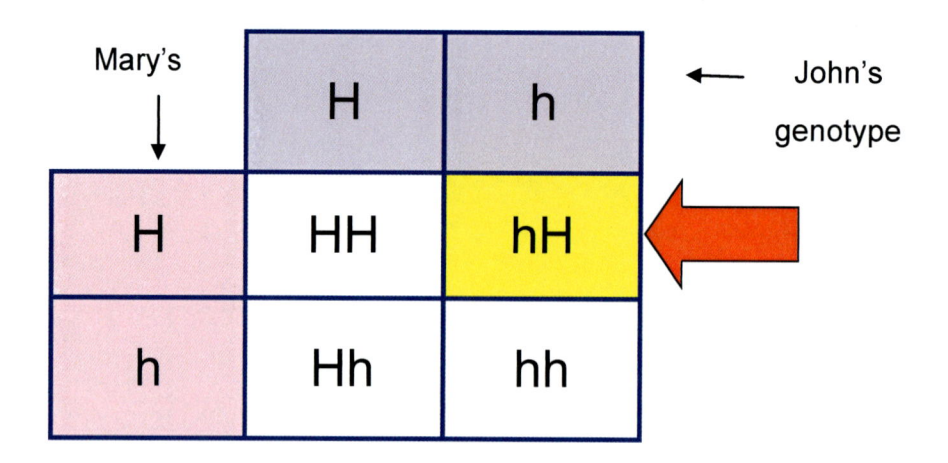

	H	h
H	HH	**hH** ←
h	Hh	hh

Mary's ↓ / John's genotype ←

- We could have an H from John join with an h from Mary: Hh

	H	h
H	HH	hH
h	**Hh**	hh

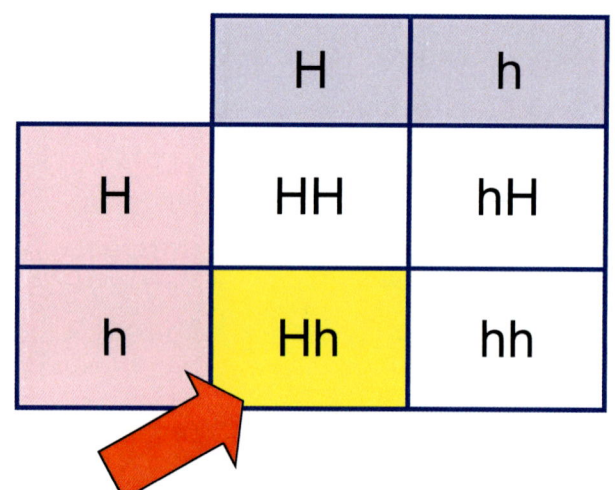

- And, finally, we could have an h from John join with an h from Mary: hh

	H	h
H	HH	hH
h	Hh	**hh** ←

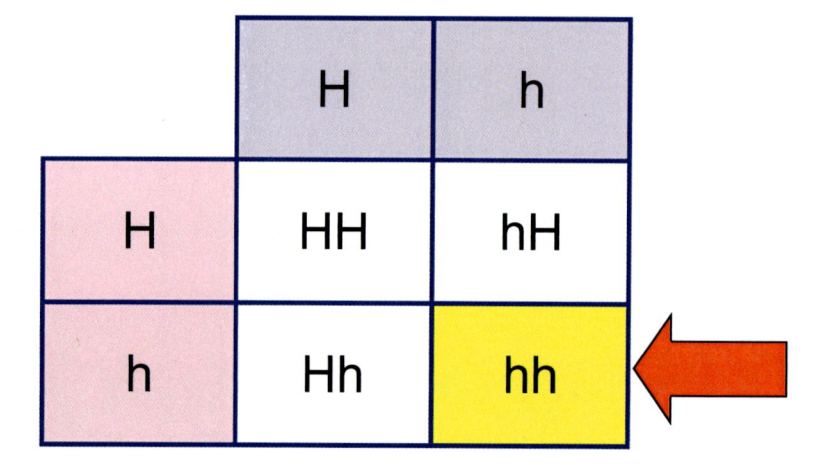

Now we have the possible genotypes for John and Mary's baby, but before we can predict the resulting phenotypes from these potential genotypes, we need to describe what hair color is "tied" to each gene. In other words, what phenotype the H represents and what phenotype the h represents.

For our example, let's pretend that H codes brown hair color and h codes for red hair color. Since H is the dominant allele (uppercase H), whenever it appears in the genotype it will always be expressed. This is a very important concept to understand, so we'll state it again. When a dominant allele appears in a genotype, it is always expressed. In other words, you will always find it in the phenotype of that individual. So, based upon these phenotypes, both John and Mary should have brown hair (each possess the dominant H allele.)

HH
Hh
hH

The (dominant) H allele is present, therefore the phenotype will be brown hair.

HH
Hh
hH

The (dominant) H allele is present, therefore the phenotype will be brown hair.

So, let's look back at the possible genotypes for John and Mary's baby. The first combination was HH. This means that the baby has two dominant alleles present for brown hair and will consequently have brown hair. The second combination was hH. Since the H allele is present, again, this baby would have brown hair. The third possible combination was Hh and, again, because the H allele is present, the baby would have brown hair. The fourth combination was hh. There is no dominant allele present; only two recessive alleles. In this case the baby would likely have red hair!

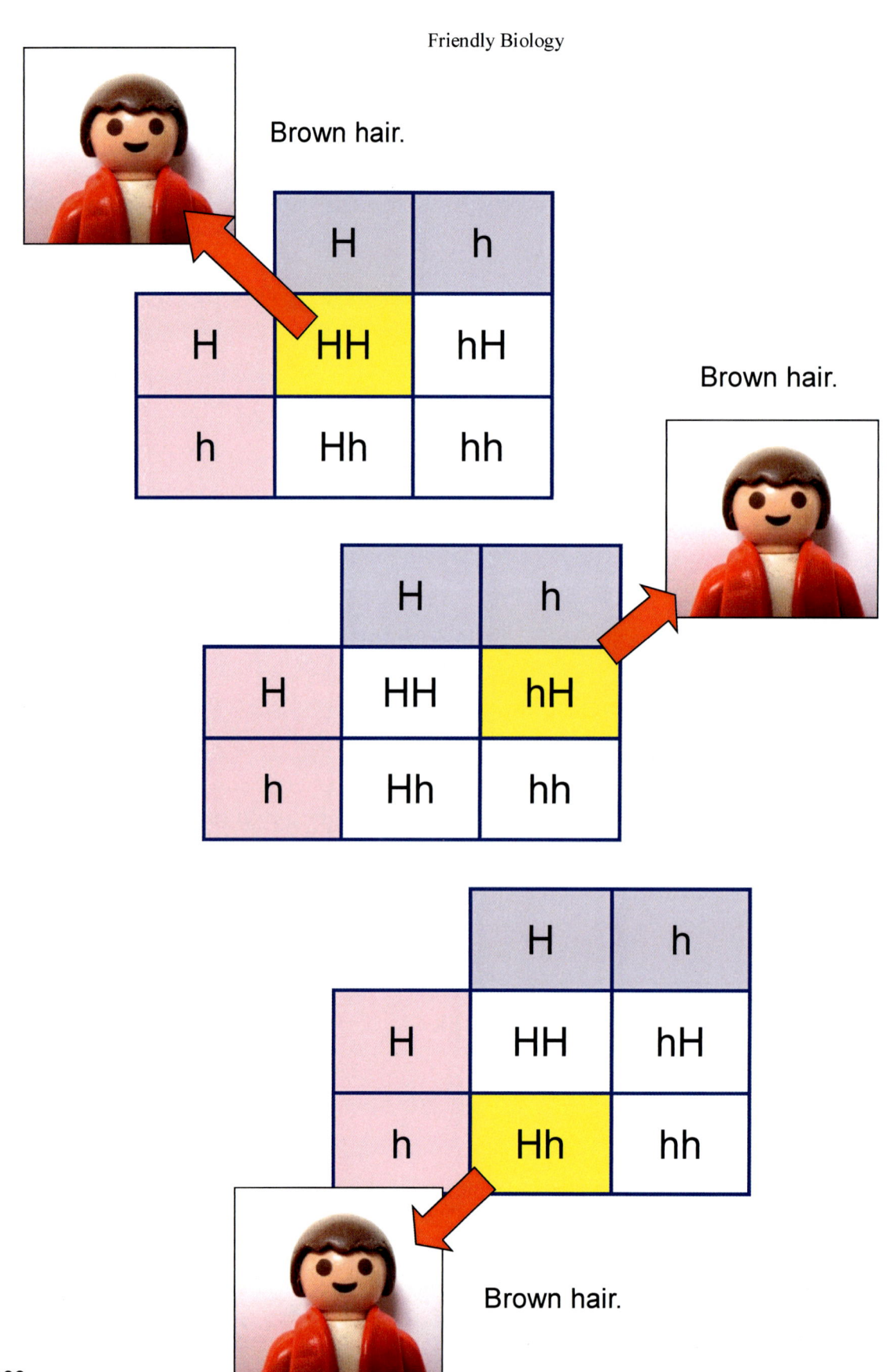

Brown hair.

	H	h
H	HH	hH
h	Hh	hh

Brown hair.

	H	h
H	HH	hH
h	Hh	hh

	H	h
H	HH	hH
h	Hh	hh

Brown hair.

	H	h
H	HH	hH
h	Hh	hh

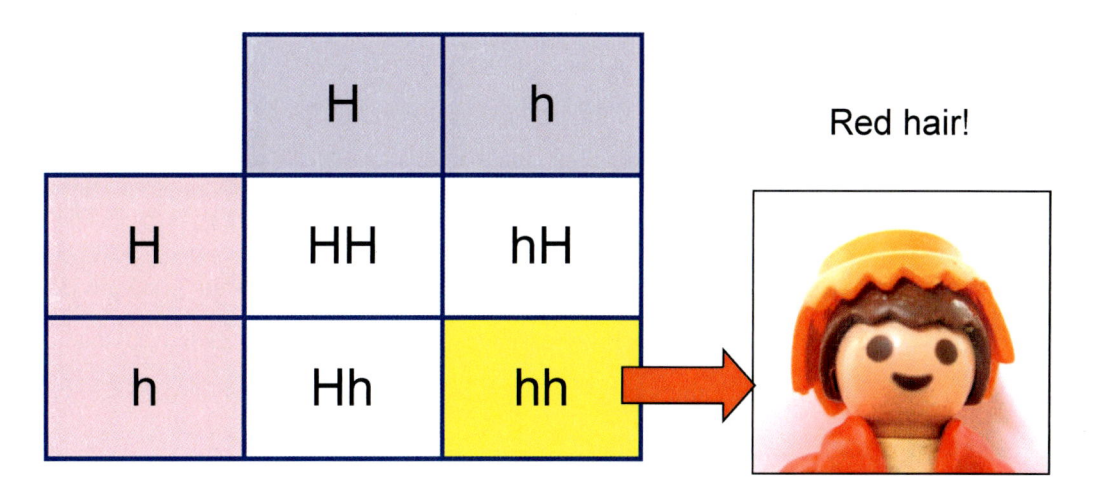

Red hair!

When an individual has a genotype for a trait in which both alleles are the same (both dominant or both recessive), we identify that individual as being homozygous. The prefix homo– meaning same and —zygous referring to the alleles in question. In our example of hair color above, the genotype HH could be termed homozygous dominant and the genotype hh could be termed homozygous recessive. The baby with the HH genotype would be identified as being homozygous dominant for brown hair color whereas the baby with the hh genotype, would be identified as being homozygous recessive for red hair color.

> Homozygous = same allele present (HH or hh)
> Heterozygous = different alleles present (Hh or hH)

> Homozygous dominant = same dominant alleles present (HH)
> Homozygous recessive = same recessive alleles present (hh)

When an individual has a genotype for a trait in which the alleles are not the same, we identify that individual as being heterozygous. Hetero– indicates opposite or different. In our hair color example, the genotypes Hh and hH are identified as being heterozygous. Because each still has a dominant allele present, we can say that the individual is heterozygous for brown hair color. Note in our example above that John's genotype was Hh. John can be identified as being heterozygous for brown hair color. Likewise, Mary with her genotype of Hh, is also heterozygous for brown hair color.

If we look at the possibilities for their baby, we see out of the four possible combinations, three combinations would result in a baby with brown hair: HH, Hh and hH. One baby could be homozygous brown (HH) and two possible babies could be heterozygous brown (Hh and hH). The fourth possible combination, hh, would be considered homozygous recessive red.

Of the four possible genetic combinations of John and Mary's baby, three combinations would result in a baby with brown hair. Mathematically, we can say that 3/4 or 75% of the time we might expect a baby to be born with brown hair given the genotypes of John and Mary. One of the four combinations results in a baby with red hair and, therefore, we could say that 1/4 or 25% of the time we might expect a baby to be born with red hair. So with these given genotypes of John and Mary, we could say the chances of their having a baby with red hair would be 1 out of 4. Now, this doesn't necessarily mean that if John and Mary have four children, three will have brown hair and one will have red. They could potentially have all brown-haired children or even all red-haired children or any combination of the two.

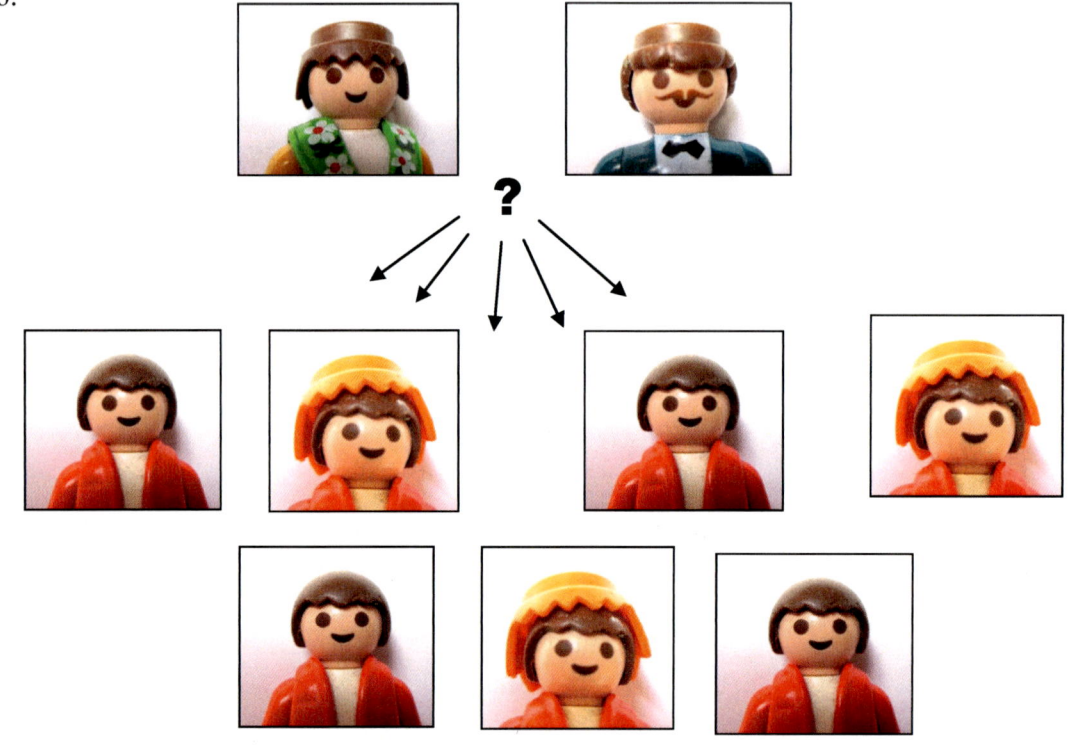

Let's change our situation slightly. Let's suppose that both John and Mary have brown hair. However, let's change their genotypes. Let's suppose that John has the genotype HH. This means that ALL of his sperm cells will have the H allele. He would considered homozygous dominant for brown hair. Let's suppose Mary still has the same genotype as she did in our first example: Hh, which would be heterozygous dominant for brown hair. This means that her ova would either be H or h. To see what the possible combinations might be for their baby, let's again set up our Punnet square. Note across the top we will put the possible genotypes for John and down the left side we will put the possible genotypes for Mary. Note for John, we put H and H. For Mary, we put H or h.

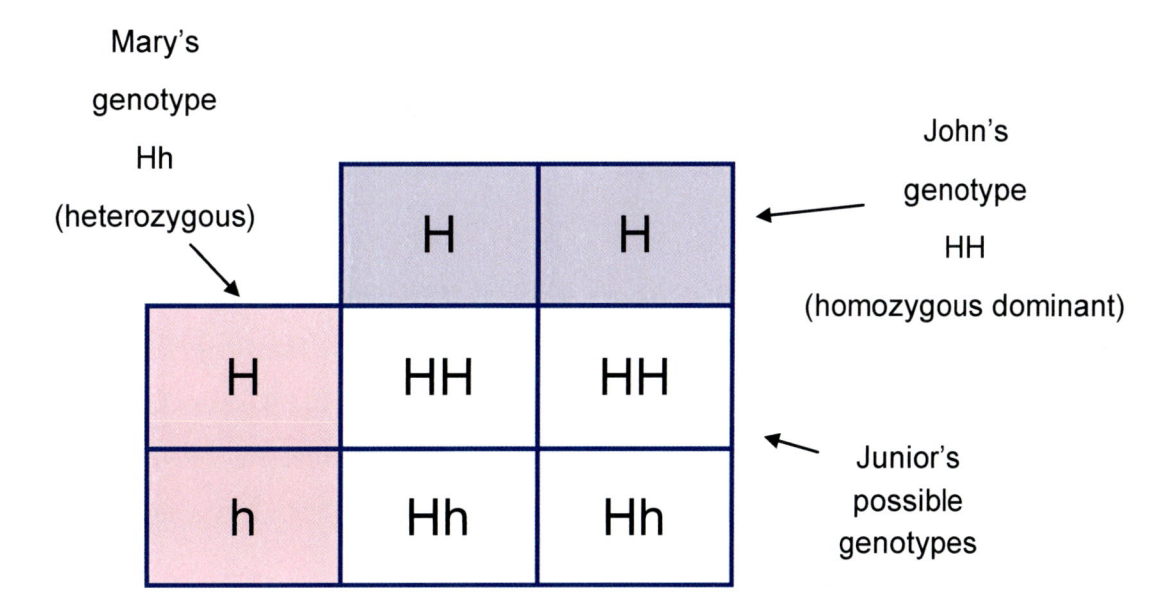

Mary's genotype Hh (heterozygous)

John's genotype HH (homozygous dominant)

	H	H
H	HH	HH
h	Hh	Hh

Junior's possible genotypes

Let's look at the possible combinations for Junior's genotype. We could have HH, HH, Hh and Hh. If you examine each combination, you'll see the dominant H allele present. Because the dominant H allele is found in all possible combinations, there is a 100% chance that all of their babies would have brown hair! Note that there is a 50% chance that the baby would be homozygous dominant for brown hair and a 50% chance that the baby would be heterozygous dominant for brown hair. Because, in this case, John does not carry the recessive h allele for red hair, there is 0% chance that any of their children would have red hair (even though Mary carries the h allele). So if John and Mary possessed this genotype, their children would all have brown hair.

Friendly Biology

	H	H
H	HH	HH
h	Hh	Hh

Mary's genotype Hh

John's genotype HH (homozygous dominant)

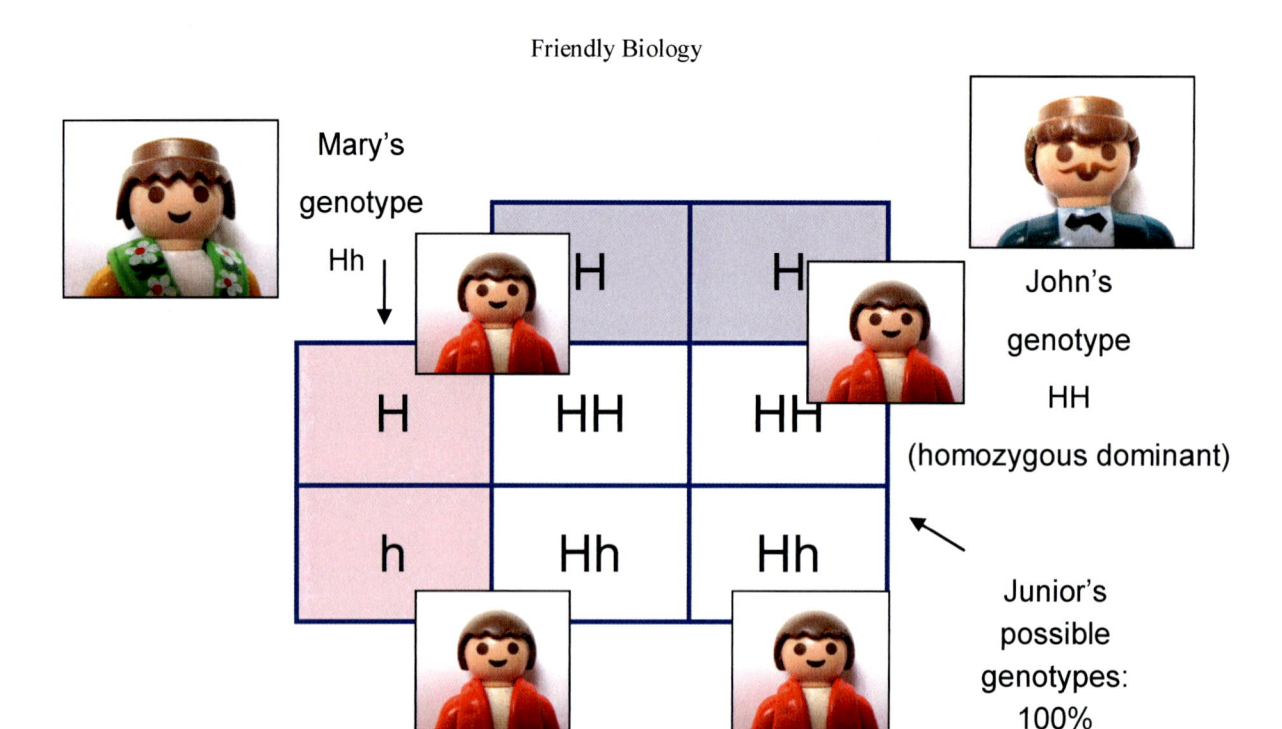

Junior's possible genotypes: 100% brown.

Let's change things a little more! Suppose John has the following genotype: Hh. What hair color would John have? If you guessed brown, you're correct. Suppose Mary has this genotype: hh. What color would her hair be? If you guessed red, you're correct, again! What might we expect the genotypes and phenotypes of their potential babies to look like? Let's create a Punnet square to find out. Start by putting John's possible alleles across the top of the square and Mary's down the left side. Note for John we have H and h. For Mary, we would have h and h.

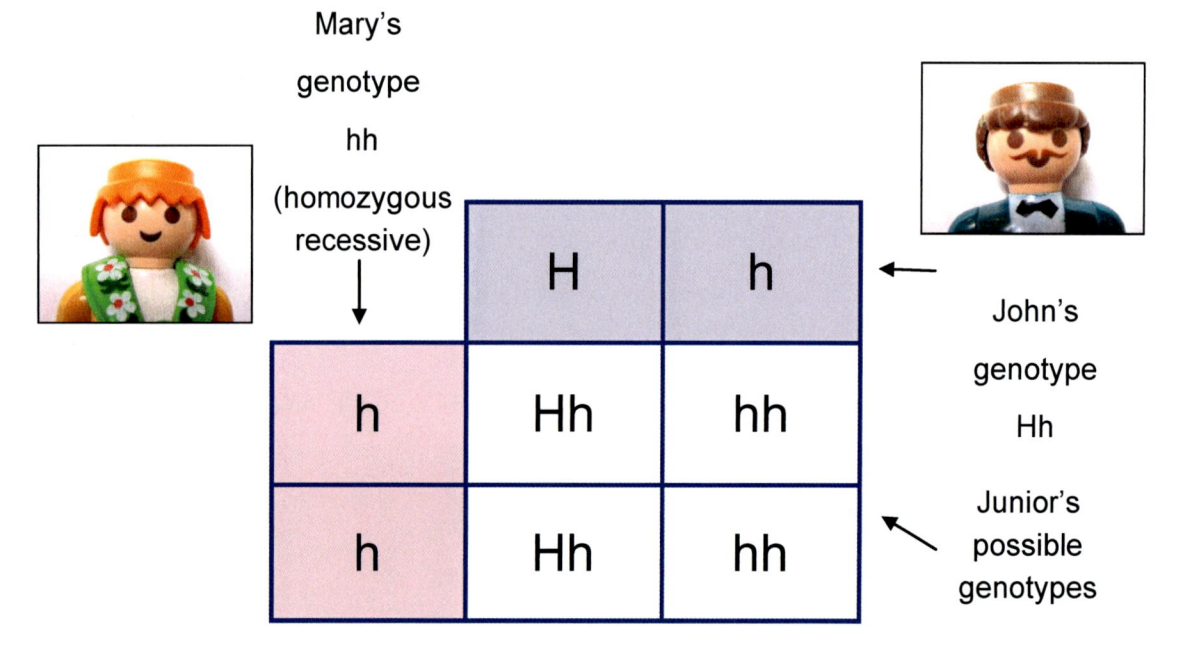

	H	h
h	Hh	hh
h	Hh	hh

Mary's genotype hh (homozygous recessive)

John's genotype Hh

Junior's possible genotypes

226

Mary's genotype hh

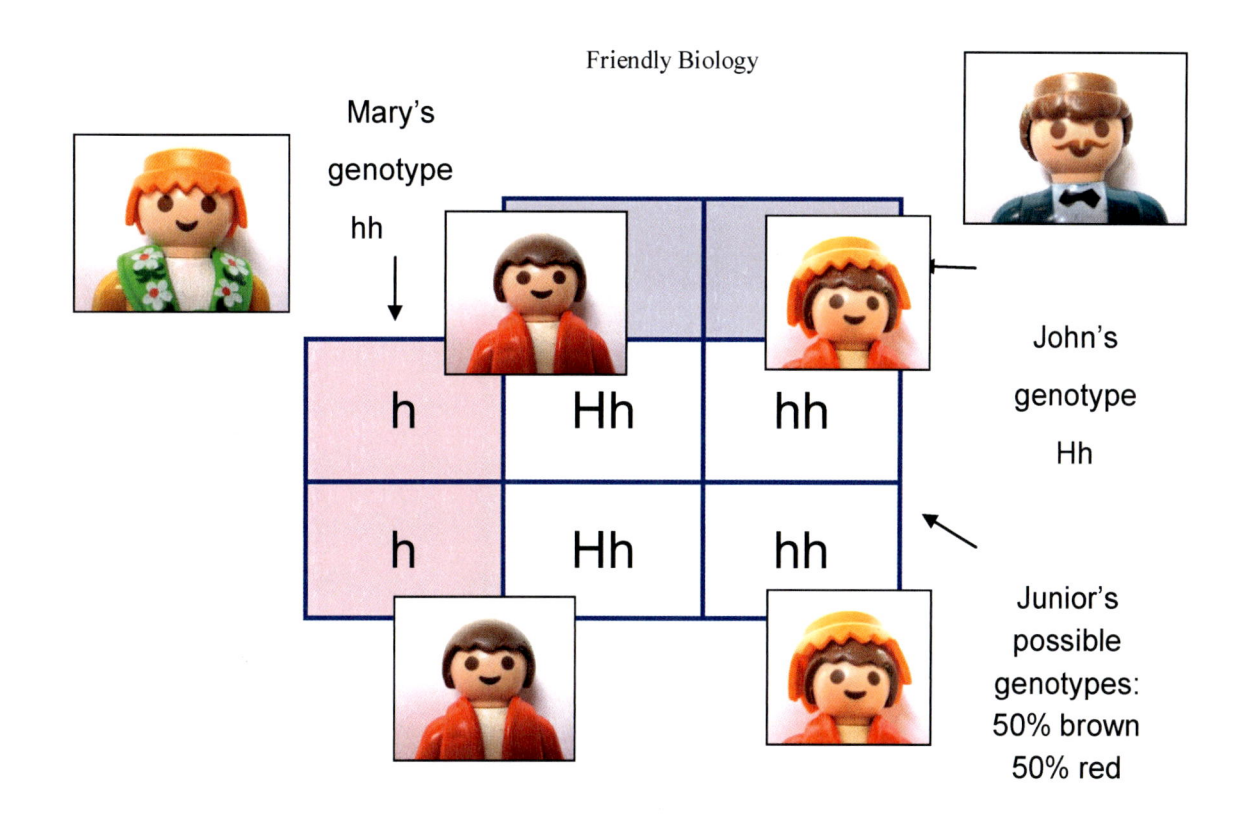

John's genotype Hh

	H	h
h	Hh	hh
h	Hh	hh

Junior's possible genotypes: 50% brown 50% red

Let's look at the possible combinations. We can see that we can get Hh, hh, Hh or hh individuals. What are the phenotypes of these individuals? Recall, that whenever the H allele appears, it is dominant and the dominant trait always appears. In this case, we see two Hh combinations which means these individuals will have brown hair. They are heterozygous dominant for brown hair (Hh). We also see two hh individuals. These two babies would have red hair (homozygous recessive, hh). So, with these genotypes of John and Mary, the chance of their having a brown-haired baby is 50% and the chance of them having a red-haired baby is 50%!

Let's look at yet another possible situation with John and Mary. Suppose John and Mary both have red hair. What are their genotypes? In order to have red hair, each of them must be homozygous recessive, right? They must both have the hh genotype. If this is the case, will they ever be able to have a baby with brown hair? Let's look at a Punnet square to answer this question. Start with John's alleles across the top and Mary's down the left side. John will provide either an h or an h genotype and Mary likewise, an h or an h.

If we examine the resulting combinations, we see that all of the babies will have an hh genotype. They will all be homozygous recessive and have the phenotype of red hair. So, the answer to our question is, no, they will not be able to have a baby with brown hair! 100% of their babies will have red hair.

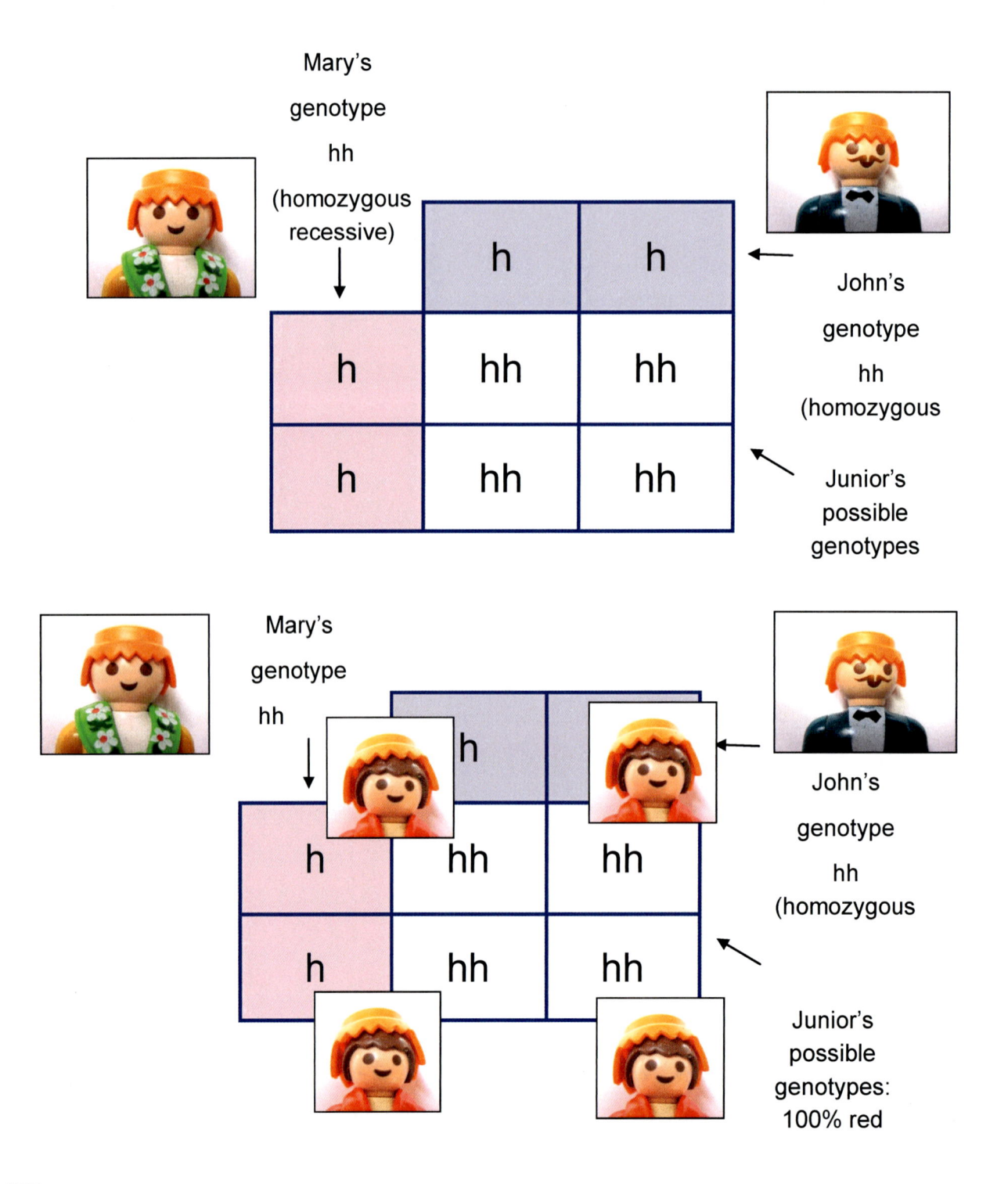

Mary's genotype hh (homozygous recessive)

John's genotype hh (homozygous

Junior's possible genotypes

	h	h
h	hh	hh
h	hh	hh

Mary's genotype hh

John's genotype hh (homozygous

Junior's possible genotypes: 100% red

	h	h
h	hh	hh
h	hh	hh

Before we leave John and Mary, let's look at one more situation. Suppose both John and Mary are homozygous dominant for brown hair. Could they ever have a child with red hair? If we examine their genotypes, recall that homozygous dominant means they have both dominant alleles in all cells. In this case, it will be HH for both John and Mary. This means that the gametes they produce will ALL be H. Consequently, all of their babies will have the HH genotype and all be brown-haired.

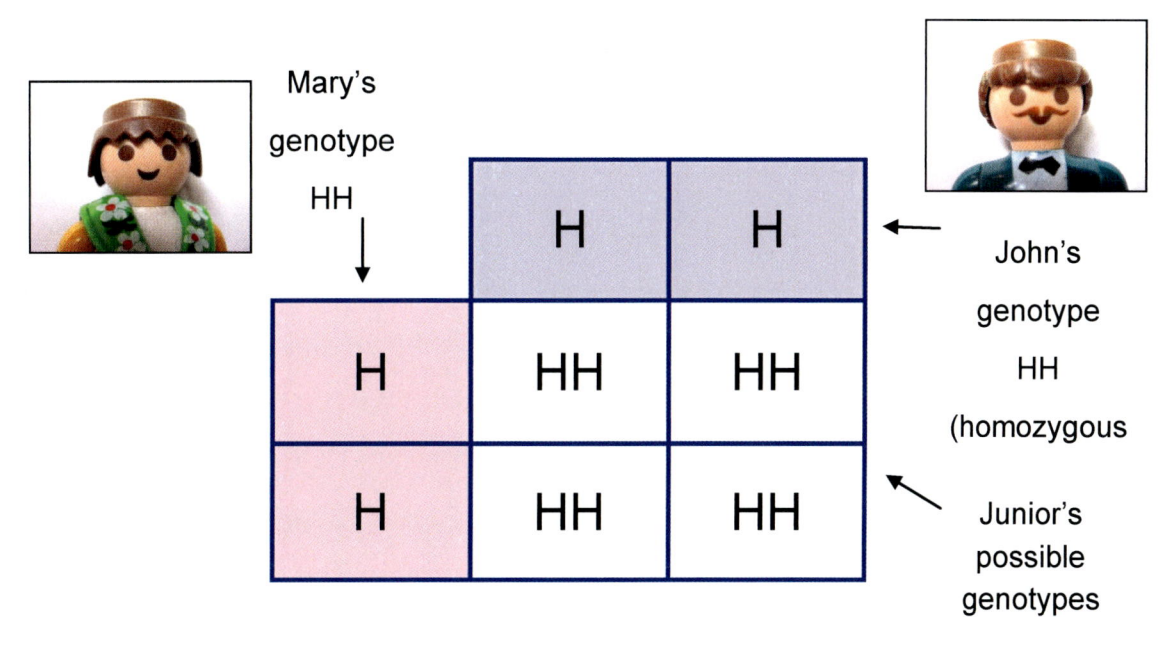

Mary's genotype HH

John's genotype HH (homozygous

Junior's possible genotypes

	H	H
H	HH	HH
H	HH	HH

It is interesting to note that at times some phenotypes may appear to skip generations. Let's look at the example of eye color. In my family, my dad had blue eyes and my mother had brown eyes. Because brown eye color appears to be dominant over blue, I have brown eyes. My genotype, because of my dad's genotype, must however, be heterozygous for brown eye color. We can say, perhaps that it is Bb. The allele B codes for brown eye color and is dominant. The allele b codes for blue eyes and is recessive. To have blue eyes, therefore, one must have a homozygous recessive genotype: bb.

My wife has brown eyes, but like me, she is heterozygous for brown eye color in that her mom has blue eyes and her dad has brown. She, too, has the genotype Bb. In our children, our oldest four have brown eyes, but our two youngest daughters have blue eyes. We can see that in this case, the phenotype for blue eyes was seen in grandparents, not seen in parents, but then seen in the children; the phenotype appears to have "skipped" a generation. The alleles were there all along, they were just not being expressed.

Lisa's genotype (brown eyes) →

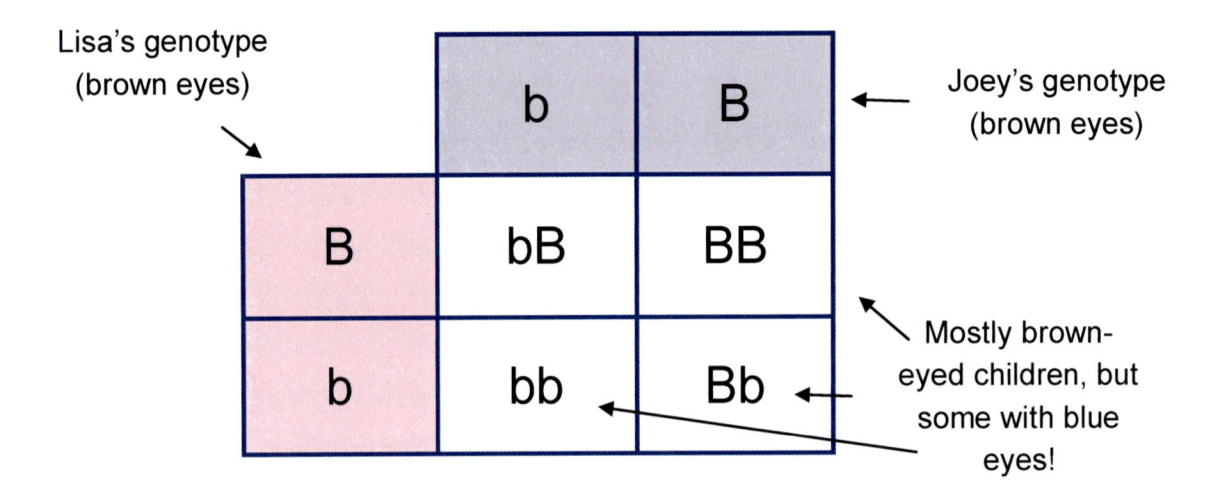

	b	B
B	bB	BB
b	bb	Bb

Joey's genotype (brown eyes) ←

Mostly brown-eyed children, but some with blue eyes!

We need to point out here, however, that not all traits being expressed in the phenotype of an individual follow the strict dominance/recessive patterns we've discussed so far. In some cases, there may not be a dominant gene, but rather a sharing of dominance between the two individuals. This situation is called codominance or incomplete dominance. The phenotype appears to be controlled by both the gene from the father parent and the mother parent.

For example, in some flowers, the color of the bloom is controlled by genes from both parent plants and the resulting flower is a blend of colors. One of the parent plants may produce one color of blooms. The other parent plant may produce another color of blooms. When pollen (male cells) from one plant is sprinkled upon the female flower parts of the other plant, the plants which eventually grow from these seeds is a blending of colors from both plants. This blending of colors is evidence that both parents contributed to the phenotype of the offspring.

There are three more topics we'll explore before we end our discussion of genetics. The first is how chromosomes determine the gender of the baby in humans. Recall that in humans, there are 23 pairs of chromosomes in all cells including the primordial sex cells. One pair of these chromosomes is identified as being the sex chromosomes as it is this pair of chromosomes which determines whether the possible baby will be a boy or girl. Within this pair of chromosomes, one chromosome is identified as being the X-chromosome and the other the Y-chromosome. Human males have one member of each possible sex chromosome: XY. However, human females have only X chromosomes: XX.

Consequently, during the spermatogenesis in males (meiosis,) approximately half of the sperm cells that are made carry the X sex chromosome while half carry the Y-chromosome. In oogenesis in females, all of the ova that are produced carry the X chromosome. So, the gender of the possible baby depends upon which sperm cell is successful at fertilizing the ova. If a sperm cell carrying the X-chromosome fertilizes the ovum, the baby will have the genotype XX and therefore be a girl. If a sperm cell carrying the Y-chromosome fertilizes the ovum the baby will have the genotype XY, and be a boy.

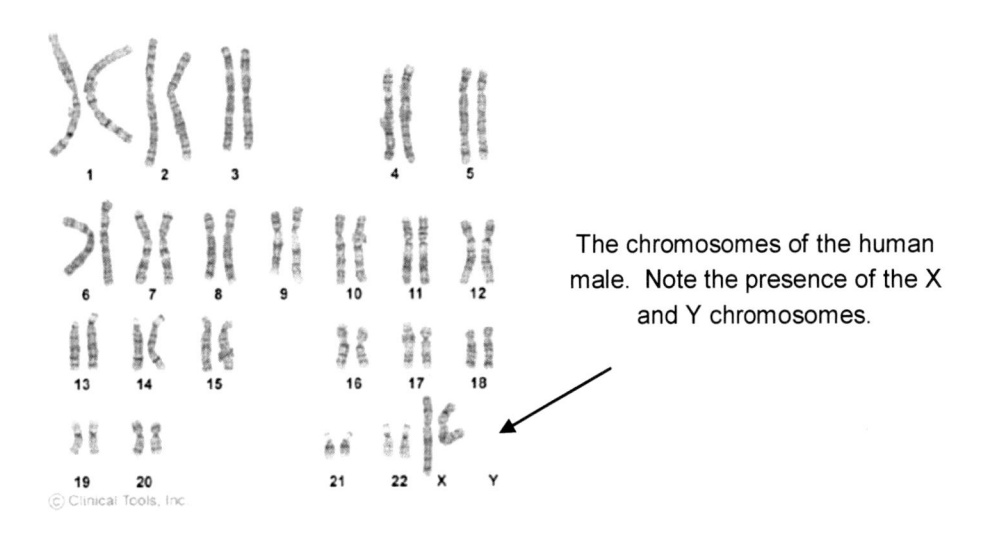

The chromosomes of the human male. Note the presence of the X and Y chromosomes.

The second concept is known as sex-linkage. A geneticist named Thomas Hunt Morgan studied fruit flies (the little gnat-like insects which are often found on very old bananas and fruits that have been allowed to sit on your kitchen counter too long) and found that certain traits of individuals were only found in the male flies and not the females. He found that these traits were linked to genes present in the male species and the traits were identified as being sex-linked traits.

A very practical use of this phenomenon is utilized by persons operating chick hatcheries. Persons who buy baby chicks often request chicks of one sex or the other. If they desire chickens for laying eggs, female chicks (known as pullets) are purchased. Male chicks may be preferred in some breeds of chickens used for producing meat. When chicks hatch they usually do not show readily visible body parts which indicate whether they are "boy" chicks or "girl" chicks. However, some breeds of chickens produce chicks which have sex-linked traits. The chicks of these breeds display obviously different color patterns in their down which are apparent right at hatching. This makes it very easy for the hatchery to sort male chicks from females. As expected, not all breeds of chickens do this and chicks of these breeds are sexed by examining their vent. So, the fact that down color is a sex-linked trait is very convenient when it comes to sexing baby chicks.

Another phenomenon which takes place during the creation of gametes which may cause variations in the phenotypes of offspring is known as crossing over. Recall that during meiosis, the chromosomes of the primordial cells make copies of themselves. During this copying process, genetic information from one chromosome from that individual's father may be "traded" with information from the chromosome of that individual's mother. Meiosis then continues with the production of gametes. However, these gametes are not strictly paternal (from the dad) or maternal (from the mom). The offspring ultimately produced from these gametes will then reveal traits from both grandparents.

The third concept we need to discuss involves an area of study and research which is very exciting, yet very controversial. This area is genetic engineering. Biologists, through extensive research and practice, have developed methods to modify the genes in many living creatures in an attempt to improve certain desirable phenotypes in those individuals. You might think of this as "tweaking" the recipe in an attempt to create an even better product. Essentially, parts of genes are removed from certain organisms (in most cases, bacterial genes) and inserted into the genes of recipient organisms. These recipients are then identified as being genetically modified organisms or GMOs.

Examples of GMOs are plants known as Roundup Ready[R] plants. In the early 1970's an herbicide known as Roundup[R] (scientifically known as glyphosate) was introduced by the Monsanto company to farmers. Roundup[R] had the ability to kill *all* kinds of plants (herbicide) both weeds and the plants the farmer was trying to grow, by inhibiting a certain enzyme required for growth in the plant. It worked great on weeds, but you had to keep it well away from your crops, trees and other plants you didn't want to harm. Geneticists, working for very large seed companies, were able to modify the genes in soybean plants which made the plants immune to the effects of Roundup[R].

This allowed farmers to spray Roundup[R] directly upon both their soybeans and weeds at the same time and only the weeds were killed! The genetically altered soybean plants were not affected by the detrimental effects of the herbicide. This resulted in marked increase in yields of soybeans for farmers all across the country because weeds could be readily controlled. The research then moved into other species of plants grown commercially like corn and alfalfa.

Corn plant susceptible to the effects of herbicide.

Soybean plants not susceptible to the effects of herbicide.

Weeds killed by the herbicide.

In addition to the Roundup Ready^R series of plants, geneticists have also developed corn plants which show greater resistance to parasites which feed on corn root systems. Like the Roundup Ready^R plants, this allowed for tremendous increases in corn yields for growers. However, not being able to control what effects these plants have on neighboring non-GMO plants as well as insects such as bees and other pollinators, not to mention what unknown effects they may have on animals and humans which consume these products, has caused skepticism to develop by many regarding the true benefits of these genetic modifications. Countries outside the United States have varying degrees of acceptance of the corn and soybeans produced from these GMOs and some countries have refused to purchase products containing GMOs. This has a marked effect upon the marketability of these crops. Plus, it has become widely documented that weeds that were once very susceptible to Roundup^R have developed resistance to its effects. Overall, geneticists may be able to realize improvements for plant and animal production through genetic engineering, however, progress and research must proceed with careful consideration of all potentially affected by these genetically modified creatures.

Let's pause here and review what we've learned in this lesson. We learned that:

- Gregor Mendel was the biologist who studied the inheritance patterns of living things and developed the basic "laws" of genetics;

- Genotype refers to the record of genes in an individual and phenotype refers to the visual or observable results of those genes;

- In organisms which are diploid (2N), gametes are produced through meiosis which are haploid (1N). The male gamete is the sperm cell and the female gamete is the ovum. The uniting of these two cells is known as fertilization;

- Variations in genes are known as alleles. Dominant alleles appear to be expressed all the time while recessive alleles, while present, are not always expressed. To be expressed, the recessive gene must be inherited from both parents;

- The Punnet square is a useful tool to help predict possible outcomes from various combinations of genotypes;

- Homozygous refers to the situation where an individual possesses the same allele for a certain gene whether it be dominant or recessive. Heterozygous refers to the situation where an individual possesses two different alleles for the same gene;

- Co-dominance or incomplete dominance is the situation where one allele is not strictly overpowering the other and there is a sharing of dominance;

- Certain visual traits are known to be sex-linked, meaning that only the male or only the female will possess those traits. In other words, the trait is linked to the sex of the individual;

- Crossing-over may occur during meiosis which results in a mixing of traits found in the gametes of an individual;

- Within the 23 pairs of chromosomes that humans have, one pair is known as the sex chromosomes. Human males have the XY genotype while human females have the XX genotype. The sex of the baby is dependent upon which sperm (X or Y) fertilizes the ovum;

- GMOs are genetically modified organisms. Care must be taken so that the benefits of these organisms can be maximized while minimizing the detrimental side effects.

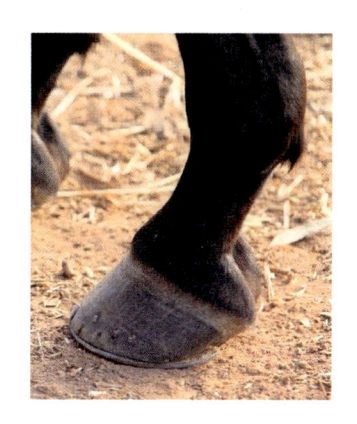

Lesson 14: Taxonomy

In the thirteen lessons we've completed so far, we've explored many features that living things have in common. We learned that living things move, living things develop and grow, living things require an energy source, living things reproduce and living things respond to their environment. In addition, we've learned that living things are composed of cells and that within the nucleus of those cells, we could find the "cookbook" or chromosomes which told all of the "recipes" or genes as to how that living thing developed and functioned.

We learned that the genetic record itself could be identified as being the genotype of that individual whereas what those genes expressed or revealed was known as the phenotype of that individual. The sum total of expression of all of the genes present in an individual, whether it be an earthworm or an elephant, results in what we can observe in that creature. The study of what we can observe in living things is known as morphology where the prefix morph- refers to form and the suffix –ology refers to the study of. Morphology, therefore, is the study of form and, more specifically, how living things are similar or different from each other. From these lists of observations of similarities and differences, we can form sets or groups of living creatures. The creation and study of these sets or groups of living creatures is known as taxonomy.

> Morph = form
>
> -Ology = the study of
>
> Morphology = the study of form (how things are similar and different)

> Taxonomy = the study of grouping living things based on morphology

Historically, there were many scientists who realized the importance of grouping living things into various sets. The primary reason for this effort to group living things, which still exists very much today, is to enable clear communication between persons studying these living things in various places around the world. For example, suppose a plant researcher found that a certain, rare plant produced a poisonous substance when its leaves were crushed. It would be extremely important that he would be able to clearly communicate this finding to other researchers all around the world. The most vital piece of information that would need to be communicated would be the exact identification of the plant. Being able to clearly communicate the identity of various living things with fellow persons can prevent serious consequences.

As a result of this need to clearly communicate, a complex system of organizing living things into various groups and subgroups, based upon visible morphology, has been developed. While records show that attempts to group living things goes back to Arisotle in 300 BC, the work of the taxonomist Carl Linneas in the 1700s is what our current system of organization is based upon. Very recent work by the partnership of two organizations known as Species 2000 and the Integrated Taxonomic Information System has resulted in approximately 1.4 million of the estimated 1.9 million different living creatures known in the world to be placed into various categories and groups. The goal of this partnership is to develop a comprehensive database of as many living organisms possible. This database is known as the Catalogue of Life and we'll explore this database later in this lesson.

An important point, however, needs to be understood here before we continue our discussion of taxonomy. We need to realize that this study of grouping organisms based upon similarities and differences is an ever-changing process. Discovery of new features of living things results in individuals being moved from one group to another. And discovery of new techniques to enable us to better examine living things has resulted in changes in how we group living things. Whole organizations of scientists exist to study these similarities and differences where extensive debates take place as to "where" and "how" living things should be grouped. Keep in mind, though, that regardless of how a living thing finds itself grouped, the big idea has always been to better communicate with others what has been discovered about living creatures in our world.

Let's begin our exploration of taxonomy by performing a search in the Catalogue of Life database. Begin by going to the Catalogue of Life website: www.catalogueoflife.org. On the opening page you will see a set of frequently asked questions regarding the Catalogue of Life database along with a choice of two options for searching the database: the dynamic checklist and the annual checklist. The dynamic checklist is the list which is constantly being updated and changed according to new discoveries regarding living creatures which may result in changes being made in their placement into the database groupings. The annual checklist is the list which gets updated only a few times a year and is intended for use when referencing the database in some sort of written report or document. For our study here, choose the annual checklist option.

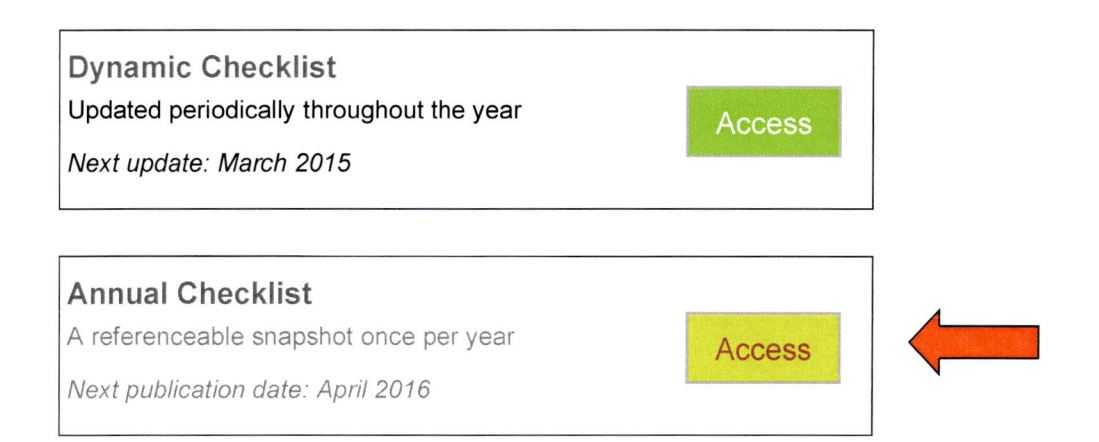

Dynamic Checklist
Updated periodically throughout the year
Next update: March 2015

Access

Annual Checklist
A referenceable snapshot once per year
Next publication date: April 2016

Access

The next page that opens includes a box which allows you to search for the living thing you are interested in knowing more about. In the search box, type 'domestic cat.'

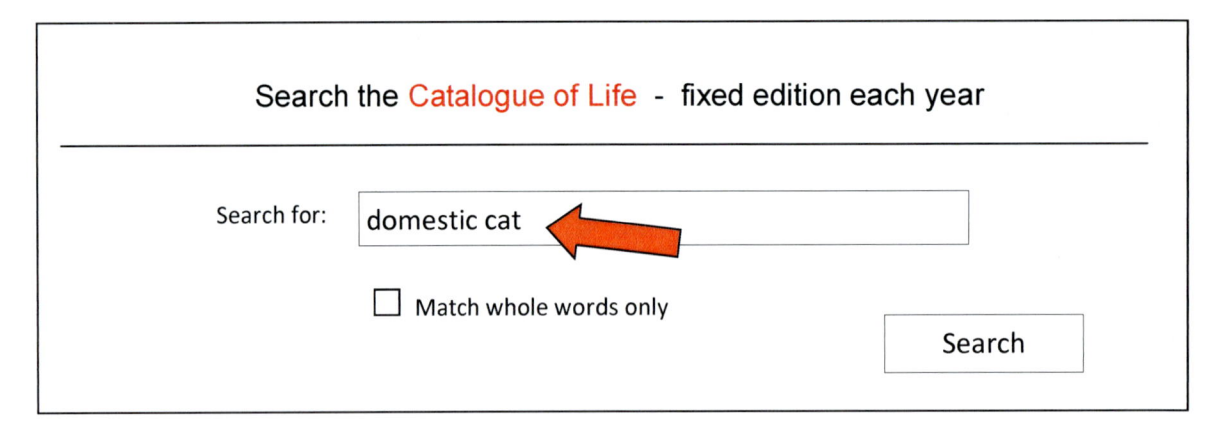

Search the Catalogue of Life - fixed edition each year

Search for: domestic cat

☐ Match whole words only

Search

The next window that opens gives you your search results which includes two options that have all or a portion of your search terms. You should see an entry for domestic cat and domestic cattle. Click on the entry for domestic cat.

Search all names—Results for "domestic cat"

Export search results / New Search   Records per page 20 Update

Records found: 2

Name	Rank	Name Status	Group	Source Database
Domestic cat (English)	Species	Common name for *Felis catus* Linnaeus, 1758	Animalia	ITIS
Domestic cattle (feral) (English)	Species	Common name for *Bos taurus* Linnaeus, 1758	Animalia	ITIS

Export search results / New Search

The screen that opens next shows the information we are seeking about the domestic cat. The first entry you see is the accepted scientific name for the domestic cat followed by synonyms and common names for the domestic cat. We'll come back to the accepted scientific name information later in our discussion.

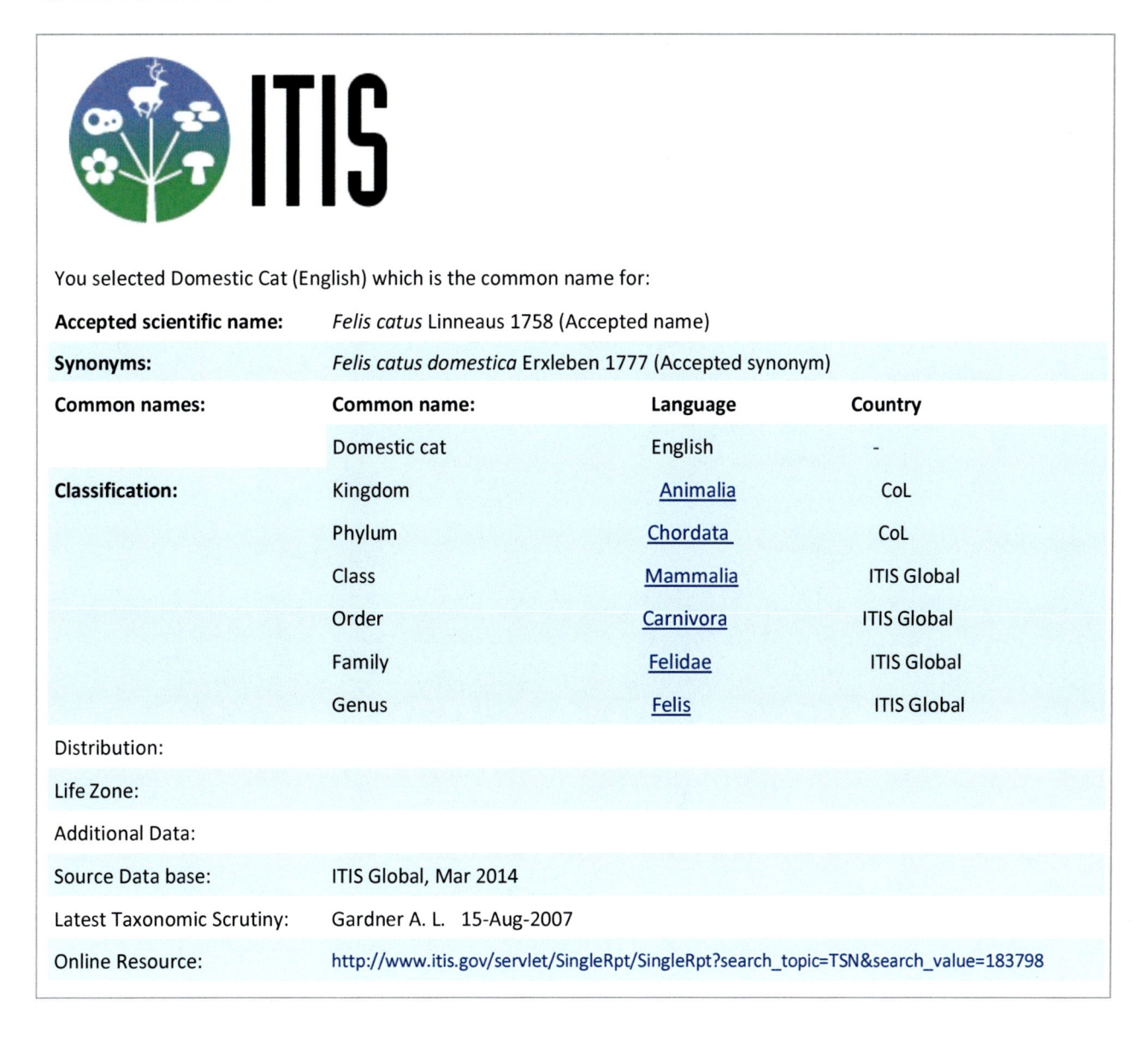

ITIS

You selected Domestic Cat (English) which is the common name for:

Accepted scientific name:	*Felis catus* Linneaus 1758 (Accepted name)		
Synonyms:	*Felis catus domestica* Erxleben 1777 (Accepted synonym)		
Common names:	**Common name:**	**Language**	**Country**
	Domestic cat	English	-
Classification:	Kingdom	Animalia	CoL
	Phylum	Chordata	CoL
	Class	Mammalia	ITIS Global
	Order	Carnivora	ITIS Global
	Family	Felidae	ITIS Global
	Genus	Felis	ITIS Global
Distribution:			
Life Zone:			
Additional Data:			
Source Data base:	ITIS Global, Mar 2014		
Latest Taxonomic Scrutiny:	Gardner A. L. 15-Aug-2007		
Online Resource:	http://www.itis.gov/servlet/SingleRpt/SingleRpt?search_topic=TSN&search_value=183798		

Look further down the entry at the section titled classification. Within this section you can read across to see that the domestic cat falls into the main group or kingdom known as Animalia. The group or kingdom Animalia refers to the animal kingdom. The animal kingdom is one of five kingdoms that serve as the first degree of classification of living things. In other words, all living things fall into one of these first five groups.

In addition to the animal kingdom, there is the plant kingdom (Plantae), the fungi kingdom (Fungi), the moneran (Monera) kingdom (which includes bacteria and other microscopic creatures) and the protozoal kingdom (Protista) (which includes microscopic organisms which are primarily found in aquatic environments). We'll discuss each of these groups in much greater detail later in this course. For now, let's go back to the search we've begun for the domestic cat.

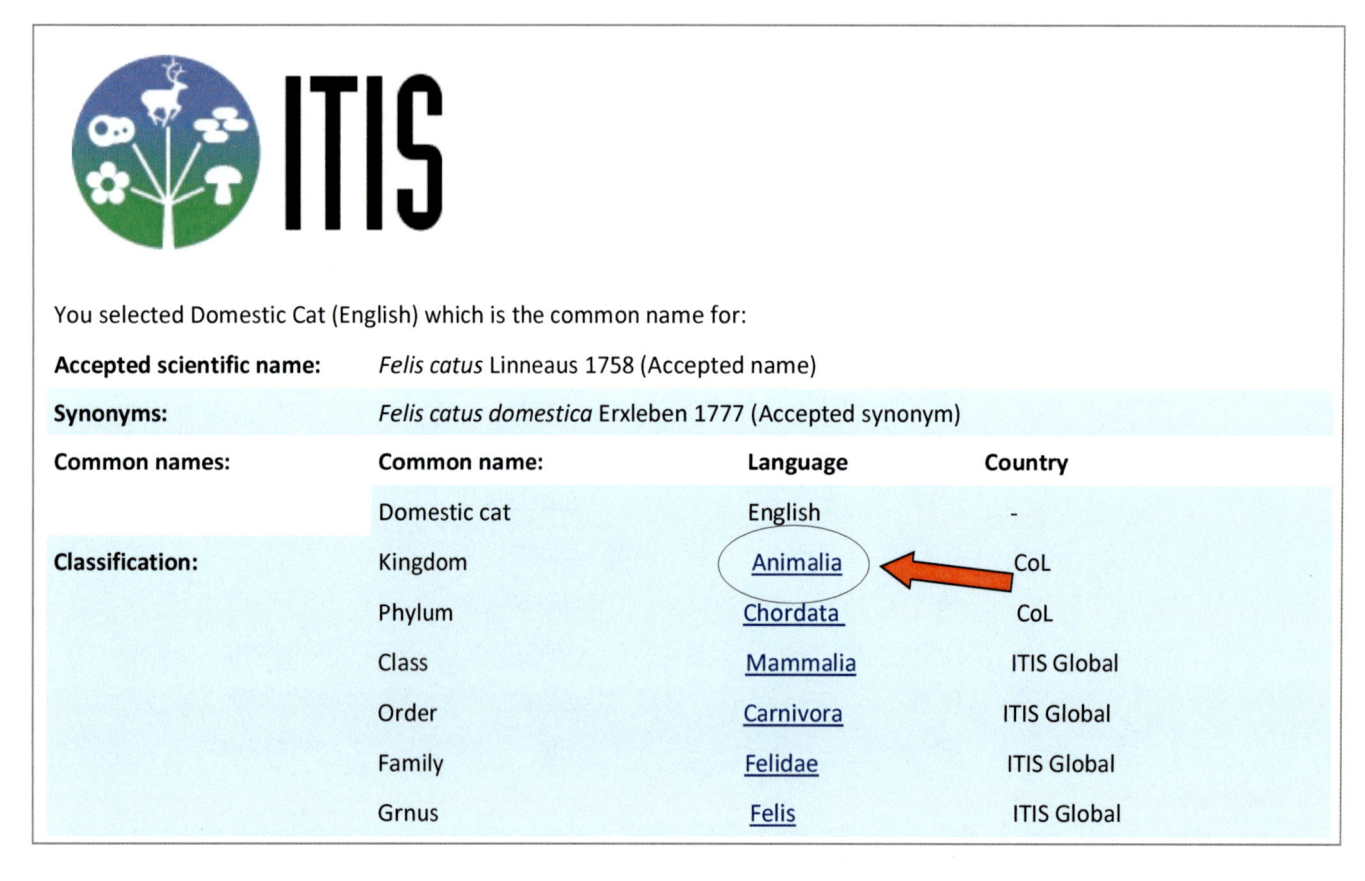

You selected Domestic Cat (English) which is the common name for:

Accepted scientific name: *Felis catus* Linneaus 1758 (Accepted name)

Synonyms: *Felis catus domestica* Erxleben 1777 (Accepted synonym)

Common names:

Common name:	Language	Country
Domestic cat	English	-

Classification:

	Common name	Language	Country
	Kingdom	Animalia	CoL
	Phylum	Chordata	CoL
	Class	Mammalia	ITIS Global
	Order	Carnivora	ITIS Global
	Family	Felidae	ITIS Global
	Grnus	Felis	ITIS Global

Beneath the kingdom level of organization, you should see the word "phylum." Each kingdom is divided into smaller divisions known as phyla, which is the plural form for phylum. The phylum for the domestic cat is Chordata. Members of this phylum all have or at one time had in their development, a notochord. A notochord is the structure which, in humans and animals, eventually develops into the spinal cord. In some creatures, the notochord remains very simple, yet in others, like humans is quite complex. Obviously, a domestic cat has a spinal cord and therefore readily falls into this grouping.

Friendly Biology

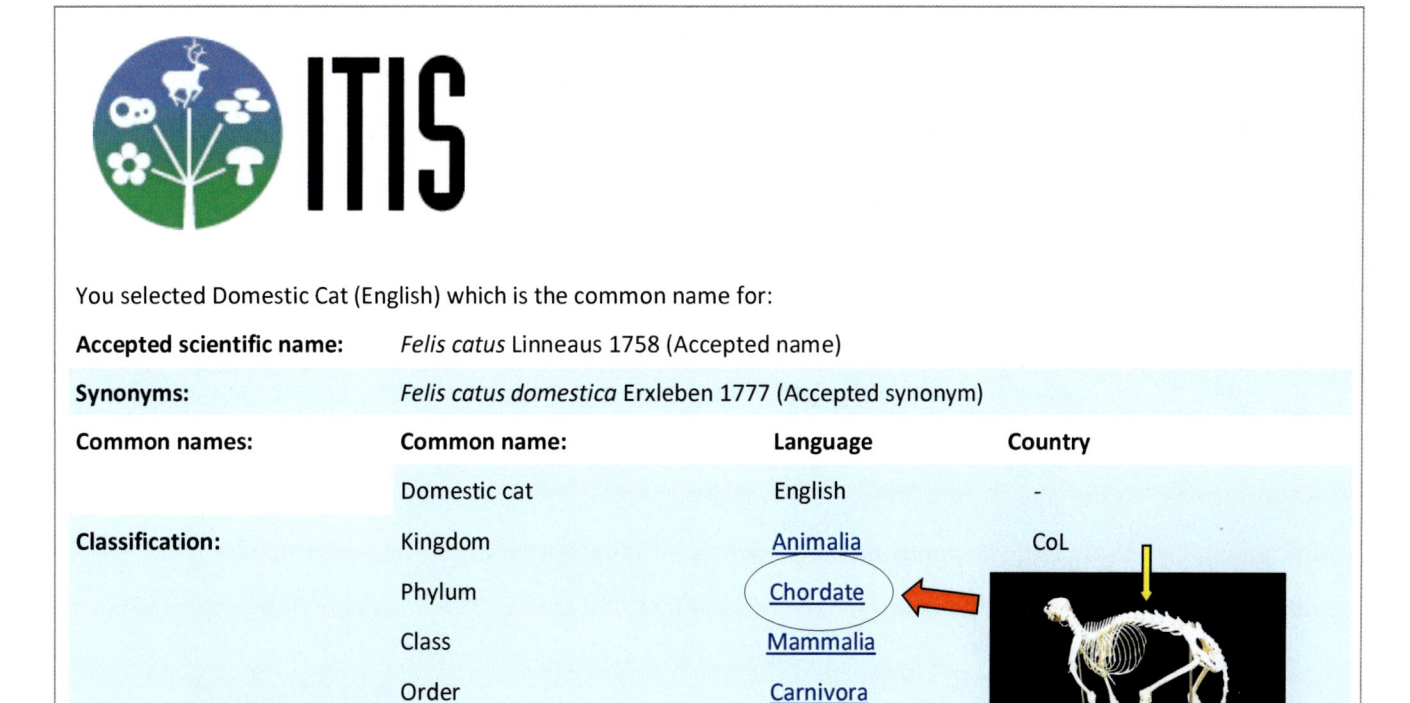

You selected Domestic Cat (English) which is the common name for:

Accepted scientific name: *Felis catus* Linneaus 1758 (Accepted name)

Synonyms: *Felis catus domestica* Erxleben 1777 (Accepted synonym)

Common names:

Common name:	Language	Country
Domestic cat	English	-

Classification:

Kingdom	Animalia	CoL
Phylum	Chordate	
Class	Mammalia	
Order	Carnivora	
Family	Felidae	
Genus	Felis	

Beneath phylum, we see the next subdivision is class. For the domestic cat, we see that this level is Mammalia or mammal. Mammals are animals which have fur, are warm-blooded and provide milk for their babies. Cats, indeed, have these characteristics.

Classification:

Kingdom	Animalia	CoL
Phylum	Chordate	Co
Class	Mammalia	ITI
Order	Carnivora	ITI
Family	Felidae	ITI
Genus	Felis	IT

Beneath the level of class, we see the division known as Carnivora. The animals which belong in this division are carnivores, the meat-eaters. Again, cats definitely fall into this category.

241

Classification:	Kingdom	Animalia	CoL
	Phylum	Chordate	CoL
	Class	Mammalia	ITIS
	Order	Carnivora	ITIS
	Family	Felidae	ITIS
	Genus	Felis	ITIS

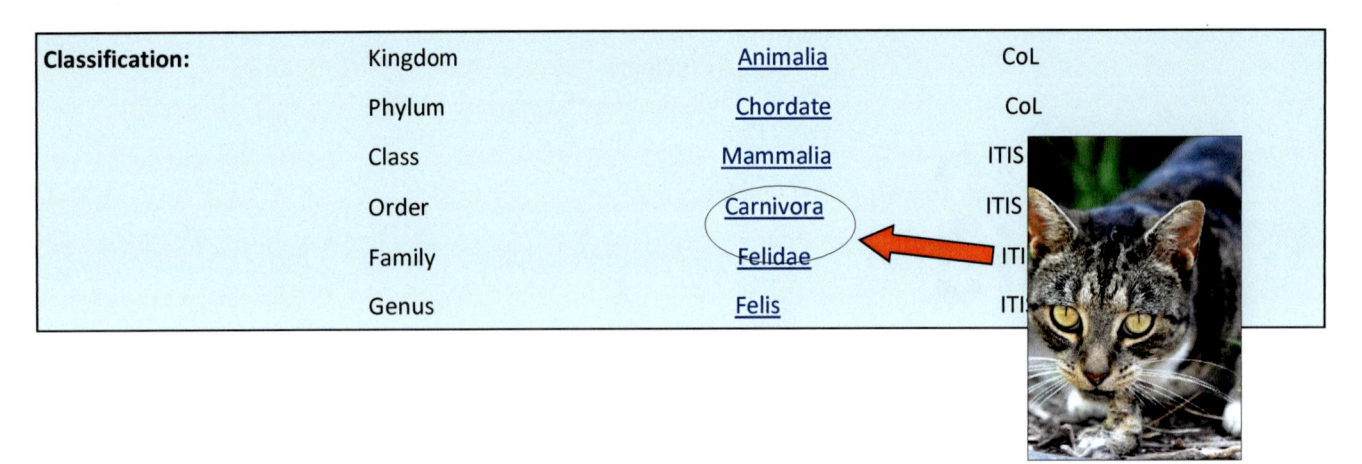

The next level down is the family which in this case is Felidae which indicates the cat family. Within the cat family we have the genus name *Felis* which refers to the small cats. Besides the domestic cat, other small cats include the ocelot, margay, caracal, jungle cat and the Chinese mountain cat. Note that the genus name is written in italicized letters. The genus name serves as part of the "official" scientific name for the domestic cat. It can be likened to the "first" name of a person.

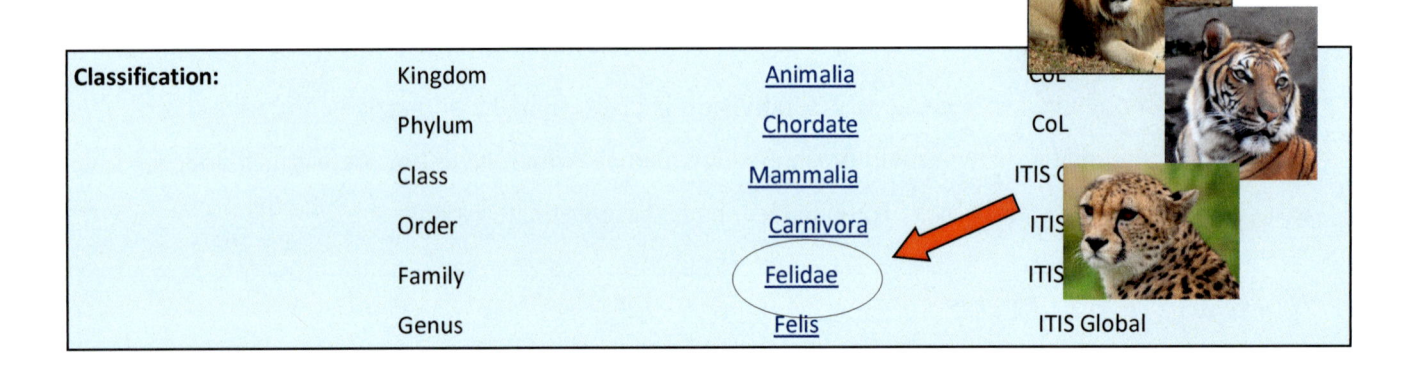

Classification:	Kingdom	Animalia	CoL
	Phylum	Chordate	CoL
	Class	Mammalia	ITIS
	Order	Carnivora	ITIS
	Family	Felidae	ITIS
	Genus	Felis	ITIS Global

Classification:	Kingdom	Animalia	
	Phylum	Chordate	
	Class	Mammalia	
	Order	Carnivora	
	Family	Felidae	ITIS Global
	Genus	Felis	ITIS Global

The final level of organization is the species name of the organism. In the case of the domestic cat, the species name is *catus*. Note that we saw earlier in our search results that the scientific name of the domestic cat is *Felis catus*. The species name, like the genus name, is always written using italicized letters. Where the genus name is analogous to a person's first name, the species name can be likened to a person's last name. Therefore, each living creature is given a scientific name which consists of the genus and species levels of organization. It is this very specific genus and species name which allows persons studying these organisms the capability of knowing that they are indeed referring to the same creature no matter where they are doing their studies. A *Felis catus* is recognized as a *Felis catus* anywhere in the world, no matter what language the person may speak!

Search all names—Results for "domestic cat"

Records per page: 20

Records found: 2

Name	Rank	Name Status	Group	Source Database
Domestic cat (English)	Species	Common name for *Felis catus* Linnaeus, 1758	Animalia	ITIS
Domestic cattle (feral) (English)	Species	Common name for *Bos taurus* Linnaeus, 1758	Animalia	ITIS

Let's now look at another member of the cat family, the tiger and see how its taxonomic organization is similar and different from that of the domestic house cat. Before we enter the name of tiger into the search window on the Catalogue of Life website, make a guess as to the following levels of organization. Note that these levels of classification are referred to as the taxonomic hierarchy of that organism. Let's make some guesses:

Kingdom: Animalia, Plantae, Fungi, Moneran or Protista? Hopefully, you've guessed Animalia.

Phylum: Does it have a spinal cord? Yes, again, therefore its phylum must be Chordata.

Order: Are tigers considered meat-eaters? Yes, therefore the order must Carnivora.

Class: Are tigers warm-blooded, have fur and give milk to their cubs? Yes, therefore the class has to be Mammalia.

Family: We said the domestic cat "fell" into the family of Felidae which were the cats. So far, so good.

Genus: The genus for the domestic cat was *Felis* indicating small cats. The tiger is definitely not a small cat, so the genus name for the tiger will not be *Felis*.

Let's enter the name of tiger now into the Catalogue of Life database to determine exactly the genus name for the tiger.

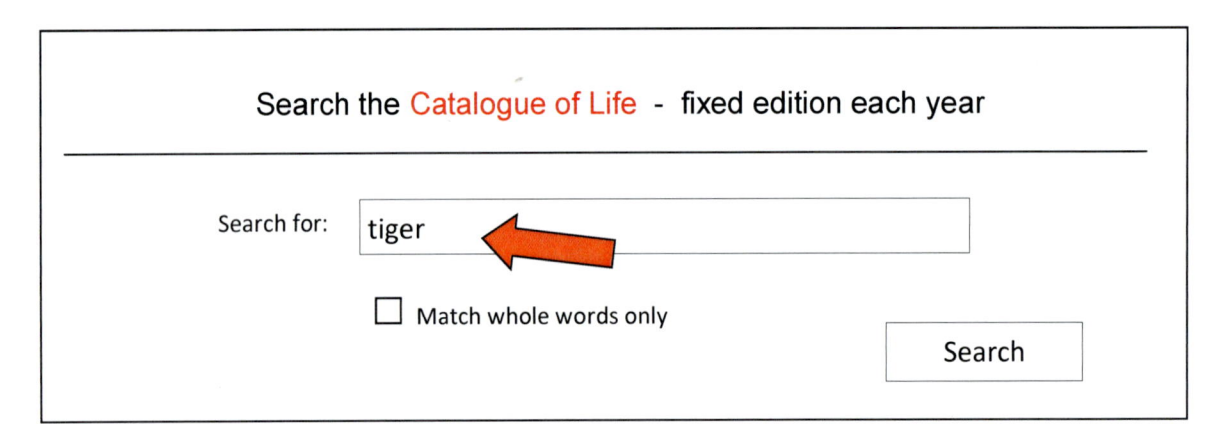

Search the Catalogue of Life - fixed edition each year

Search for: tiger

☐ Match whole words only

Search

After entering "tiger," you'll see quite a selection of living things which have tiger as part of a common name. To find the entry you're looking for, look at the classification levels "above" the genus and species levels. For example, click on the first entry which Betanodavirus: tiger puffer nervous necrosis virus. Note that the group to which that organism belongs is the viruses and not the "tiger" we are seeking.

Try the second entry. This entry is a tiger which *does* belong to the Animalia kingdom. Upon clicking this entry we find that common names which go along with the "tiger" common name for this organism are muskie and pike—which are names of fish! This is not the tiger entry we are looking for! The third entry also belongs to the Animalia kingdom so let's try it.

Friendly Biology

		Search all names—Results for "tiger"		
		Export search results / New Search	Records per page	20 Update
Records found: 2				
Name	Rank	Name Status	Group	Source Database
Betanodavirus: tiger puffer nervous necrosis virus ICTV	Species	accepted name	Viruses	ICTV
Tiger (English)	Species	common name for *Esox masquinongy* Mitchill, 1824	Animalia	FishBase
Tiger (English)	Species	common name for *Panthera tigris* (Linnaeus, 1758)	Animalia	ITIS
		Export search results / New Search		

Upon clicking this entry, we see that this "tiger" is indeed the member of the cat family that we are searching for. If we look down the levels of classifications we find that the tiger is a member of the phylum Chordata, the class Mammalia, order Carnivora, family Felidae and genus *Panthera*. Note that the species name for the tiger is *tigris* and that, once again, the scientific name is written using italicized letters: *Panthera tigris*.

You selected Tiger (English) which is the common name for:

Accepted scientific name: *Panthera tigris* (Linnaeus, 1758) (accepted name)

Synonyms: -

Common names:	**Common name:**	Language	Country
	Tiger	English	-
Classification:	-	Animalia	CoL
	Phylum	Chordata	CoL
	Class	Mammalia	ITIS
	Order	Carnivora	ITIS
	Family	Felidae	ITIS
	Genus	Panthera	ITIS Global

So the domestic cat and tiger share many levels of the taxonomic hierarchy. However, we see at the level of genus, they are separated into the small cats and the large cats. We can see that size was the primary dividing factor at this level for members of the cat family.

While we are looking at the scientific name for the tiger, you might note a person's name and a year which follows the scientific name. In the case of the tiger, you can see the name Linnaeus and the year, 1758. This indicates that Carl Linnaeus was the taxonomist who placed the tiger into this classification in the year 1758. Take a moment and look back at the search you did for the domestic cat. Who was the taxonomist who classified the cat? _____ And in what year? _____ Search now for the lion's hierarchy. Who classified the lion? _____ And in what year? _____

You selected Tiger (English) which is the common name for:			
Accepted scientific name:	*Panthera tigris* (Linnaeus, 1758) (accepted name)		
Synonyms:	-		
Common names:	Common name:	Language	Country
	Tiger	English	-
Classification:	Kingdom	Animalia	CoL
	Phylum	Chordata	C
	Class	Mammalia	ITIS
	Order	Carnivora	I
	Family	Felidae	IT
	Genus	Panthera	ITIS Global

Let's take a look now at two other common animals and compare their taxonomic hierarchies: the domestic horse and domestic cow (or cattle). Before we enter these names into the database search box, let's make some guesses as to their taxonomic hierarchies.

Let's begin with the horse. First, write the kingdom to which the horse belongs.

Kingdom:_____

For the phylum, look back at the taxonomic hierarchy we found for the cat. Note the deciding feature here was whether the creature had a backbone or not. Obviously, like the cat, the horse does have a backbone and therefore would fall into the same phylum. Write its name here.

Phylum: _____

The next division is the class. Refer again to the cat's taxonomic hierarchy where we found that the cat was a mammal (had hair, nursed babies and was warm-blooded). Does a horse fall into the same class? Yes, so write that class name below.

Class: _____

We're now down to the division of order. Look back at the cat hierarchy once again. At this point we found that the cat fell into the order Carnivora based upon the fact that it was primarily a meat-eater. What about the horse? Would the horse fall into the same category? Hopefully, you're saying, "No, horses don't eat meat; horses are plant-eaters!" Consequently, there must be another order for horses. Go to the Catalog of Life database now and search for the hierarchy for the domestic horse. Below is what you should find.

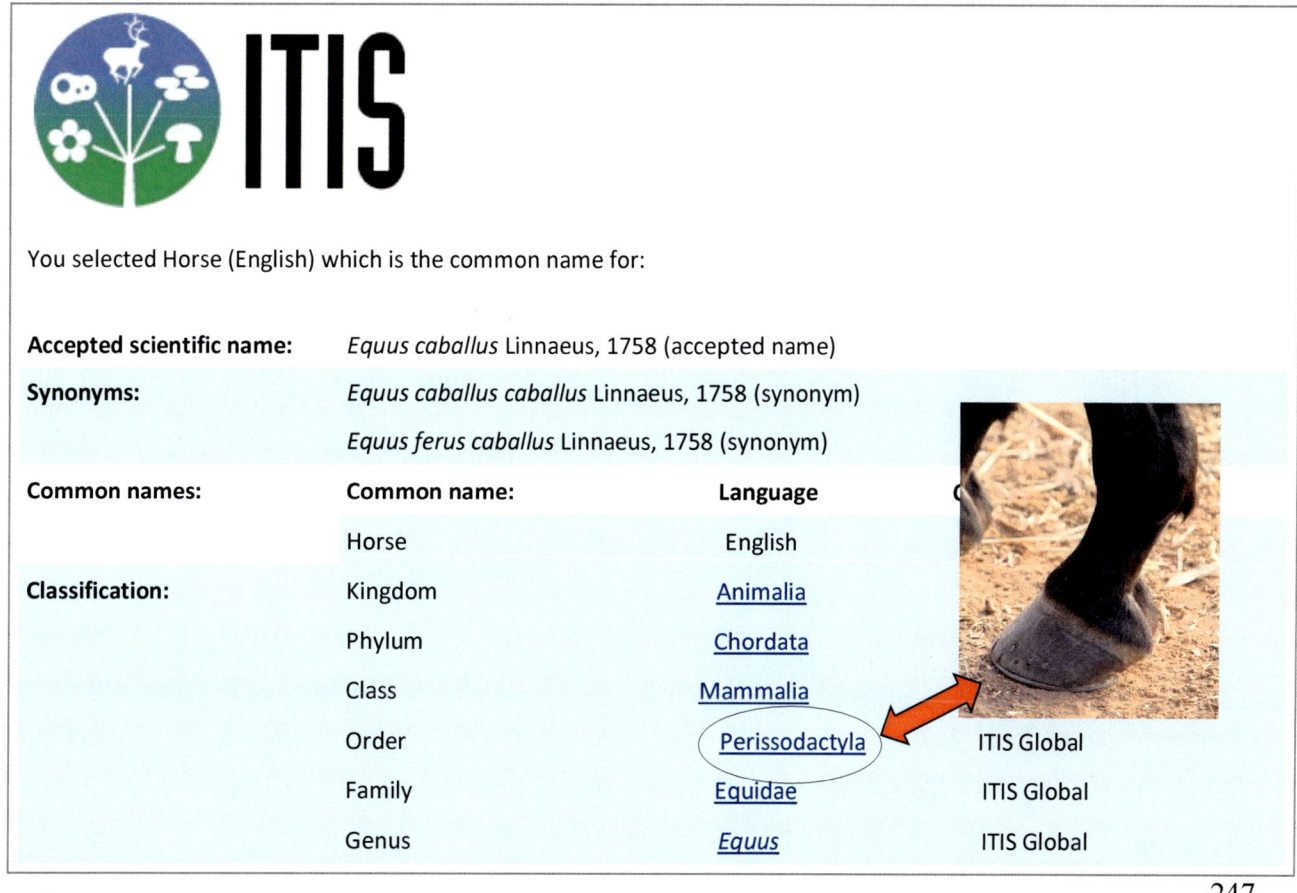

You selected Horse (English) which is the common name for:

Accepted scientific name: *Equus caballus* Linnaeus, 1758 (accepted name)

Synonyms: *Equus caballus caballus* Linnaeus, 1758 (synonym)

Equus ferus caballus Linnaeus, 1758 (synonym)

Common names:

Common name:	Language
Horse	English

Classification:

Kingdom	Animalia	
Phylum	Chordata	
Class	Mammalia	
Order	Perissodactyla	ITIS Global
Family	Equidae	ITIS Global
Genus	*Equus*	ITIS Global

For the horse we see that the order is Perissodactyla. Now, you might have been guessing that the order should indicate that this group of animals is not consisting of meat-eaters and, instead, herbivores. However, we see that taxonomists are using a different feature for this group of animals. This odd-looking word, Perissodactyla, describes the number of toes found on individuals in this classification. In this case, perisso– refers to odd-numbered and the root word --dactyla refers to toe or finger. Therefore, members of the order Perissodactyla have an odd number of toes or hooves. In the case of the horse, we see that it has one main toe (hoof) which touches the ground and two other smaller, relatively dysfunctional hooves or dewclaws which do not touch the ground. The horse, then, has three toes, an odd number. Other members of this order include rhinoceroses, tapirs, donkeys and zebras.

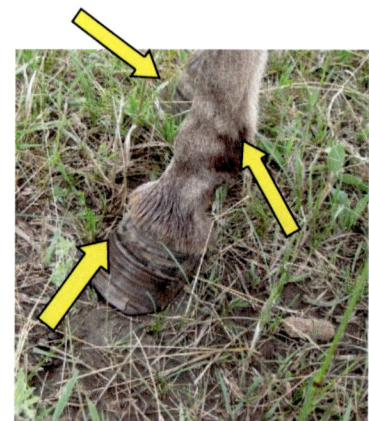

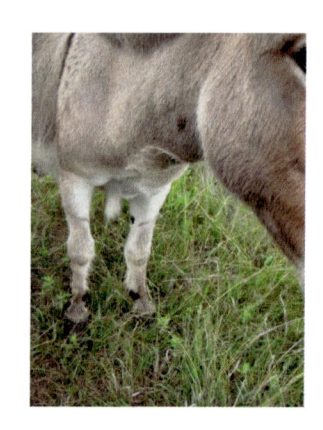

Donkeys have an odd number of toes on each foot.

Let's look at the taxonomic hierarchy now for the cow. Before you search on the database, see how many of the following classifications you can complete.

Kingdom: _____

Phylum: _____ (Hint: Backbone present?)

Class: _____ (Give milk, hairy, warm blooded?)

Now, we come to the order level with the cow. With the horse, we found that the order was based upon the number of toes found on the creature. Off-hand, do you know how many toes you would find on a cow? Cows have two toes which touch the ground and then two smaller toes referred to as dewclaws which are up higher on the foot and do not touch the ground. All-together, cows have four toes on each foot. Knowing this, do you think cows fall into the same order as horses? Recall that horses are found in the "odd-toed" order of Perissodactyla. Search the database now to find out the order for the cow.

You selected Domesticated Cattle (English) which is the common name for:

Accepted scientific name:	*Bos taurus* Linnaeus, 1758 (accepted name)		
Synonyms:	*Bos indicus* Linnaeus, 1758 (synonym)		
	Bos primigenius Bojanus, 1827 (synonym)		
Common names:	Common name:	Language	Country
	aurochs	English	
	domestic cattle (feral)	English	
	domesticated cattle	English	
Classification:		Animalia	
	Phylum	Chordate	
	Class	Mammalia	
	Order	Artiodactyla	ITIS Global
	Family	Bovidae	ITIS Global
	Genus	*Bos*	ITIS Global

We see that cows and horses share the same classifications down to the level of the order where we find the cow belongs to the order Artiodactyla. Can you guess what the meaning of artiodactyla might be? If you said even-numbered toes, you are correct! Cows have two toes which touch the ground (sometimes referred to as cloven-hooved) and then two dew claws, one on the inside of the foot and a second on the outside surface of the foot. Cows, then, have a total of four toes on each foot, an even-number of toes. Other members of the order Artiodactyla include sheep, goats, deer, pigs, camels, llamas and giraffe.

Sheep have an even number of toes on each foot.

So, with members of the cat family we discussed earlier, we found that size was utilized as a characteristic upon which divisions were made. In the case of horses and cows, we see that number of toes was utilized as a characteristic upon which taxonomic divisions are made. These are just two examples of the thousands of characteristics that taxonomists use to place living creatures in one group or another. In addition, a very current approach to determining categorization is examination of DNA and RNA samples of organisms. By observing similarities and differences, taxonomists are able to place organisms in appropriate orders, families, etc.

Up to this point in our discussion on taxonomy, we've been targeting a specific organism and then investigating its taxonomic hierarchy (the various phyla, class, order, etc. to which it belongs). Suppose, however, that you know an organism's hierarchy, but would also like to know which *other* organisms share a particular level in the hierarchy. For example, from our discussions above, we've learned that tigers are members of the genus *Panthera*. Suppose, now, you would like to find a list of all other known large cats which are also members of this genus.

To accomplish this, direct your browser back to the home page of the Catalogue of Life homepage (www.catalogueoflife.org). Once there, click, again, to access the annual checklist.

Earlier we clicked the "search" option in the box which appears next. This time, click the "browse" option.

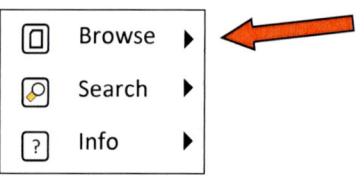

Two options appear next: taxonomic tree and taxonomic classification. Choose taxonomic classification.

Next, a box will appear with several drop-down menus. The first menu allows you to choose a specific group or kingdom. Because we know that the tiger is a member of the kingdom Animalia, begin with that menu.

Browse taxonomic classification

Top level group	Animalia
Phylum	
Class	
Order	
Superfamily	
Family	
Genus	
Species	
Infraspecies taxon	

Clear Form   Search

The next menu down asks for the phylum we are interested in. Recall that tigers are members of the phylum Chordata (have backbone). Choose Chordata in this drop-down menu.

Browse taxonomic classification

Top level group	Animalia
Phylum	Chordata
Class	

251

The next level "down" is the class level of organization. Open this list of choices and you should recognize the class for tigers which is Mammalia. Choose this option.

Browse taxonomic classification	
Top level group	Animalia
Phylum	Chordata
Class	Mammalia
Order	

The next level is order. Open this list of options. Do you recognize the order to choose? Hopefully, you recall that tigers are carnivores so find Carnivora on the list. If you didn't remember this level from our earlier discussion (or if you're investigating an organism other than a tiger) you may want to "get" the taxonomic hierarchy, first. Copy or print it and the refer to it as you complete the drop-down menus here.

Browse taxonomic classification	
Top level group	Animalia
Phylum	Chordata
Class	Mammalia
Order	Carnivora
Superfamily	

After choosing the order Carnivora, the next level down is superfamily. Go ahead and click this drop-down menu. Note that the message "no matching results found" appears. This means that "below" the order carnivora the level of superfamily does not exist. A superfamily can be considered as being an intermediate grouping between orders and families. In this case, a superfamily does not exist between orders and families of carnivores. Continue by moving on down to the family drop-down menu.

Recall that tigers are members of the Felidae or large cat family. Choose that option.

Browse taxonomic classification

Top level group	Animalia
Phylum	Chordata
Class	Mammalia
Order	Carnivora
Superfamily	No matching results found
Family	Felidae
Genus	

We're now down to the genus level of organization. Click that menu and find *Panthera*.

Browse taxonomic classification

Top level group	Animalia
Phylum	Chordata
Class	Mammalia
Order	Carnivora
Superfamily	
Family	Felidae
Genus	Panthera
Species	

At this point, because we are interested in which other cats also belong to the genus *Panthera*, we're not going to choose a species, but instead click the "search" button at the bottom of the box.

253

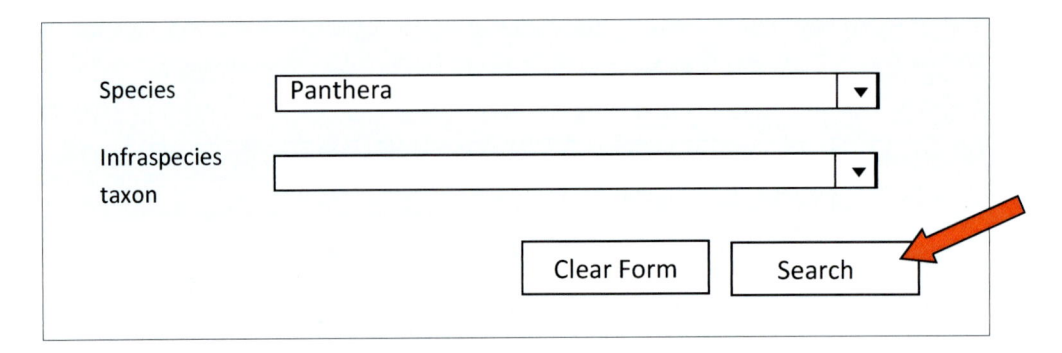

A page should now open with your results which lists all members of the genus *Panthera*. At the top of the list you'll see that there are four species found with 36 infraspecific taxa found. This means there are four members of the *Panthera* genus and of those four there are 36 subspecies "beneath" the species level of organization. For our example, we are interested in knowing these four species. Note that each member is listed using its scientific name (genus and species) which may still prove puzzling. Scan down through the listing to see if you can recognize any of the names.

Search results for taxonomic classification

Records found: 40 (4 species, 36 infraspecific taxa)   Records per page   20   Update

Scientific name	Name status	Group	Source database
Panthera leo (Linnaeus, 1758)	Accepted name	Animalia	ITIS
Panthera onca (Linnaeus, 1758)	Accepted name	Animalia	ITIS
Panthera pardus (Linnaeus, 1758)	Accepted name	Animalia	ITIS
Panthera tigris (Linnaeus, 1758)	Accepted name	Animalia	ITIS

Hopefully, you recognize *Panthera leo* to be the scientific name of the lion. Confirm this by clicking on the scientific name *Panthera leo*. The data for *Panthera leo* should appear next. Scan down through the information and find the common name data. Note, indeed, that *Panthera leo* is the scientific name for the lion. Note all of the other subspecies (infraspecific taxon) of the lion.

Accepted scientific name: *Panthera leo* ((Linnaeus, 1758) (accepted name)

Synonyms:

Infraspecific taxon

Panthera leo azandica (J. A. Allen 1924

Panthera leo bleyenberghi (Lönnberg, 1914)

Panthera leo krugeri (Roberts, 1929)

Panthera leo leo (Linnaeus, 1758)

Panthera leo melanochaita (C. E. H. Smith, 1858)

Panthera leo nubica (de Blainville, 1843)

Panthera leo persica (Meyer, 1826)

Panthera leo senegalensis (J. N. von Meyer, 1826)

Go back now to your listing of members of the genus *Panthera* (click your back arrow on your browser). Scan down through the list to find *Panthera onca*. This scientific name may not be familiar to you so go ahead and click its name to take you to the classification data. Scan down through the information to the common name. Note that *Panthera onca* is the jaguar. So far, we now have identified two of the four members of genus *Panthera*.

Return to your listing of members of the genus *Panthera*. Scan down until you find *Panthera pardus*. Click this name to reveal its common name. *Panthera pardus* is the scientific name for the leopard. We now have three of the four members of the genus *Panthera*: lions, jaguars, leopards. And, of course, our fourth is the tiger, *Panthera tigris*!

Let's pause now and review what we've learned in this lesson. We have learned that:

- Many individuals have worked and continue to work on studying and then placing living organisms into various groups and categories based upon the phenotypes and more recently, genotypes, of these creatures. The study of this placement is known as taxonomy;

- The placement of organisms into various groups is an ever changing process;

- The Catalogue of Life is a recent effort to develop a comprehensive database of taxonomic information available for online use;

- There are several "layers" of classification with the most frequently used being: kingdom, phylum, class order, family, genus and species;

- The genus and species serve as the official scientific name of a living creature. The genus portion is always capitalized while the species portion is written in lower case. Both terms are typed in italics or underlined when hand-written.

Lesson 15: The Animal Kingdom

The purpose of this lesson and the four which follow is to provide a survey of the major phyla, classes and orders of living things within each kingdom. With each lesson, you will be presented with an enormous amount of information. You will not be required to memorize any of these facts for the lesson test. Instead, utilize these lessons to gain a greater appreciation for the vast diversity of living things which exist in our world today. Note that there are practice pages at the end of each lesson as well as a lesson test. It is our recommendation that the Lesson Test be taken as an open-book test.

We'll begin by looking at each phylum within the animal kingdom. To help you gain a perspective of "where" we are in the kingdom, we have provided a flowchart-style diagram. As we move from phylum to phylum and class to class, we'll refer you back to the diagram to help you keep things relatively organized in your mind. Take a look at the diagram below and note where we are beginning our discussion.

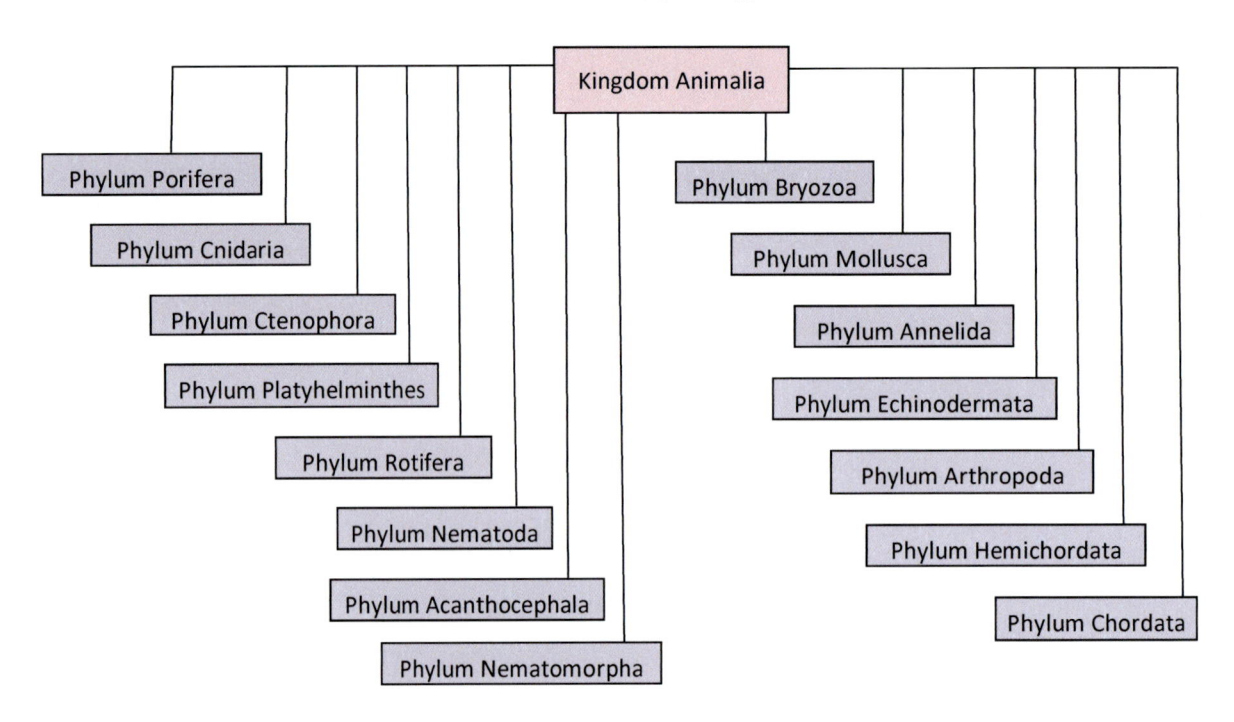

```
                            Kingdom Animalia
    │
    ├── Phylum Porifera
    ├── Phylum Cnidaria
    ├── Phylum Ctenophora
    ├── Phylum Platyhelminthes
    ├── Phylum Rotifera
    ├── Phylum Nematoda
    ├── Phylum Acanthocephala
    ├── Phylum Nematomorpha
    ├── Phylum Bryozoa
    ├── Phylum Mollusca
    ├── Phylum Annelida
    ├── Phylum Echinodermata
    ├── Phylum Arthropoda
    ├── Phylum Hemichordata
    └── Phylum Chordata
```

Recall from our earlier discussions that members of the Animal Kingdom are capable of moving from one place to another and require a source of energy provided by some other organism. This means that, unlike plants, animals are incapable of producing their own food supply. Also, unlike plants, the cells of animals have no cell walls. Animals are multicellular (consist of many, many cells) and these cells are organized into specialized tissues and complex organ systems. Most animals reproduce by sexual reproduction. The cells of animals are eukaryotic (have a nuclear membrane).

Let's look now at the first phylum within the Animal Kingdom: Phylum Porifera. Reflected in its name, the members of this phylum are characterized by having many, many pores or holes. These are the sponges. The body wall of sponges consists of two layers of cells which are penetrated by thousands and thousands of pores. Because sponges live underwater, these pores allow water carrying food products for the sponge to freely move in and out. Sponges are generally found to be sessile which means they are "stuck" in one place.

Within this phylum we find three classes. These classes are based upon chemical compounds which make up the "skeleton" or supporting structure within the sponge. The first class, known as Class Calcarea, has a "skeleton" made up of calcium carbonate and is found in shallow water. Calcium carbonate is the same substance which makes up the shells of many other sea creatures. The second class of Porifera is the Class Hexactinella. The "skeleton" of this class is made up of silicon dioxide and these are found in deep ocean waters. The third class is Class Desmospongiae whose "skeleton" is a mixture of a substance known as spongin and siliceous material. These sponges can be found both in marine or freshwater and are used as bath sponges. The final class of this phylum is Class Sclerospongiae which has a "skeleton" made up of calcium, silica and spongin materials.

Sponges can reproduce asexually through budding or sexually through the formation of free-swimming sperm cells which travel to neighboring sponges. Most sponges are considered to be hermaphroditic which means individual specimens can produce both egg and sperm cells. Self-fertilization rarely occurs. Sponges also have the amazing capability to regenerate missing parts and continue living successfully when placed into a suitable environment.

A natural sponge from the Phylum Porifera.

The next phylum of Animalia we will discuss is called Cnidaria (ni-dare-ee-a) and is also known as Ceolenterata. These animals are all found in water environments, both freshwater and saltwater. Their distinguishing characteristic is that they all possess cnidocytes (ni-do-sites) or nematocysts which are stinging cells which they use to capture prey.

There are three classes of the phylum Cnidaria. Class Hydrozoa includes the *Hydra* and *Physalia* (which is commonly known as the Portuguese Man of War or Floating Terror). The Portuguese Man of War is commonly found floating on the surface of the Atlantic, Pacific and Indian Oceans. The second class is Class Scyphozoa which includes 200 recognized species of what are known as "true" jellyfish. These are strictly marine animals and can range in size from one-half inch to over six feet across. The third class is Class Anthozoa which includes sea anemones, corals and sea fans.

Members of the Phylum Cnidaria. Note the presence of tentacles with stinging cells.

The next phylum in the Kingdom Animalia is Phylum Ctenophora (ten-offer-uh). The members of this phylum are commonly known as comb jellyfish. They possess cilia which are hair-like projections they use to capture prey. Unlike the "true" jellyfish we discussed above, these jelly fish do not have any stinging cells.

The phylum we will discuss next is Phylum Platyhelminthes. The term platy– refers to "flat" as in plate or the Platte River (which was named for being very flat). Helminth refers to worms and therefore these organisms are known as the flatworms. There are three classes within this phylum which include Class Turbellaria, Class Trematoda and Class Cestoda.

Class Turbellaria is the class to which the planaria belongs. Recall from our discussion of reproduction in Lesson 12 that we learned how the planaria has the ability to regenerate from broken parts of its body.

Class Trematoda includes organisms commonly known as flukes. Flukes are parasitic to animals— including humans—and include liver flukes found in sheep and cattle. They spend part of their life cycle in moist grasses and marshes and are ingested by grazing cattle and sheep. They then migrate to the liver where they "rob" nutrients from their host.

The third class of Phylum Platyhelminthes is Class Cestoda. These flat worms are commonly known as tapeworms. Like the flukes, the members of this class are parasitic animals which are found in the intestines of many warm-blooded hosts, including humans. They are characterized by a mouth

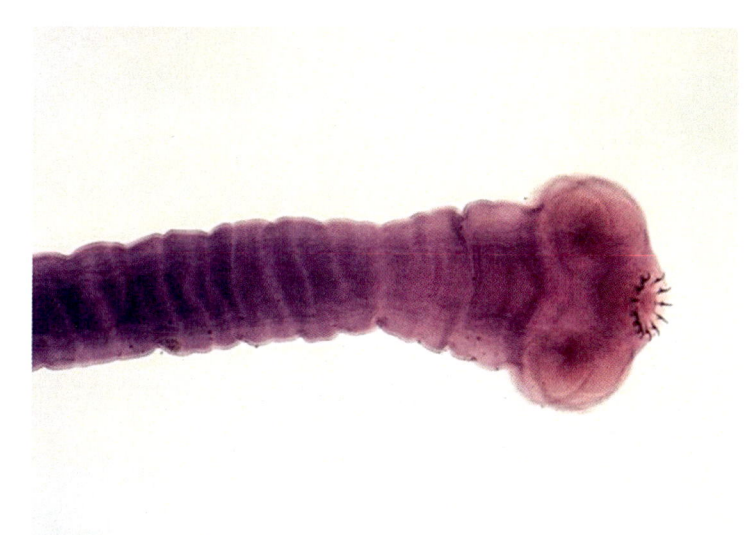

The scolex (head) of a tapeworm.

scolex with a mouth with small hooks on it called a scolex that they use to attach to the wall of the intestine. Their body is then composed of many segments and can gain great length. They are a common parasite of dogs and cats.

Segments of a tapeworm body, known as proglottids.

The next phylum in Animalia we will look at is Phylum Rotifera. Commonly known as rotifers, these animals are tiny creatures which live in freshwater or in thin layers of water on land. They have a crown of rotating cilia (hair-like projections) which make them appear to have little rotating wheels on top of their heads.

Phylum Nematoda is the next phylum we will discuss in our look at the Animal Kingdom. This phylum includes what are commonly referred to as "round" worms and includes many parasitic species. A common member of this phylum is the hookworm often found in puppies. These voracious parasites affix their mouths to the intestinal lining of young puppies and "steal" a tremendous amount of blood leaving the puppy weak and anemic which can result in death if not treated quickly. Roundworms are also a problem for cattle and swine producers. Pinworms (found in humans) are also a member of this phylum. While not a dangerous parasite, pinworms gain access to the digestive tract by ingestion. At night, the female pinworm exits the infected person through the anus to lay her eggs on the surrounding skin. This results in itchy discomfort around the anus.

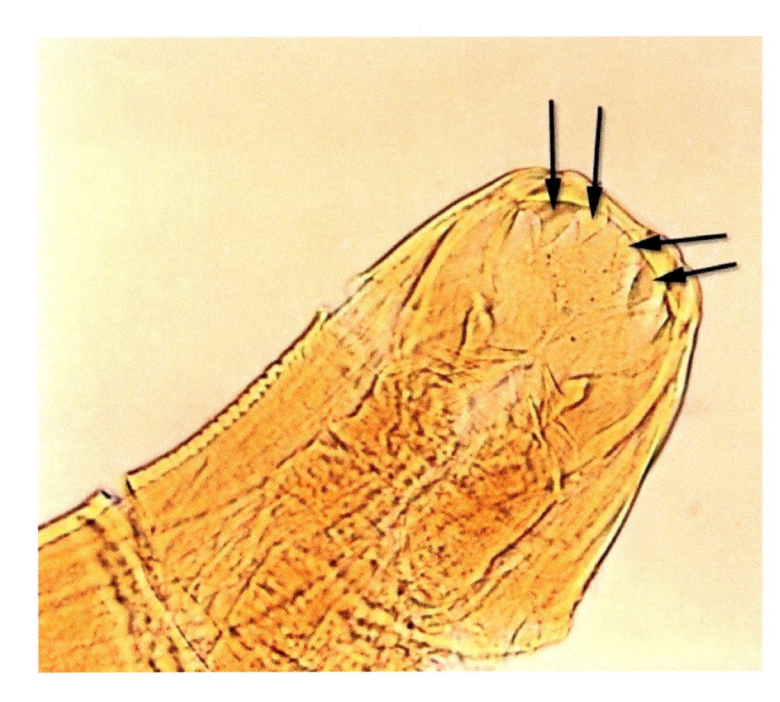

The arrows indicate the teeth by which the hookworm attaches itself to the lining of the intestine.

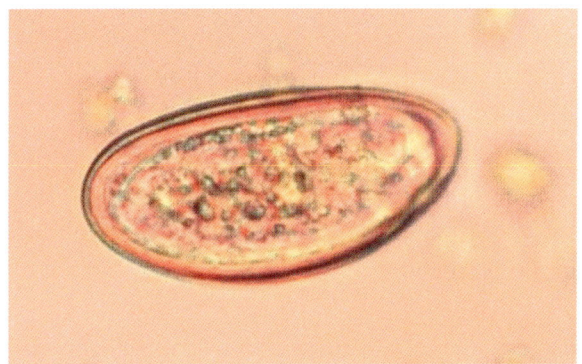

The egg of the pinworm.

The next phylum we'll look at is Phylum Nematomorpha which are long, hair-like worms. Their larval (immature) stage is parasitic to insects but as adults they are found in freshwater environments. They are commonly known as horsehair worms. Horsehair worms are commonly found in the bodies of crickets and grasshoppers and may be up to a foot long. When ready to leave their host, they modify nerve transmitters in the brain of the cricket or grasshopper which cause it to dive into water. This allows the worm to burrow out of its host and into the watery environment where it lays eggs and begins the process again.

The next phylum is named Phylum Acanthocephala. The prefix acantho– means horn or spine and the root word -cephala refers to head. These are commonly known as the spiny-headed or thorny-headed worms. Like most of the other worms we've discussed so far, this worm is parasitic. A portion of its lifecycle is found in crustaceans of watery environments which eventually end up in the digestive tracts of birds, primarily ducks. Humans, too, have been know to be infested with these parasites. Some recent studies have revealed that these organisms are actually only a portion of the Phylum Rotifera and should not be classified separately.

The next phylum we'll look at is the Phylum Bryozoa. These organisms are very short (only 0.5 mm in height) tube-shaped animals which affix themselves to underwater rocks. They are commonly referred to as moss animals. They are filter feeders, taking in water through their mouths.

The Phylum Brachiopoda is the next phylum we'll examine. This phylum, which mainly consists in fossil forms, does exist today in marine environments and members are commonly referred to as lamp shells. These organisms resemble other marine creatures which are made up of a system of two shells (bivalves). However, unlike others, these organisms have an upper and lower shell versus a right and left shell (valve).

The next phylum we'll examine is the Phylum Mollusca. Many of the members of this phylum have a mantle which secretes substances which create a shell. They also have a "foot" portion known as the radula. Let's look more specifically at classes of this phylum.

Snails (above and top next page) are members of the Phylum Gastropoda.

The Class Gastropoda includes snails, slugs and whelks. The name gastropoda literally means stomach and foot. They have a coiled shell (except slugs), a body with a distinct head and eyes with tentacles.

The Class Bivalvia is another class found in the Phylum Mollusca. They are characterized by having two halves of shells, a mantle with gills, a head with eyes, but no tentacles. Members of this class include clams, oysters and scallops.

Oysters are members of Class Bivalvia of Phylum Mollusca.

The Class Cephalopoda is yet another class of the Phylum Mollusca. The name cephalopod literally means head and foot. These animals live in ocean waters and members include the squid, octopus and chambered nautilus. The "foot" portion of these creatures is modified into many grasping tentacles.

Suction-cup appendages enable the octopus to capture prey.

Chitons are members of the next class of the Phylum Mollusca known as Class Polyplacophora. Chitons are sea creatures which are distinguished by have a shell made of eight valves known as plates. These plates are joined together which allow the chiton to flex. This enables it to move about as well roll up into a ball when being harmed.

The final class of Phylum Mollusca is the Class Scaphopoda. These animals, too, are sea creatures and have a shell shaped like a tooth. They are commonly known as tooth shells or tusk shells as the shell often curves to one side like the tusk of an elephant.

So far in our discussion of members of the Animal Kingdom, we've looked at eleven different phyla. We have learned:

Phylum Porifera: sponges that are found in marine environments;

Phylum Cnidaria: sea creatures with tentacles and stinging cell, jellyfish;

Phylum Ctenophora: comb jellyfish which move by cilia;

Phylum Platyhelminthes: flat worms, tapeworms;

Phylum Rotifera: worms with crown of cilia—whirling wheel on its "head";

Phylum Nematoda: round worms (hookworms) and pin worms;

Phylum Nematomorpha: horsehair worms which infect crickets;

Phylum Acanthocephala: spiny-headed worms found in crustaceans;

Phylum Bryozoa: moss animals or sea moss;

Phylum Brachiopoda: sea animal with upper and lower valves;

Phylum Mollusca: snails, clams, scallops, squids and octopi.

The next phylum found in the Animal Kingdom is Phylum Annelida. Phylum Annelida consists of worms which have segments in their bodies. The best examples of this phylum are the earthworms that you may have used as fishing bait. Some members of this phylum also have setea or bristles on their underneath surface which allow them to move across surfaces. There are three classes within this phylum including the sandworms, the earthworms and leeches. Leeches are most often found in fresh-water streams and lakes and have suckers on both ends of their bodies. They use these suckers to attach to rocks and other structures. If you've ever waded in a river or lake you may have had an encounter with a leech that has attached itself to you!

The earthworm (right) and leech (below) are both segmented worms belonging to the Phylum Annelida.

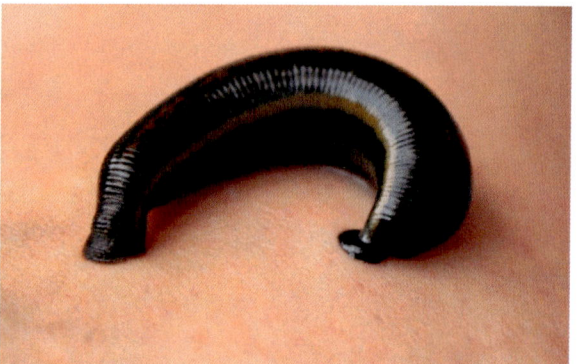

The next phylum we will discuss is the phylum which has the greatest number of members. This phylum is Phylum Arthopoda and consists of animals which have segmented body parts and jointed legs. They also have an external skeleton (exoskeleton) which functions like our skeleton, but is on the outside of their bodies. This is what makes the crunching sound when you step on an insect or crayfish. Members can be found in the air, on land or in the water. Because this is a very large phylum, it is first divided into four subphyla based upon whether the member has antennae or not.

Members which have no antennae belong to the Subphylum Chelicerata (chi-liss-er-ah-ta). These animals have chelicerae which are front legs specialized to work like pincers or fangs. Within the Subphylum Chelicerata we find the Class Arachnida (ar-rack-nid-uh). A member of this class has its head and thorax fused into one part. They have four pairs of legs (eight in all). Spiders, scorpions, ticks and mites are members of this class.

Note the chelicerae on this spider which work to grasp prey.

The red hourglass on the abdomen of this spider identifies it as the poisonous black widow spider.

The scorpion has pincers to secure prey.

The second subphylum of Phylum Arthropoda is Subphylum Crustacea. The members of this subphylum *do* have antennae as well as mandibles (jaws). In fact, members of this group have two pairs of antennae. Like the Subphylum Arachnida, these animals also have their head and thorax fused into one body part. However, instead of 4 pairs of legs, the animals of this subphylum have five pairs of legs or ten in all. The legs of this group are jointed. Most members of this subphylum live in aquatic environments and include the crayfish, lobster, crab, shrimp, barnacle and water flea. The sowbug or pillbug is a land-dwelling member of this subphylum.

Members of the Subphylum Crustacea have two pairs of antennae and ten legs. Crayfish (left) and shrimp (below) are members of this subphylum.

The third subphylum is Subphylum Trilobita. Members of this subphylum were marine creatures and are now extinct.

The fourth subphylum of Phylum Arthropoda is Subphylum Uniramia. The members of this group have antennae and mandibles, however their legs have only one branch. There are three classes within this subphylum. The first is known as Class Chilopoda which includes centipedes. Centipedes are characterized by many body segments with one pair of legs per segment.

The trilobite is extinct.

The centipede (left) has one pair of legs per body segment while the millipede (below) has two pairs of legs per body segment.

The second class within this subphylum is Class Diplopoda. Where the members of Class Chilopoda have one pair of legs per body segment, members of Diplopoda have two pairs of legs per body segment. Millipedes are members of this class.

Friendly Biology

The third class within Subphylum Uniramia is Class Insecta. This is the class commonly known as the insects. The bodies of members of this class are in three distinct parts: the head, thorax and abdomen. They have one pair of antennae and two pairs of wings.

Because you are likely familiar with insects we will also explore the Orders of this class. There are sixteen orders of insects which include:

Order Thysanura: the silverfish

Order Ephemeroptera: the mayflies

Order Odonata: the dragonflies and damsel flies

Order Orthoptera: grasshoppers (left), crickets, cockroaches, walking sticks and mantis (far right)

Order Isoptera: termites

Order Dermaptera: the earwigs

Order Mallophaga: chicken lice

Order Anoplura: human body lice

Order Hemiptera (referred to as the "true bugs": waterbugs, water striders (below left), back swimmers, bedbugs (below right), squash bugs and stinkbugs

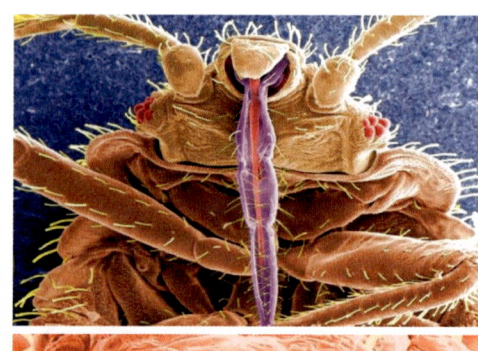

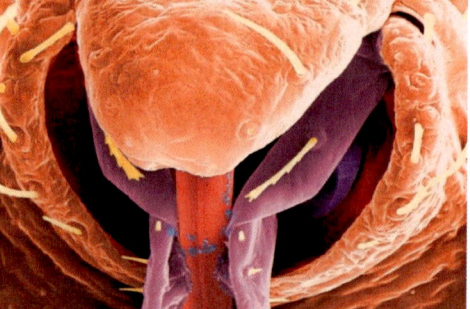

Photo courtesy Janice Haney Carr

Order Homoptera: the cicadas (left), aphids (right), leaf hoppers and scale insects

Order Neuroptera: the Dobson fly, aphis lion and lacewing (below)

Order Coleoptera: beetles including the tiger beetle (left), ladybugs (right), fireflies and boll weevils

Order Lepidoptera: butterflies (left), moths (right) and skippers

Order Diptera: houseflies (left), bot flies, blow flies, crane flies, mosquitoes (right), midges and gnats

Order Siphonaptera: fleas (right)

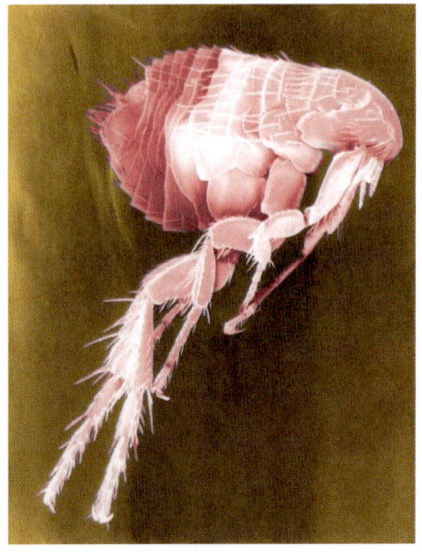

Photo courtesy Janice Haney Carr

Order Hymenoptera: bees (left), ants, wasps (right), hornets and ichneumon fly.

This concludes the Phylum Arthropoda. The next phylum we will examine is Phylum Echinodermata. The members of this phylum are said to be pentaradially symmetrical. The term pentaradial can be broken down into two parts: penta- meaning five (as in pentagons being shapes with five sides) and -radially meaning from the center out (as in radius of a circle). Symmetrical means the same or a mirrored image. The familiar member of this phylum is the starfish.

There are five classes within the Phylum Echinodermata which include:

Class Crinoidia: the sea lily and feather star (below)

Class Asteroidia: starfish (right)

Class Ophiuroidia: brittle stars (below) and basket stars

Class Echinoidea: sea urchins (left) and sand dollars (right)

Class Holothuroidea: the sea cucumber (below)

Photo courtesy Teddy Fotiou

The next phylum in the Animal Kingdom we will discuss is the Phylum Hemichordata. This is a very small phylum which includes marine worms known as acorn worms or tongue worms (right).

Photo courtesy Dennis Tappe & Dietrich W. Büttner

The last phylum in the Kingdom Animalia is Phylum Chordata. Because members of this phylum are very familiar (you and I are included), we will delve deeper into the various levels of classifications of Phylum Chordata in Lesson 16.

Lesson 16: The Animal Kingdom Phylum Chordata

In Lesson 15, we began our look at the many phyla of the Animal Kingdom. We'll continue now with the last phylum, Phylum Chordata. The members of this phylum have, or had at one time in their development, a notochord. A notochord is the early developmental stages of the spinal column.

For humans, one of the first things that developed after we were conceived was the notochord. Through a series of many cell divisions, the notochord became our spinal column including our spinal cord, brain and vertebra. In addition to the presence of a notochord, members of Phylum Chordata also possess (or at one time have possessed) pharyngeal gill slits and what is referred to as a post-anal tail. In humans and mammals these structures are present during very early fetal development but are later replaced by other body structures.

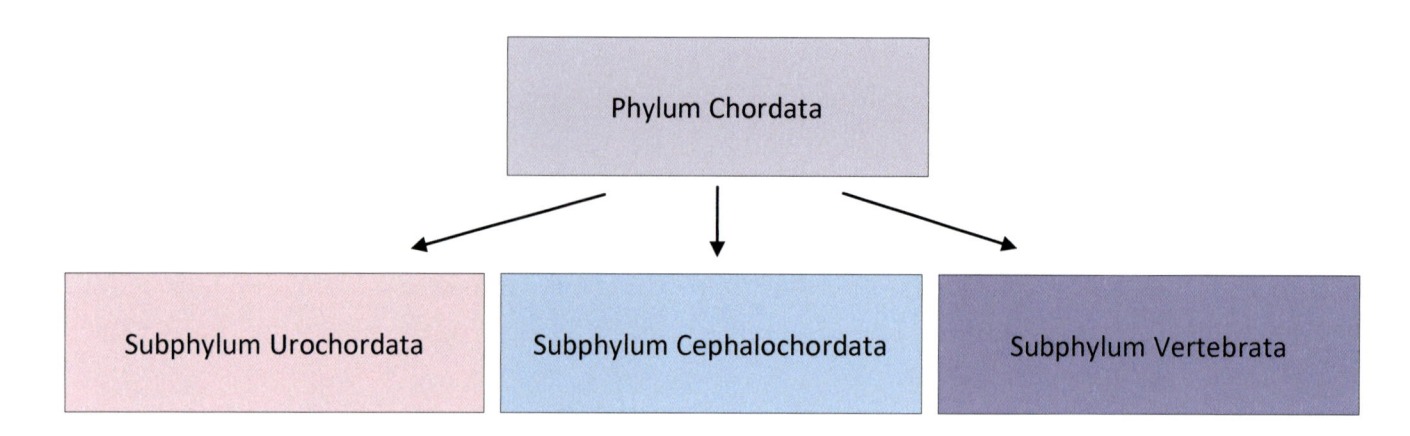

```
                    ┌─────────────────┐
                    │ Phylum Chordata │
                    └─────────────────┘
                   ↙         ↓         ↘
┌──────────────────────┐ ┌──────────────────────────┐ ┌──────────────────────┐
│ Subphylum Urochordata│ │ Subphylum Cephalochordata│ │ Subphylum Vertebrata │
└──────────────────────┘ └──────────────────────────┘ └──────────────────────┘
```

The Phylum Chordata is divided into three subphyla: Subphylum Urochordata (also known as Tunicata), Subphylum Cephalochordata and Subphylum Vertebrata.

Let's take a look at Subphylum Urochordata first. The most abundant members of this subphylum are the sea squirts. As adults, these marine animals live fixed to the ocean floor but in the early stages of development, they were free-swimming tadpole-like larvae which had a notochord, gill slits and a post-anal tail. Once they attach to the ocean floor, their nervous system largely disintegrates.

Blue-bell tunicates of the Subphylum Urochordata

Photo courtesy of Nick Hobgood

The next subphylum is Subphylum Cephalochordata. There are about 25 species of animals which belong to this subphylum and are known as lancelets. Lancelets are sword-shaped marine creatures which resemble an eel. They live in shallow tropical waters spending most of their time buried in the sand.

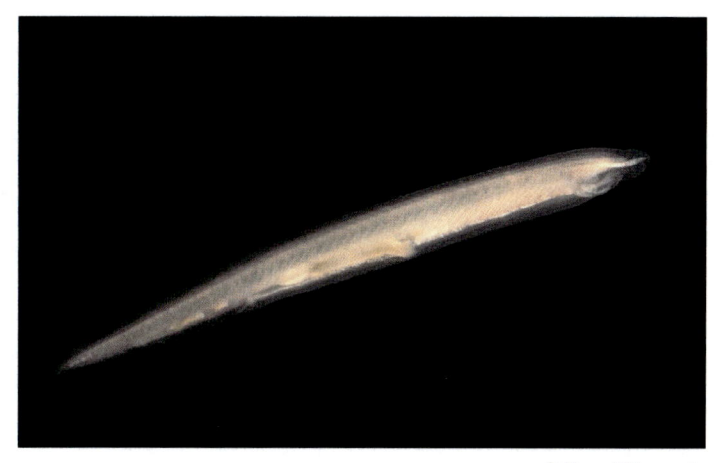

The lancelet is approximately two inches in length.

© Hans Hillewaert

The third subphylum, where we will spend most of our study in this lesson, is the Subphylum Vertebrata. This group of animals is commonly known as the vertebrates. As we mentioned earlier, members of this subphylum meet the "requirements" of being in Phylum Chordata in that at one time in development each had a notochord, gill slits and a post-anal tail. The notochord in vertebrates develops into the spinal column which further develops into the vertebral bones and skull which protect the spinal cord and brain.

The Subphylum Vertebrata is divided into eight classes. These classes include categories for fish, amphibians, reptiles, birds and mammals. The first class we will discuss is Class Ostracodermi. This class of vertebrates is extinct. Members of this class were marine animals which had no jaw but did have a covering of scales. They were known as ostrocoderms.

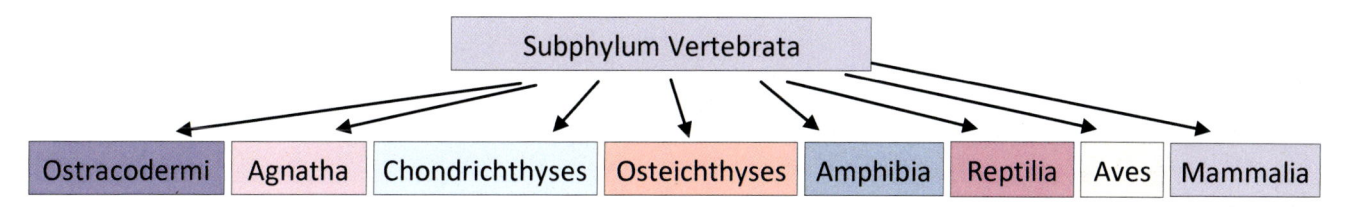

Subphylum Vertebrata → Ostracodermi, Agnatha, Chondrichthyses, Osteichthyses, Amphibia, Reptilia, Aves, Mammalia

The second class found in Subphylum Vertebrata also includes marine animals. This is the Class Agnatha. The root word -gnatha (nath-uh) refers to the jaw and when coupled with the prefix a- which means without, we know these animals do not have true jaws. They do not have scales nor paired fins, but do have a skeleton made of cartilage. Members of this subphylulm are the lamprey and hagfish.

The lamprey can be parasitic to other fishes.

Photo courtesy Tiit Hunt

The third and fourth classes we will explore are also marine animals which *do* have true jaws. These are the fishes and are divided according to whether they have skeletons made up of cartilage or skeletons made up of bones.

The Class Chondrichthyes (kon-drick-these) is the class of fishes that has skeletons made up of cartilage. The prefix chondri- refers to cartilage while the root word -ichtheyes refers to fish. They do have paired fins but no air bladder. Members of this class include the sharks, rays and skates.

The Class Osteichthyes (ahs-tee-ik-these) is the class of fishes that has bony skeletons The prefix oste– refers to bones Again, -ichtheyes refers to fish. Like the Class Chondrichthyes, members of this class have paired fins. However, members of this class *do* have air bladders. Air bladders are internal organs which enable fish to maintain desired depths within the water without having to expend energy swimming. Members of this class are include trout, bass, perch, pike and catfish.

The next class we will look at is Class Amphibia which includes the amphibians. These animals have gills at one stage in their lives which, in some orders, are replaced by lungs. This means that these animals are aquatic at some point in their lives and then become terrestrial (land dweller). Their skin is slimy and lacks any protective outgrowths and their feet lack claws. They have a

The shark (left) has a skeleton composed of cartilage and is classified within the Class Chondrochthyes while the sockeye salmon (right) has a skeleton made of bone and is classified within the Class Osteichthyes.

three-chambered heart as an adult and fertilization of their numerous eggs usually takes place externally. Many go through the process of metamorphosis during which their body undergoes major changes in form. The best example of this is the frog. Initially, the fertilized egg develops into the aquatic tadpole. The tadpole grows four legs and slowly loses its tail. Its gills are replaced by lungs which enable it to utilize oxygen in the air. This major change in body form is known as metamorphosis.

There are four orders in the Class Amphibia we will discuss. The first order is Order Apoda. As their name implies, members of this order have no legs (the prefix a- means without and poda refers to feet). Caecilians, also known as blind worms, are the primary member of this order.

The caecilian has no legs and is in the Order Apoda.

The second order of Class Amphibia is Order Urodela. These amphibians do have legs and retain a tail throughout their lives. The primary members of this order are the newts and salamanders.

The tiger salamander belongs to the Order Urodela.

The third order of Class Amphibia is Order Anura. The name anura literally means without tail so members of this order do not retain their tail during development. Frogs, toads and tree frogs are the primary members of this order. Fertilization of eggs laid by the female is done externally with the hatching tadpoles having gills to obtain oxygen from their watery habitat. Through metamorphosis, legs grow, the tail is resorbed and lungs develop allowing them to utilize air for their source of oxygen.

The frog (left) is characterized by a moist skin and legs built for leaping long distances. The toad (right) has a dry, leathery skin and short legs built for taking short hops.

The fourth and final order of Class Amphibia is the Order Trachystoma. The term trachystoma literally means rough mouth. These amphibians are eel-like in appearance yet have two very small forelimbs and no hindlimbs. They have no teeth yet do have a roughened area on their palate. As adults they utilize gills but do have the capability of walking out onto land if necessary. Members of this order include the sirens and mud eels.

The siren has two very small forelimbs and no hindlimbs.

This completes our look at the Class Amphibia. Let's pause briefly and review where we are in the classification scheme for the Kingdom Animalia. We are currently in the Phylum Chordata (animals with notocords) and the Subphylum Vertebrata (animals with developed spinal columns). We've looked at five classes now which included the fishes and amphibians. We'll now move onto the next class which is Class Reptilia.

As its name reveals, Class Reptilia includes the reptiles. While some of the reptiles may resemble members of the Class Amphibia, there are notable differences which place them into a separate class. Unlike the amphibians which have gills during a part of their lives, reptiles only have lungs. Their bodies are covered with scales or plates rather than a slimy skin and while they also lay eggs like the amphibians, fertilization of the eggs takes place internally. This means that the sperm cells are deposited by the male inside the body of the female. The eggs are covered with a leathery-like shell. Finally, like the Class Amphibia, reptiles have a three-chambered heart. There are four orders within the Class Reptilia. Let's look at each of these now.

The first order we'll look at in Class Reptilia is Order Rhynchocephalia or the "beak-headed" lizards. The tuatara, of which there are only two species, is the only surviving member of this order. While tuataras resemble lizards like the iguana, their internal anatomy requires that they be placed in a different classification. Tuataras are only found in New Zealand.

The tuatara can remain reproductively active to over 100 years of age. The sex of the offspring is dependent upon the temperature at which the eggs are incubated.

The second order of Class Reptilia is the Order Chelonia. Members of the Order Chelonia include turtles and tortoises. Turtles and tortoises are well-known for taking their "home" with them in the form of an upper shell known as the carapace and the lower component known as the plastron. There are approximately 300 different species of turtles in the world.

The diet of the tortoise (left) is mainly plant material while the turtle (right) is omnivorous.

The third order of Class Reptilia is the Order Crocodilia. As you might guess from its name, this order contains the crocodiles as well as the alligators, gavials and caimans. Most members of this order are carnivorous, meaning that they are meat-eaters. Unlike other reptiles, members of the Order Crocodilia care for their young.

The teeth and mouthparts of crocodiles and alligators are well-suited to their meat diet.

The fourth order of Class Reptilia is the Order Squamata. The term squamos refers to the flat scales found on the skin of members of this group of animals. Snakes and lizards of over 7000 different species make up this order. An interesting characteristic of members of this group is the capability of dislocating their jaw to enable the eating of large prey. There is quite the range of size with the

The iguana (left) and garter snake (right) are members of Order Squamata.

smallest dwarf gecko measuring slightly more than one-half inch to the green anaconda reaching lengths of seventeen feet. While the majority of reproduction in snakes and lizards is through sexual reproduction, a process known as parthenogenesis is known to occur. In this method, there is no male gamete involved. Instead the female ovum divides to form two ova when then rejoin to create a new viable living organism. All offspring produced are female. The komodo dragon and cottonmouth snake are two members of the Order Squamata which are capable of carrying out this alternative means of reproducing.

Now that we've completed all four orders within the Class Reptilia, let's focus on the next class of animals which is the Class Aves. This class is home to all birds. Two of the most obvious features of birds are that their skin is covered with feathers and their forelimbs are modified into wings. Unlike the amphibians and reptiles, the heart of birds is four-chambered and birds are said to be endotherms. The prefix endo- in this term refers to the idea of within and the root word -therm refers to heat or energy. Endotherms are those animals which produce their own heat from their glucose sources and are commonly referred to as being warm-blooded. On the other hand, amphibians and reptiles are known as ectotherms meaning they are dependent upon outside sources of energy to heat their bodies. While they are commonly referred to as being cold-blooded, these animals do produce heat from their glucose sources, just at very low levels.

There are seventeen different orders of birds and these different classifications are based on many features. Examples include body shape, mode of getting food, whether they sit or perch or time of day they are most active. Below is a listing of these orders with a representative species of each.

Order Gaviiformes: Loons

Order Pelecaniformes: White or brown pelicans, cormorants

Order Ciconiiformes: Long-legged wading birds such as herons, bitterns, spoonbills and the ibises

Order Anseriformes: Short-legged goose-like birds such as ducks, geese and swans

Order Falconiformes: Large birds of prey such as hawks, falcons, eagles, kites and vultures

Order Galliformes: Fowl-like birds which often feed on the ground such as chickens, turkeys, pheasants, quail and grouse

Order Gruiformes: Crane-like birds such as the cranes, coots, gallinules and rails

Order Charadriiformes: Shore birds such snipes, sandpipers, plovers, gulls, terns, auks and puffins

Order Columbiformes: Pigeons and doves

Order Psittaciformes: Parrots, parakeets and macaws

Order Cuculiformes: Cuckoos and roadrunners

Order Strigiformes: Nocturnal (night-time) birds of prey such as owls

Order Caprimulgiformes: Goatsuckers, whip-poor-wills, chuck-will's-widow and nighthawks

Order Apodiformes: Swifts and hummingbirds

Order Coraciiformes: Fishing birds such as kingfishers

Order Piciformes: Woodpeckers, sap-suckers and flickers

Order Passeriformes: Perching birds such as robins, bluebirds, jays, sparrows, warblers, cardinals and thrushes

Before we move to the final class of animals in the Subphylum Vertebrata, let's pause briefly and review the classes we've considered so far. We learned that:

- Class Ostracodermi were extinct, jawless marine animals;

- Class Agnatha are also jawless marine animals with cartilaginous skeletons;

- Class Chondrichthyes which are fish with jaws and cartilaginous skeletons;

- Class Osteichthyes includes all fishes with jaws and having bony skeletons;

- Class Amphibia are the slimy-skinned animals, such as frogs, salamanders and newts, with the ability to change from utilizing gills to utilizing lungs to obtain oxygen;

- Class Reptilia includes the animals with lungs that are covered with scales or plates, utilized only lungs and includes alligators, crocodiles and lizards;

- And Class Aves which includes the many orders of birds.

We will now examine the final class of Subphylum Vertebrata of the Phylum Chordata. You are probably famliar with Class Mammalia, the mammals. Members of this class share the following features:

- They have hair or fur on some part of their body;

- They provide milk to nourish their young;

- They are endotherms, meaning they produce heat themselves to maintain specific body temperatures and are commonly termed warm-blooded;

- They have four-chambered hearts and utilize lungs to get oxygen;

- And many have a highly developed brain.

Let's look now at the sixteen different orders of Class Mammalia.

The first order is Order Monotremata. The prefix mono– refers to one and the root word –treme indicates opening. These animals have one opening to eliminate urine, feces and eggs in the case of the female. In addition, these mammals are characterized by the fact that they are oviparous which means they are egg-layers. The duck-billed platypus and spiny anteater or echidna are the two members of this order. While these animals are considered endotherms, they do not maintain as strict a range of body temperature as other mammals.

The echidna (above) and platypus (right) are egg-laying mammals.

The next order of mammals is the Order Marsupialia which are the pouched mammals. Once the offspring are born from these mammals, they migrate up to the pouch and find a nipple to get their milk supply and continue growing and developing. The Virginia opossum is the only marsupial found in North America while the kangaroo, wallaby and koala are marsupials found in Australia.

Kangaroo (left) and opossum (right) carry and nurse their young in a pouch.

Order Insectivora is the next order of mammals we will examine. Can you guess what the desired food is for this set of animals? Insects is correct! These mammals, which include the moles and shrews, are known for having a very high metabolic rate. This means that they utilize their food supply at a very fast rate which requires that they must eat very frequently.

The next order is Order Chiroptera (ky-rop-ter-uh). The name chiroptera means "hands of wings." This order consists of bats which are the only true flying mammals. The forelimb of bats is modified to form a wing which allows it to fly in order to catch its food supply which is insects for most bats. Bats utilize a unique means of sensing known as echolocation in which they emit a sound and wait for the sound to "bounce" back to them.

The mole hunts for insects beneath the soil surface.

The webbed forelimb of the bat allows it to fly.

The next order of mammals share the feature of being toothless or having peg-like teeth. Their class name reflects this characteristic: Order Edentata, where dentate refers to teeth. Members of this order include the armadillo, sloth and great anteater.

Armadillo (left) and sloth (right) have peg-like teeth.

Photo courtesy of Valerius Tygart

Another group of mammals which are also anteater-like but have scale-like skin make up the next order of mammals. This order is the Order Pholidota. The pangolins are members of this order.

The next order of mammals makes up the largest order and is the Order Rodentia, commonly known as the rodents. This group of mammals is known as the gnawing mammals in that they use their teeth to chew and gnaw on woody foods. Unlike most other mammals which have four incisors on each jaw, rodents only have two. Incisors are the scissor-like teeth which we refer to as our "front" teeth. Unlike our incisors, these teeth continue to grow throughout the life of the rodent. Members of this order include squirrels, mice, rats, chipmunks, marmots, gophers and porcupines. The guinea pig is a rodent commonly used as a pet.

The tree pangolin is covered with scales.

The chipmunk is a rodent.

299

The gray squirrel is a common member of the Order Rodentia.

The next order of mammals is similar to the Order Rodentia in that they share a common incisor tooth feature. This order is the Order Lagomorpha which includes rabbits, hares and the pikas. On the upper jaw of lagomorphs there are four incisors with one large pair being forward and a smaller pair behind. Like the rodent's incisors, these incisors continue to grow throughout their lives.

Marine mammals make up the next order of mammals. This order is Order Cetacea and includes whales, porpoises and dolphins. These mammals—of which there are ninety different species—are known for having a highly developed brain.

The black-tailed jackrabbit is a member of the Order Lagomorpha.

The killer whale (left) and porpoise (right) are known for their ability to learn complex tasks.

The next order of Class Mammalia is also an aquatic mammal. This order is Order Sirenia. Members of this order include the manatees and dugongs.

Nose shape is the common feature of the next order of mammals and is so indicated by its name: Order Proboscidae. Proboscis means nose. This order includes animals with a trunk, including elephants, the extinct mammoth and mastodon.

The manatee, also known as the sea cow, can be up to thirteen feet long and weigh over a ton.

Elephants are known for having a prehensile trunk.

The next order is Order Carnivora, commonly known as the carnivores. These mammals are the flesh- or meat-eaters and are able to do so by having teeth and claws which allow them to hunt and kill their prey. This group of mammals includes the domestic dog and cat as well as many wild meat-eaters like lions, tigers, bears and wolves. There are also many small carnivores such as weasels, mink, skunks and otters.

Photo courtesy Lisa McMillan Theye

The coyote (above) and mink (right) are members of the Order Carnivora.

The next order of Class Mammalia is the Order Pinnipedia which literally means having "fins as feet." The members of this order are the flesh-eating marine mammals which include seals, sea lions and walruses.

The next two orders of mammals are orders that we discussed back in Lesson 14 where we were introduced to taxonomy. Recall the example where we looked at how the number of toes present on these hoofed animals allowed for a divi-

Fins as feet, as seen in this sea lion, is the distinguishing feature of members of the Order Pinnipedia.

sion in their grouping. We looked at the Order Perissodactyla which had an odd number of toes and the Order Artiodactyla which had an even number of toes on each foot. Members of Order Perissodactyla include the horse, donkey, ass, rhinoceros and tapir.

The donkey (left) and tapir (right) are members of the Order Perissodactyla having an odd number of toes.

Members of the Order Artiodactyla include the cattle, sheep, deer, giraffe, hippopotamus and goats. Members of both of these groups are considered to be herbivores meaning they utilize plants as their main food source.

Sheep (left) and cattle (right) are members of the Order Artiodactyla having an even number of toes on each foot.

The final order of Class Mammalia is the Order Primates. This order includes mammals which walk in an upright or erect position. Members of this order include the monkeys, lemurs, gibbons, orangutans, chimpanzee, gorillas and humans. Members of this order are considered omnivores meaning that they utilize both plant and animal sources of food. It is also readily apparent that members of this order have the most developed brain of all mammals.

The chimpanzee (left) and baboon (right) are members of the Order Primates.

Let's pause here and review the orders which make up Class Mammalia. We learned that:

- The Order Monotremata includes the egg-laying mammals which were the platypus and echidna;

- The Order Marsupialia which were the mammals who had a pouch utilized for further development of their babies;

- The Order Insectivora which are the insect-eating mammals like moles and shrews;

- The Order Chiroptera which are the true flying mammals, the bats;

- The Order Edentaria which are the toothless or peg-like toothed mammals;

- The Order Pholidota which are the scale-like skinned pangolins;

- The Order Rodentia which are the rodents;

- The Order Lagomorpha which are the rabbits, hares and picas;

- The Order Cetacea which are the marine mammals including the whales, dolphins and porpoises;

- The Order Sirenia which includes the manatees;

- The Order Proboscidae which includes the elephants;
- The Order Carnivora which are the meat eaters;
- The Order Pinnipedia which are the sea-dwelling meat eaters including the seals, sea lions and walruses;
- The Order Perissodactyla which are the odd-toed hoofed animals;
- The Order Artiodactyla which are the even-toed hoofed animals, and finally
- The Order Primates which are the upright-walking mammals which includes monkeys, apes and humans.

This ends our discussion of the Kingdom Animalia. Know that there is ongoing debate regarding placement of various species within each level of the taxonomic system. There are two practice pages which come next in your study. Refer back to pages in this lesson for help in completing these practice pages. The test which accompanies this lesson has also been designed to be an "open book" test meaning you are welcome to use the lesson pages as a reference for completion.

Friendly Biology

Lesson 17: The Plant Kingdom

In our last two lessons we examined the many phyla and corresponding classes and orders of the Animal Kingdom. In this lesson we will examine members of the Kingdom Plantae, the plants. Before we begin looking at the various groups, let's look first at general characteristics of all members of this kingdom.

First, plants are multicellular organisms, meaning they are composed of many cells. In a later lesson, we will study organisms which have similar features of plants, however, they are unicellular, consisting of a single cell. Because plants are composed of many cells, cells with similar functions or jobs make up tissues within the plant organism. These tissues, in turn, work together to make "organs" of plants much like tissues do in our bodies. These "organs" then complete jobs for the plant in order for it to carry on life processes that we learned about in previous lessons in this book.

One such job is the job of photosynthesis where carbon dioxide, water and sunlight are converted into glucose which is the "fuel" for the plant. Recall that this process of photosynthesis takes place utilizing chlorophyll found in the chloroplasts of plant cells (Lesson 8). Because plants can make their own fuel source, they are said to be autotrophic. The prefix auto- refers to self and the root word, -trophic refers to feeder. Plants, therefore are self-feeders or can feed themselves. Animals and humans, on the other hand are said to be heterotrophic where the prefix hetero- refers to unlike self or different. Our bodies and incapable of making our own food supplies. We, instead, rely upon the work of other organisms to provide our food.

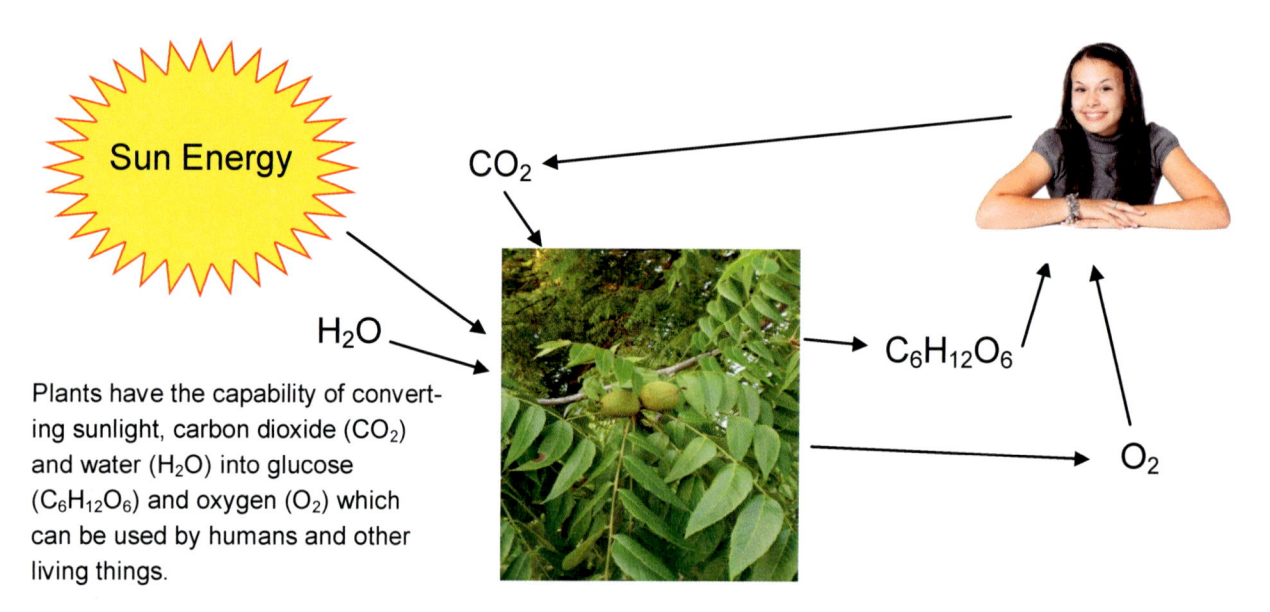

Plants have the capability of converting sunlight, carbon dioxide (CO_2) and water (H_2O) into glucose ($C_6H_{12}O_6$) and oxygen (O_2) which can be used by humans and other living things.

A second characteristic of members of the Kingdom Plantae is that plant cells contain cellulose. Recall from Lesson 3 that we introduced the structure of cellulose and then its function as a structural material in plant cells. We said that cellulose made the crunchiness of carrots and celery as well as wood used for our homes, furniture and buildings. Plant cells have cell walls made up of cellulose. We'll find similar organisms that have features like plants, but these organisms will lack cellulose.

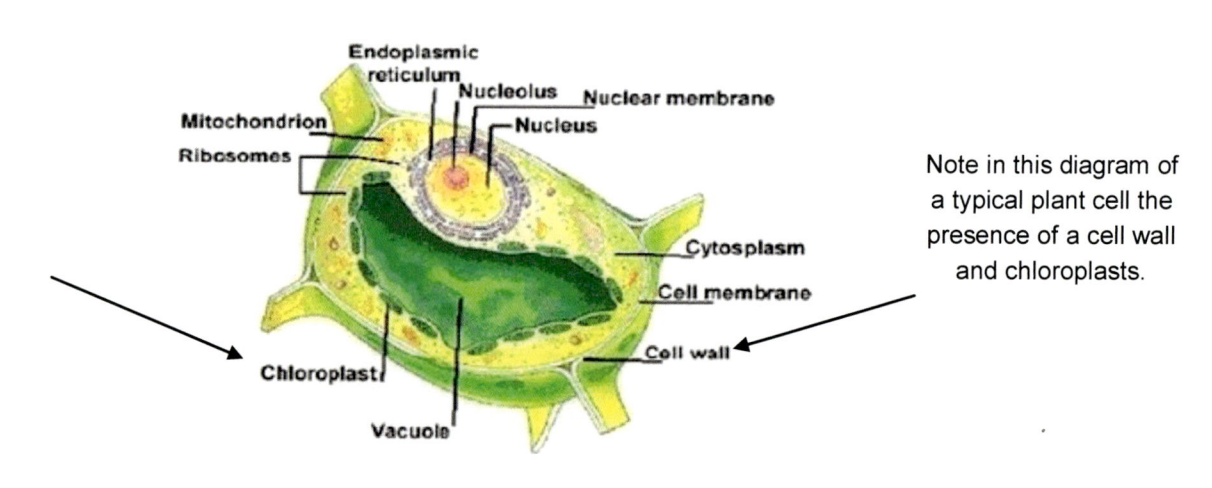

Note in this diagram of a typical plant cell the presence of a cell wall and chloroplasts.

A third characteristic of members of the Kingdom Plantae is the fact that plants have the capability of storing glucose in the form of amylose or starch (Lesson 3).

A fourth characteristic which places organisms into the Kingdom Plantae is that the parts of plants which create the sex cells (gametes) of plants are multicellular.

So, we've said that members of the Kingdom Plantae:
- Are multicellular;
- Are autotrophic;
- Have cell walls made up of cellulose;
- Can convert glucose into amylose (starch) for storage;
- And, have multicellular organs which create sex cells (gametes) for the plant.

Now that we've discussed the features all plants have in common, let's begin our look at the various groups of organisms which make up the plant kingdom. Taxonomists have divided all plants into two groups based upon the presence of vascular tissue. Vascular tissue in plants is analogous to our circulatory system. Like blood in our bodies, substances within the plant are moved from place to place through its vascular tissue. This vascular tissue can be identified as being xylem (zi-lum) which carries primarily water and phloem (flow-um) which carries primarily "food" products such as glucose and amylose through the plant.

So, whether a plant has vascular tissue or not determines into which of these first two groups it falls. Now, we also need to know that unlike the animal kingdom (where classifications are by phyla) the plant kingdom's "first" levels of classifications are known as divisions. We have ten different divisions within the Kingdom Plantae with only one division being the non-vascular plants. The remaining nine divisions all include variations of vascular plants.

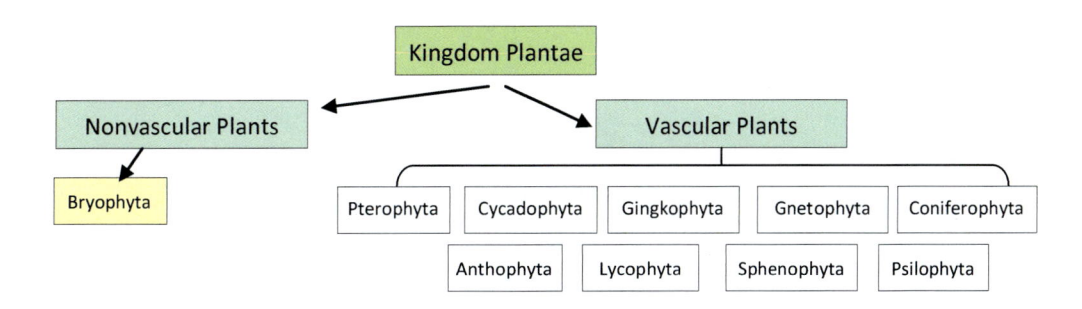

```
                        Kingdom Plantae
                       ↙              ↘
        Nonvascular Plants              Vascular Plants
               ↓
          Bryophyta        Pterophyta  Cycadophyta  Gingkophyta  Gnetophyta  Coniferophyta
                           Anthophyta  Lycophyta    Sphenophyta  Psilophyta
```

One other feature upon which plant divisions are based is which phase of generation is dominant. Recall from Lesson 12 that we discussed how plants have two forms of generations: the gametophyte and sporophyte forms. As we move through each division of plants here in this lesson, we'll point out the dominant form for that division. We'll look now at the first division which includes the non-vascular plants.

The division of Kingdom Plantae which contains non-vascular plants is the Division Bryophyta. These plants are usually found in moist growing conditions and because they lack vascular tissue, they are very limited in how high they can move substances. Because of this, they are very small, low-growing plants, yet multicellular. All members of this division have the gametophyte as the dominant generation. Members of this division can be divided into three classes: Class Hepaticae which are the liverworts; Class Anthocerotae which are the hornworts; and the Class Musci which include mosses like Sphagnum moss.

Liverworts (top left), hornworts (top right) and mosses are all low-growing, non-vascular plants.

The next nine divisions include the vascular plants. Because these plants have vascular tissue they are capable of growing much higher than the non-vascular plants of Division Bryophyta. We'll start with the most simple vascular plants first.

The first division is Division Psilophyta. These plants are fern-like having no roots or leaves. They have short branches on which sporangia form. The whisk fern found in tropical and sub-tropical regions is a member.

The next division of the plant kingdom is Division Lycophyta. These small vascular plants, commonly known as the club mosses, have simple leaves with the sporophyte generation being the dominant generation above ground. The gametophyte generation is found underground.

The next division is Division Sphenophyta which includes plants known as horsetails. The sporophyte stage is the dominant generation. Stems of these plants are hollow and jointed and contain silica.

Ferns and tree ferns are members of the next division of Kingdom Plantae which is the Division Pterophyta. Members of this division have well-developed fronds and underground rhizomes, however, they do not produce any flowers. The sporophyte generation is the dominant generation.

The horsetail (above right) is a member of the Division Sphenophyta. The fern (right) is a member of the Division Pterophyta.

The next division of the plant kingdom is the Division Cycadophyta. These plants are palm-like plants which have male and female cones on different trees. The seeds are said to be naked, meaning they are not enclosed in a fruit. Plants which produce seeds that are naked are also referred to as gymnosperms. Cycads and sago palms are members of this division.

There is only one species found in the next division of vascular plants which is Division Ginkgophyta. This lone member is the ginkgo tree. All other members of this division are extinct. While the seeds of this tree are naked, there is a fruit-like structure present on female trees which is exceptionally odorous and messy. For this reason, male trees are often used in landscaping projects.

The cycad (top right) has seeds which are not enclosed in a fruit. The gingko (leaf, middle right and fruit, lower right) does have its seeds enclosed in a fruit.

The next division is Division Gnetophyta which includes three genera that have naked seeds yet have a more complicated vascular structure. One genera of this division is the strange plant, the welwitschia found only in the Namib Desert of Africa. This plant has only two very large leaves which can grow up to nine feet in length upon a very short trunk just above the ground. It has a long tap root but primarily gets moisture from fog.

The welwitschia plant belongs to the Division Gnetophyta.

The next two divisions include plants which are very familiar to most people. These divisions are Division Coniferophyta and Division Anthophyta. Division Coniferophyta are the cone-bearing plants, commonly referred to as conifers or evergreens. The seeds of this division are naked and leaves are formed as needles or scales. Common members include pines, firs, spruces, larches and yews.

The conifers have needle-like leaves and seeds produced in cones.

The flowers of grasses are not as showy as other more colorful flowers

The Division Anthophyta is the final division we'll examine and includes the very familiar flowering plants. The seeds of members of this division are not naked, but rather enclosed in an ovary which ripens into a fruit. This ovary is part of a whorl of leaves or petals known as the flower. Note that not all flowers resemble the stereotype flower with petals. Some flowers are small and rather inconspicuous.

The Division Anthophyta is divided into two classes based upon the anatomy of seeds produced by these plants. The first class is Class Monocotyledonae and the second is Class Dicotyledonae. Let's look first at the anatomy of a seed and then we'll address the differences found between these two classes.

While many may think that a seed is the very earliest stage of the development of a plant, it is actually made up of many, many cells and the plant, at this stage, is a small embryo waiting on appropriate conditions to continue development. Recall from Lesson 3 where we discussed how plants have the capability of packing a "lunch" for their offspring in their seeds in the form of stored glucose known as amylose. Look at the photo here of a seed. Note that it is covered by a tough outer covering known as the seed coat. Beneath the seed coat we find the storage area of amylose and then the location of the plant embryo.

Photography by Curtis Clark Copyright Curtis Clark

An opened seed of the gingko revealing parts of a seed.

It is the design of the storage area of amylose by which members of Anthophyta are divided into classes. If we look at the class names for this division, we can find within the name the word cotyledon. The cotyledon is the location of amylose storage and, depending upon the number of cotyledons present, plants are either classified as being monocotyledons (having a single cotyledon) or dicotyledons (having two cotyledons). The best way to better understand this difference is to examine two great examples of this means of classification: the wheat seed and the bean seed.

Wheat and its relatives, the grasses, are all members of the Class Monocotyledonae. The wheat seed is made up of one section with the wheat embryo down near the lower, pointed portion of the seed. Look at the diagram here. Note that the amylose is also referred to as the endosperm of the seed. Members of this class are commonly known as the monocots.

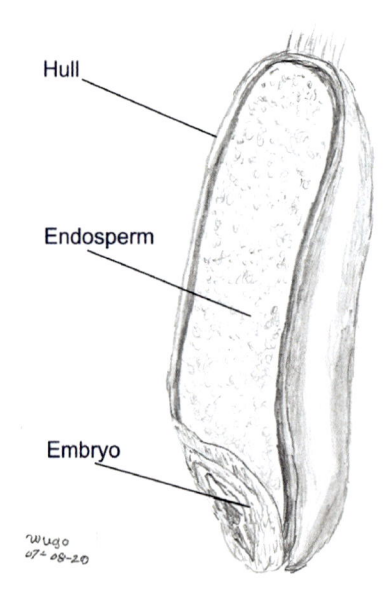

Bean seeds, on the other hand, are formed in two halves. There are two cotyledons which each contribute to supplying the one embryo with glucose. Look at the photo here to see the parts of a bean seed.

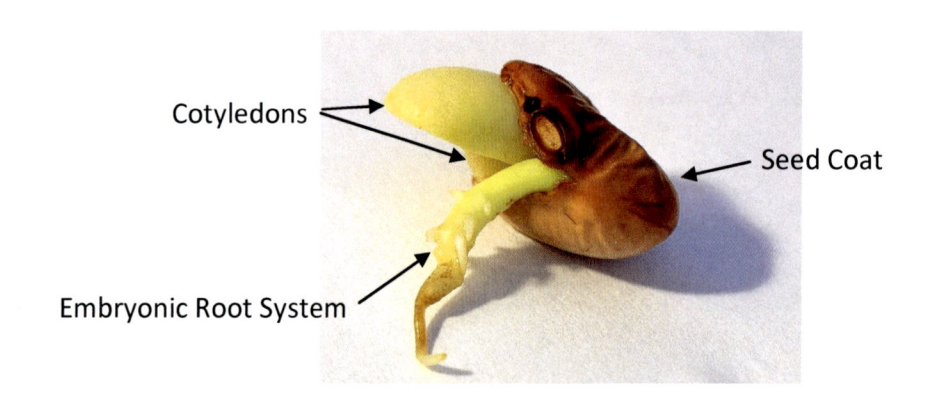

In addition to differences in seed anatomy, there are also differences in the way the vascular tissue (xylem and phloem) is arranged within the stems of monocots and dicots. In monocots, the xylem and phloem, known as a vascular bundles, are scattered throughout the stem in no apparent arrangement. In dicots, however, the vascular bundles are arranged in a circle within the stem.

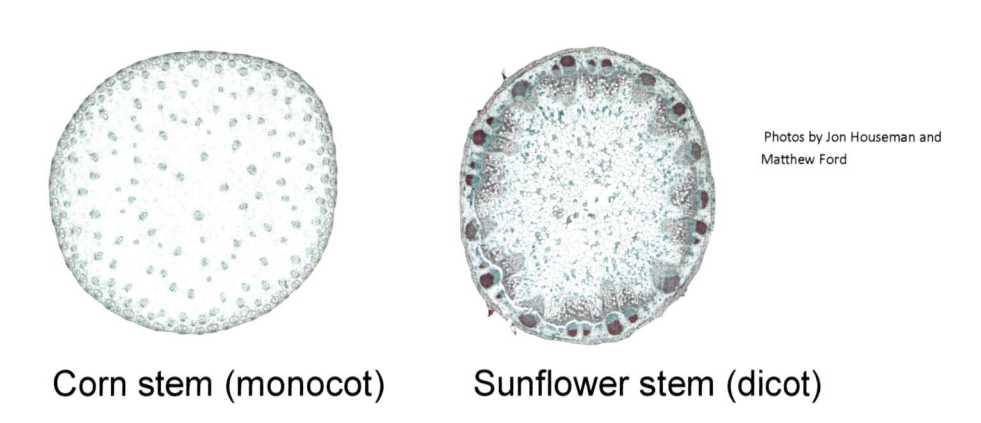

Photos by Jon Houseman and Matthew Ford

Corn stem (monocot)   Sunflower stem (dicot)

Yet another difference in monocots and dicots is the way the veins of the leaves are arranged. In monocots, the veins of the leaves run parallel to each other from the base of the leaf to the tip. In dicots, the veins of the leaves run in various configurations other than parallel. Common monocots include corn, the grasses and common garden flowers like irises, tulips and daylilies. Common dicots include trees like maples, oaks, elms and hackberries and shrubs like lilacs, azaleas and roses. Most fruit trees are dicots.

The veins in a monocot leaf (left) of the corn plant run parallel to each other while the veins of a dicot leaf (right), a maple, are not parallel, but branched in various configurations.

Lesson 17 Lab Activity: Parts of a Flower

The purpose of the lab activity is to become acquainted with the major parts of a flower. Materials you will need include:

- A flower. You can search your yard or garden or go to a local florist, garden center or grocery store which sells flowers. Many times the caretaker of the flowers is willing to share any spent or faded flowers at no cost. A lily or hibiscus flower is ideal.

- Paper plate or placemat on which to work.

- Sharp knife or X-acto knife. Be careful with this as a sharp blade can cut you very easily!

- Magnifying glass

The first thing to note in learning about the parts of a flower is that not all flowers will have all parts we describe here. Some flowers may only have the female structures while some may only hold the male structures. These flowers are known as imperfect flowers, not that they are deficient in any way, but only have one set of the gamete-producing organs. Flowers which have both male and female parts present in the same flower are known as perfect flowers. The diagram and the lily flowers pictured below both show perfect flowers.

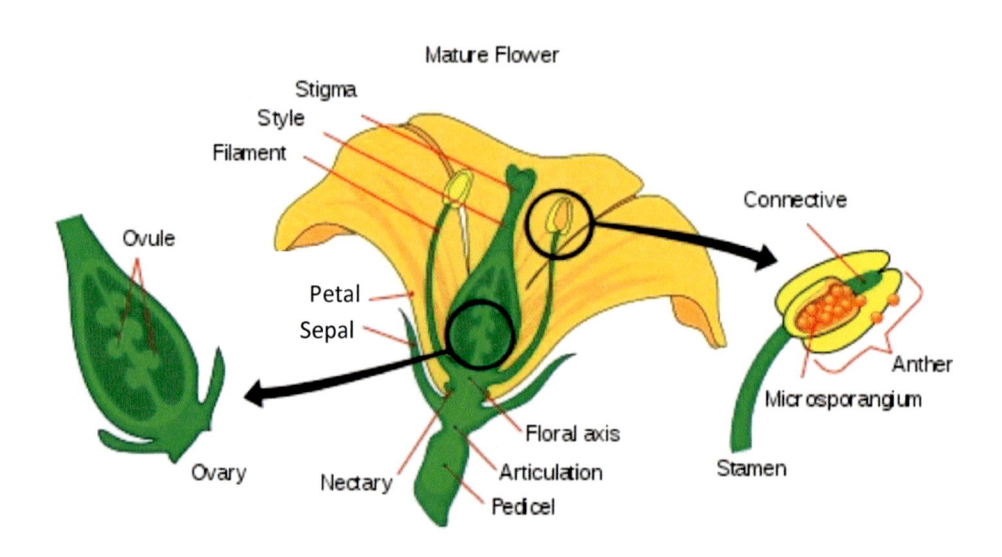

Begin your exploration by finding the stem which holds the flower. Refer to the diagram to find the pedicel and articulation ("joint" between stem and flower). Note that the first structures of the flower are the sepals which had at one time enclosed the flower before it opened. Gently peel the sepals back from the flower. Inside, you may or may not be able to find the nectary which are the glands which secrete nectar. Nectar serves to attract feeding insects, birds and bats. The actions of these animals enhances the chances of pollination.

The next structures you will find are the petals. Like the sepals, the petals are actually modified leaves and encircle the internal parts of the flower. They serve to protect as well as attract pollinators.

Gently pull away the petals of the flower to expose the ovary or female portion of the flower. Note the long tube-shaped structure on the upper portion of the ovary which is known as the pistil. The pistil can be separated into two basic parts: the style and the stigma. The stigma is the location where pollen cells (the male gametes) enter the ovary and is usually sticky to touch.

Surrounding the ovary you will also find the male structures of the flower known as the stamen.

The stamen is the male part of the flower and includes the pollen-producing anther supported by the filament.

There may be several of these depending upon the type of flower you are using. The stamen has two main components: the filament and the anther. The filament is the thin, stem-like structure which supports the anther. The anther is the location where pollen, the tiny structures which carry the male gametes, is formed. Depending upon the age of your flower, you may or may not find pollen on the anthers of your flower. Pollen is moved by the action of the wind or pollinating animals as we discussed above. Use your magnifying glass to observe the pollen on the anther.

Note that the pollen grains themselves are not the actual male gametes, but that they carry cells of two types: one type of cell will eventually divide and form the pollen tube while the other type will divide into two sperm cells. When a grain of pollen lands upon the sticky stigma of the pistil, the cells which form the pollen tube begin to burrow down into the stigma. The burrow creates a tube that the sperm cells move through on their way to the ovules.

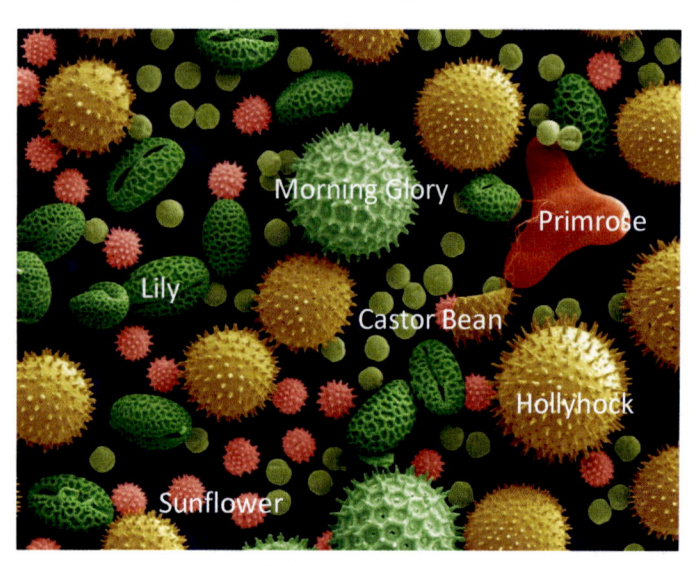

Dartmouth Electron Microscope Facility, Dartmouth College

Above is a scanning electron microscope view (colorized) of pollen granules of several varieties of common garden flowers.

Photo by J. J. Harrison

Above is a close-up view of pollen grains on the anther of a tulip flower.

Above is a cross section of a developing rose hip. Note the ovary below housing the ovules. Note also the developing pistils, stamens and petals above all enclosed by the sepals.

Now turn your attention back to the ovary of your flower. Using your knife, open the side of the ovary. Again, depending upon the age of the flower, you should find developing ovules inside the ovary. The ovules eventually form the seeds of the plant. Note that in most plants, the seeds are actually plant embryos meaning they have undergone fertilization and are now many, many cell divisions past that stage. Development slows as the seed becomes dormant waiting for appropriate conditions to continue growth and development.

This concludes this lab activity. Discard your flower or consider pressing or drying it to enjoy again later.

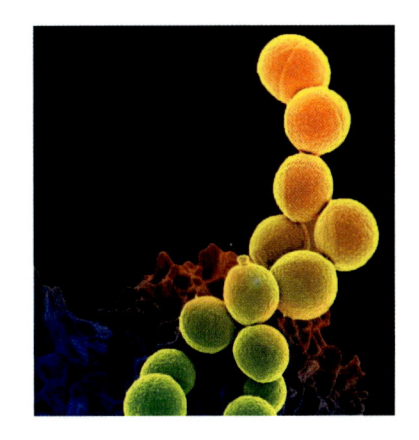

Lesson 18: Kingdom Monera and the Viruses

    Think back to Lesson 7 where you were introduced to the various parts of cells. When we learned about the nucleus of the cell we emphasized the point that some living things had a nucleus which was enclosed in a nuclear membrane while others did not. You learned that organisms that did not have a nuclear membrane present were identified as prokaryotes while those that did were known as eukaryotes (pp. 126-127). In addition to lacking a nuclear membrane, prokaryotes also do not have organelles bound by membranes. These are the organisms which belong to the Kingdom Monera which will be our focus in this lesson.

    Members of the Kingdom Monera are unicellular organisms and most members of this kingdom obtain nutrients by absorbing them straight across their cell membranes. However, some can carry on photosynthesis or chemosynthesis. Some may have a cell wall which is composed of a polysaccharide with polypeptide cross links. In Lesson 12 you learned that reproduction by members of this kingdom is mainly through binary fission (splitting into two new cells) or by budding. Let's take a closer look at members of the Kingdom Monera now.

There are four phyla in this kingdom: Phylum Archaebacteria (some feel these organisms belong in a kingdom to themselves), Phylum Schizophyta, Phylum Cyanophyta and Phylum Prochlorophyta. Let's look at some characteristics of Phylum Archaebacteria first.

Members of the Phylum Archaebacteria are all anaerobic bacteria meaning they carry on life activities with very little to no oxygen in their environment. These bacteria have unique transfer RNA unlike all other bacteria, as well as a unique cell wall. They are found in very harsh conditions. For example, some members of this group are known as methanogens meaning they utilize methane gas for their source of energy. Consequently, they live in the digestive tracts of cattle and sheep as well as in the bottom of sewage treatment ponds, bogs and lakes.

Another group within this phylum is known as the extreme halophiles. The prefix halo- refers to salt and the root word phile refers to loving (recall hydrophilia). These bacteria love very salty conditions and consequently are found in the Great Salt Lake of Utah and the Dead Sea. Yet another group

Members of the Archaebacteria thrive in very extreme habitats. These shown here impart the yellow color to the water at Yellowstone National Park.

of strange bacteria within this phylum is the thermoacidophiles. The prefix thermo- refers to heat and acido- refers to acid. Therefore these bacteria love very hot, acidic conditions. They are found in the very hot water of hot springs which may have a pH of less than 2! Yet, they are capable of maintaining an internal pH near neutral (7). Some thermoacidophiles are found living near volcanic vents deep within the ocean where they are thought to utilize gases being emitted by the volcano as a nutrient source.

Let's move to the second phylum of Kingdom Monera, Phylum Schizophyta. The members of this phylum are ones you are likely familiar with as these are the bacteria which cause diseases. They are usually saprophytic (live off of dead living matter) or parasitic (live off of other living organisms at the expense of the host). Like all monerans, they are prokaryotic and reproduce by binary fission. However, some reproduce by forming endospores. Many of these bacteria also produce endotoxins which are poisonous substances which wreak havoc for cells of the host organism. Food poisoning is a very common example of a disease caused by endotoxins of bacteria.

Members of this phylum include the staphylococcus bacteria (known to cause staph infections), streptococcus bacteria which cause "strep" throat, actinomycetes which cause lumpy jaw in cattle and

Examples of Phylum Schizophyta include the spirochetes (below) which cause Lyme disease transmitted by ticks (inset), *E. coli* (upper right) and *Staph aureus* (lower right).

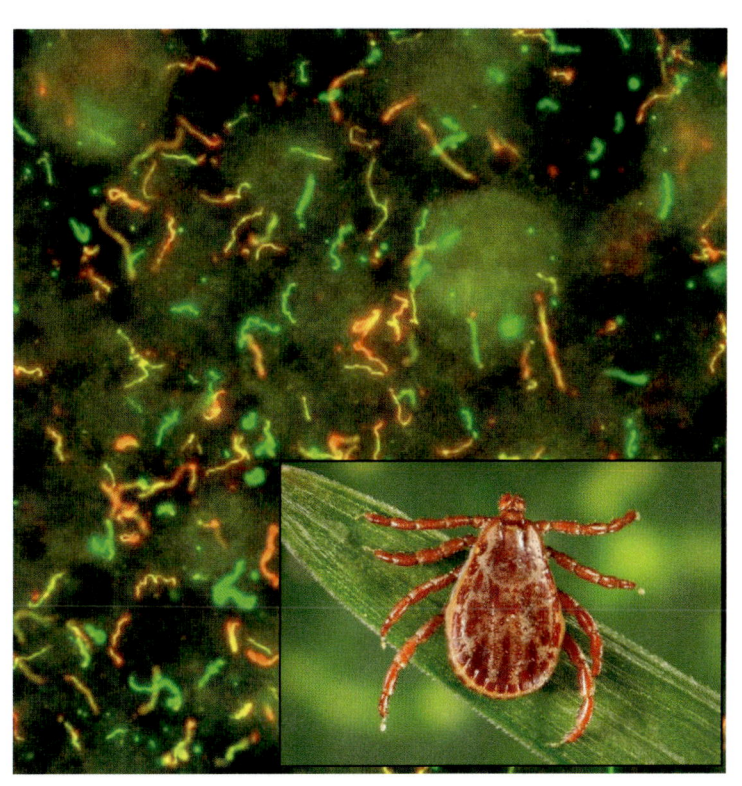

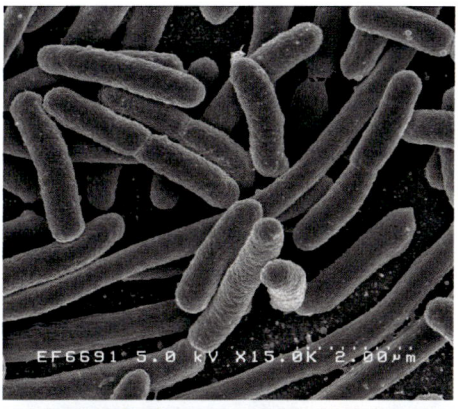

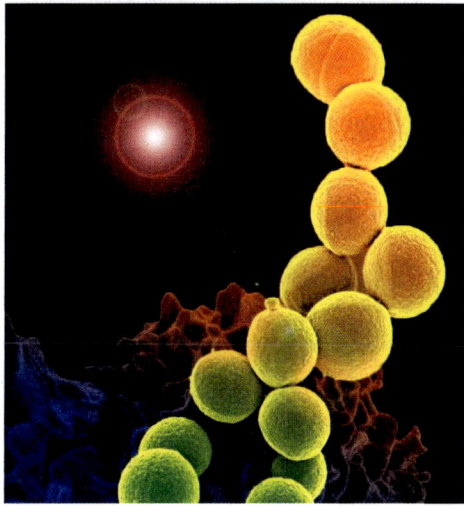

tuberculosis, the debilitating lung infection of humans and animals. Other members of Phylum Schizophyta include the rickettsiae which cause typhus and Rocky Mountain Spotted Fever. Ticks transmit these diseases to humans and pets. Finally, there are the spirochetes which, as their name implies, have a spiral shape. These bacteria are known to cause Lyme disease in humans and pets as well as syphilis, a sexually transmitted disease of humans.

Let's move on to the third phylum of the Kingdom Monera now: Phylum Cyanophyta. Members of this group of bacteria are known as the blue green bacteria. The prefix cyano- refers to the color blue. These bacteria have chlorophyll and other pigments, but, unlike plants, the chlorophyll in these bacteria is not bound into chloroplasts. Because of the presence of chlorophyll, members of this phylum are capable of undergoing photosynthesis to produce their food source. (Don't confuse these organisms with bluegreen algae which are eukaryotic organisms.)

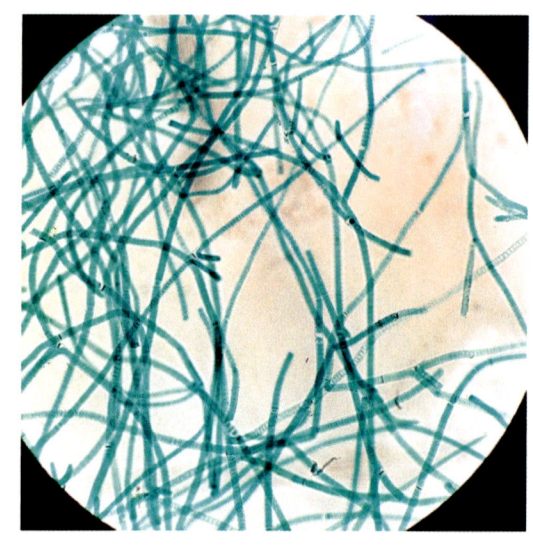

Blue green bacteria of the Phylum Cyanophyta are capable of photosynthesis.

The fourth phylum of the Kingdom Monera is the Phylum Prochlorophyta. Like the Phylum Cyanophyta, members of this phylum also contain chlorophyll. However, they do not contain another pigment known as phycobilin. They are thought to carry out a unique form of photosynthesis and for this reason and others, researchers feel that they should be placed into their own phylum. However, others feel they are actually members of the Phylum Cyanobacteria. Members of this phylum make up some of the tiniest forms of plankton organisms.

Let's pause to review what we've learned in our discussion of members of the Kingdom Monera. We learned that:
- Members of the Kingdom Monera are unicellular prokaryotes;
- They reproduce by fission or budding;
- Some members are capable of conducting various forms of photosynthesis and;

- Some members cause serious diseases.

Now we'll move our discussion to another set of tiny "living" things: the viruses. We have put the term living in quotes since biologists feel that because viruses do not display the features of living things, they aren't actually living things. Some biologists think that viruses are on the "fringe" of being a living thing. Nonetheless, they do cause significant problems for living things so we feel it is important to include them in our study of living creatures of our earth.

Let's look first at the general nature of a virus and then we'll discuss how they are placed into various groups. Viruses are defined as infectious agents capable of only living within the cells of another living thing and, therefore, they are known as being obligate intracellular agents. They consist of a portion of DNA or RNA with a covering of protein known as the capsid. Some viruses have a layer of lipids surrounding the capsid. They have no nucleus nor any other organelles, therefore they are dependent upon the cells they have infected to carry on any "life-like" activities.

Viruses work as clandestine saboteurs, meaning once they enter a cell, they commandeer the cell into doing the viruses' desired mission. In other words, the virus is capable of taking control of the cell and causing it to do what it wants. Most of the time the virus wants to replicate in order to make more viruses and this usually results in the death of the cell. We mentioned above that viruses are made up of strands of DNA or RNA. By utilizing these portions of "bad" genetic material, they either insert their DNA into the existing DNA of the host cell and cause the cell to begin producing "rogue" proteins or directly use their "bad" RNA to commandeer the cell's ribosomes to produce "rogue" proteins. Either way, the cell no longer has control of its normal behavior. The virus is then capable of reproducing itself (one of the only life-like features it possesses) and then rupturing the cell, releasing these newly formed viruses destined for further cell invasion.

Because viruses are found *within* living cells of all types of living organisms (yes, even bacteria can be infected with viruses), using antibiotic drugs to treat infection is ineffective. So, medicines like penicillin or tetracyclines that work well to fight bacterial infections are useless against a viral infection.

Instead, immunologists (biologists who specialize in the immune system) have developed vaccines to fight viral infections. Vaccines are severely "beaten-up" fragments of disease-causing cells (known as attenuated cells) or, even, killed viruses that work to stimulate the body to develop antibodies against the virus. By purposefully exposing ourselves to these small pieces of viruses, our bodies can build up an arsenal of antibodies ready to attack should the live virus ever arrive in the body at a later time. You are likely familiar with vaccinations you may have received as a child against the measles virus. Your dog or cat may have been vaccinated for the rabies virus or your puppy vaccinated for the parvovirus. The influenza virus (flu virus) is another virus for which control is attempted through vaccination.

Scientists have developed antiviral medications such as Tamiflu^R to reduce the effects of viral infections. It's interesting to note that the agent in Tamiflu^R actually has no direct effect upon the influenza virus itself, but, rather, once the drug goes to the liver to be processed for removal by the body, it's one of these metabolites made by the liver which alters proteins used on the surfaces of the body's cells. This alteration makes it difficult for viruses to enter the body's cells, which in turn reduces the effects of the virus.

Pictured here are H1N1 (influenza) virus particles.

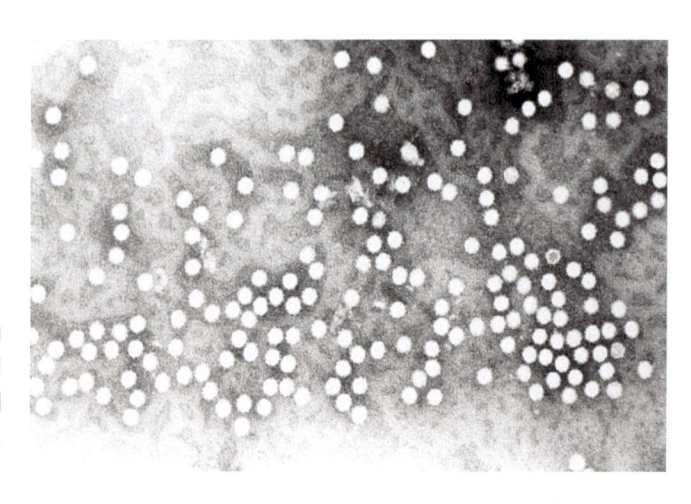

Parvovirus particles can wreak havoc for cells lining the intestines of puppies. Infection results in blood vessel damage causing hemorrhage which can lead to rapid death.

Because viruses don't really fit the traditional classification schemes created for other living things, the International Committee on Taxonomy of Viruses has developed a means of classifying viruses which currently (2013) includes seven orders which are divided into over 100 families. To augment this classification system, another method developed by the virologist, David Baltimore, known as the Baltimore Classification, utilizes categories based upon how a virus builds its mRNA. Together, these two systems allow virologists to place viruses into various taxonomic groups. There are close to 3000 classified species of viruses but many, many yet to be classified.

Let's pause now to review what we have learned about viruses in this lesson. We learned that:
- It's controversial as to whether viruses can really be considered living things;
- Viruses consist of a strand of DNA or RNA covered by a protein coat known as a capsid;
- Viruses work by entering cells and taking over the genetic "machinery" of the cell;
- Treatment with antibiotic drugs which normally kill bacteria are ineffective against viruses as viruses live within cells; and
- Control of diseases caused by viruses is mainly through the use of vaccinations.

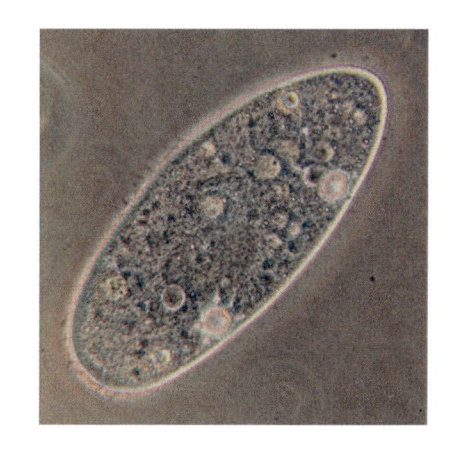

Lesson 19: Kingdom Protista

Like members of the Kingdom Monera, members of the Kingdom Protista are all unicellular organisms. However, unlike members of the Kingdom Monera, members of Kingdom Protista are eukaryotic organisms, meaning they have a nuclear membrane and organized cellular organelles where members of the Kingdom Monera do not.

The Kingdom Protista consists of fourteen different phyla grouped into three main groups: the animal-like protistas, the plant-like protistas and the fungal-like protistas. We'll look first at the animal-like protistas.

The animal-like protistas are often referred to as protozoans. The prefix proto- refers to first or early and the root word -zoan refers to animal (as in zoo or zoology). These creatures are considered by some to be the first animals. There are four phyla in this group: Phylum Sarcodina, Phylum Ciliophora, Phylum Zoomastigina and Phylum Sporozoa. The feature which distinguishes these four phyla is how they move about or means of locomotion. Let's take a closer look at each phyla.

Members of the Phylum Sarcodina move through the use of a pseudopodia. The prefix pseudo- means false and the root word -podia refers to foot. Therefore, these microscopic organisms move about with a foot-like appendage. Most live in water and reproduce by fission. Some are known to have shells. Examples of this phylum include the Ameba, Foraminifera and Radiolaria.

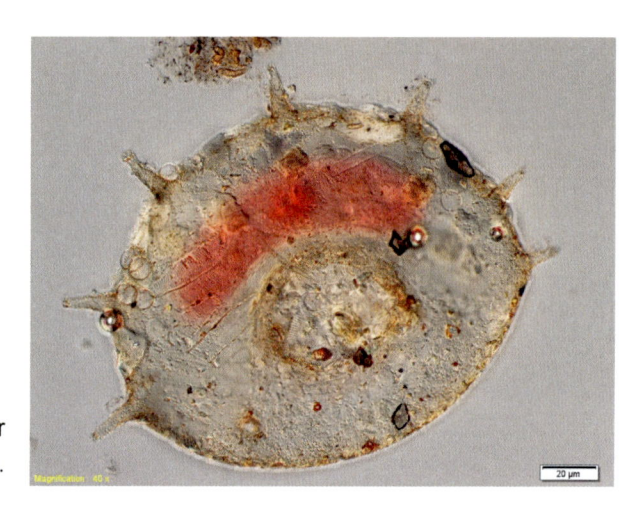

A testate (shelled) ameba is a member of the Phylum Sarcodina.

The second animal-like protista group is the Phylum Ciliophora. Members of this group move about through the use of cilia. Cilia are hairlike projections of the cell membrane which move about allowing the organism to swim. Members of this phylum also have a pellicle which is a thin skin on their outer surfaces. Some have what is referred to as a macronucleus and a micronucleus. The most familiar member of this phylum is the Paramecium, the slipper-shaped organism that can readily be found in pond water. Other members include the Stentor and Vorticella.

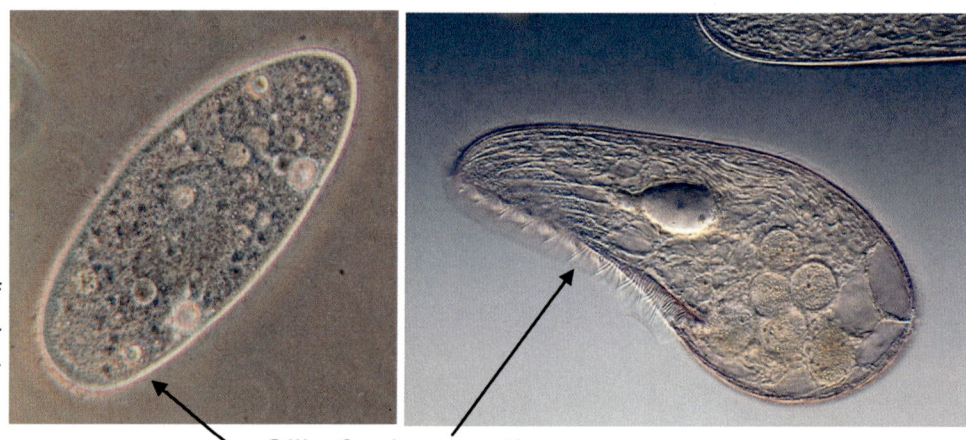

Two examples of Phylum Ciliophora are *Paramecia spp.* (right) and *Blepharisma spp.* (far right).

Cilia for locomotion.

The third animal-like protista group is the Phylum Zoomastigina. The members of this group move through the use of one or more flagella. A flagellum (singular) is a tail or whip-like structure which moves side to side allowing the organism to swim through its watery environment. All members of this group are heterotrophic meaning they depend upon another organism for their food supply. Likely the best known of all members of this group is the *Trypanosoma spp.* which cause African Sleeping Sickness and Chagas Disease. Another member of this group is the *Giardia spp.* which is readily found in the digestive tract of cattle. The organism can be found in water found in the tracks of cows made in muddy areas. If a cattle owner comes into contact with manure from infected cattle, he or she can become infected also which results in a severe diarrhea with potential dehydration.

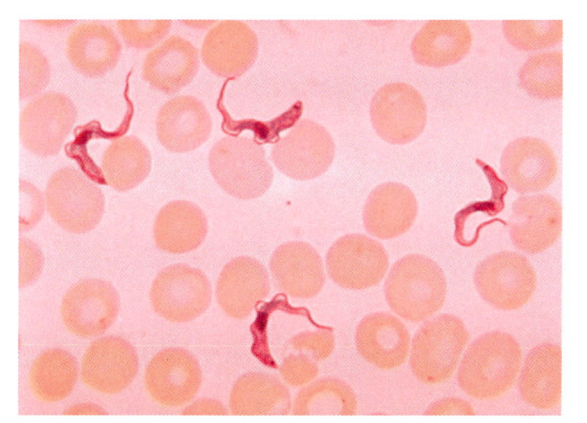

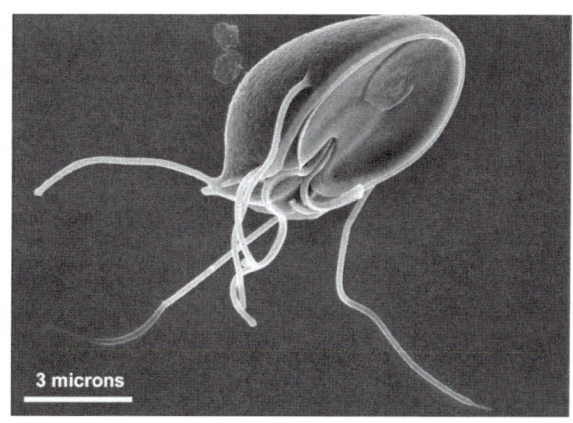

Trypanosoma spp.(left) are parasites found in the blood. *Giardia spp.* (right) can be found in the digestive system.

The fourth animal-like protista group is the Phylum Sporozoa. Unlike the other three animal-like group members, these protozoans do not move about on their own. All members of this group are considered parasitic. Two well-known examples are *Plasmodium spp.* and *Toxoplasma spp.* Plasmodium is the organism responsible for malaria. The organism spends part of its life cycle in mosquitos. Once an infected mosquito bites a human, it can transmit the organism to its new host where the organism causes red blood cells to lyse resulting in severe, and often deadly, anemia. The toxoplasma organism as we discussed in Lesson 12 spends part of its life cycle in cats causing diarrhea in newly infected cats and then brain problems in fetal kittens.

We'll now move to the next set of phyla of the Kingdom Protista which are the fungal-like protista. At one time this set of four phyla were thought to be members of the Kingdom Fungi, but taxon-

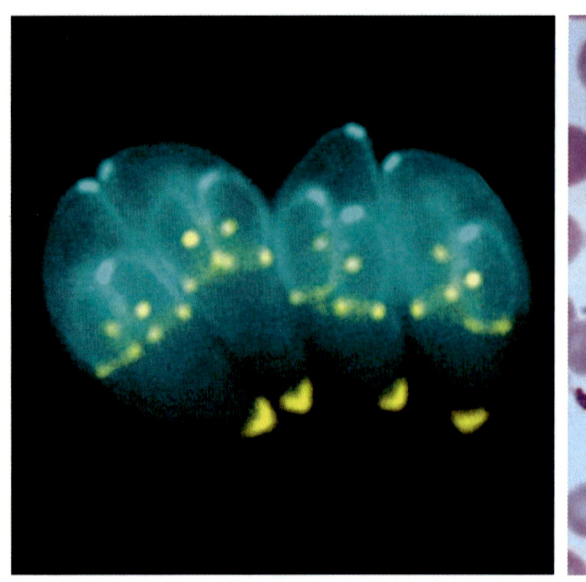

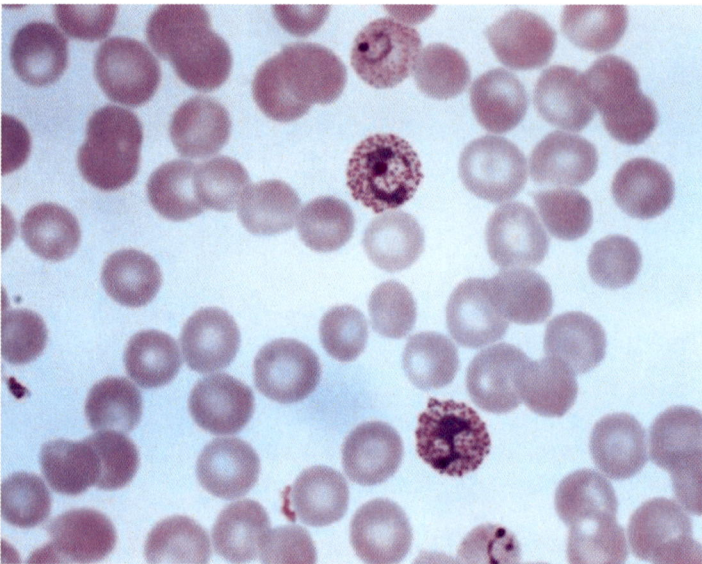

Toxoplasma spp. organisms (left) are undergoing mitosis. *Plasmodium spp.* organisms (right) which cause malaria can be seen within and among erythrocytes (red blood cells).

omists now agree that these unicellular creatures actually fit better in Kingdom Protista.

The first phylum is Phylum Acrasiomycota. There are approximately 65 species within this phylum and they are commonly known as the slime molds. They can be found on the forest floors where they slowly move about feeding upon dead and decaying leaves. During this feeding stage, they are characterized by being single cells. Members of this phylum are frequently used in research activities by biologists.

The second phylum of fungus-like protists is Phylum Myxomycota. Like the Phylum Acrasiomycota, members of this phylum are also known as slime molds. However, during the feeding stage of their life cycle, they appear to be multi-nucleate meaning they have many nuclei which all work together to move across surfaces.

The third phylum of fungus-like protists is Phylum Chytridiomycota, commonly referred to as the chytrids. These protists are primarily found in

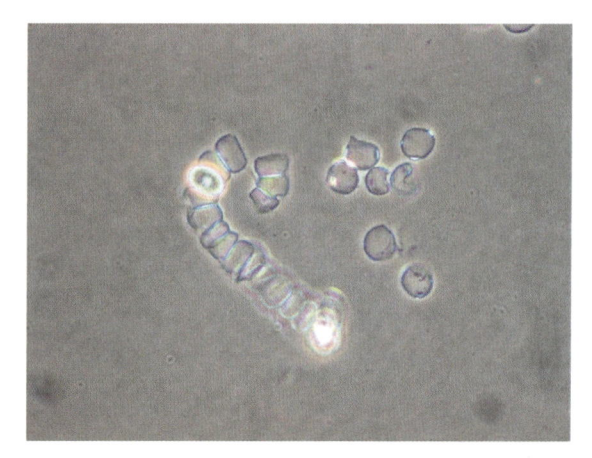

Photo by Shirley Chio
Acrasia rosea is a slime mold.

Fuligo septicum is a member of the Myxomycota Phylum, commonly known as the "dog vomit" fungus.

Potato blight is caused by *Phytophthora spp* which is a member of the Phylum Oomycota.

water and have gametes with a single flagellum. They are found to be parasitic on algae and other water plants.

The fourth phylum of fungus-like protists is the Phylum Oomycota. There are approximately 475 members of this phylum which include water molds, white rusts and downy mildews. One important member is *Phytophthora spp*. which was responsible for causing the devastating potato blight and resulting famine in Ireland in 1845-47.

The final six phyla of the Kingdom Protista are those that make up the algae. While algae are plant-like in that they make their own food supply through photosynthesis, they differ in the way they produce reproductive cells. The reproductive cells of plants are made in structures which are multicellular whereas in algae these structures are unicellular. While most algae spend their entire life cycle in the water, some do live on land, yet require water to reproduce. The criteria for making these six phyla is based upon the color of the pigments within the cells which conduct photosynthesis for the cells. All algae have chlorophyll a, however, the other phyla may also have chlorophyll b, c or d and vary in the wavelengths of light absorbed by the cells to perform photosynthesis. Another criteria for the phyla divisions is the means by which algae store energy.

The first division is Phylulm Chlorophyta. As we learned earlier in Lesson 8 about chloroplasts, the prefix chloro- refers to the color green. Members of this phylum are known as the green algae. These algae contain chlorophyll a, b and c and store their produced food as starches. They also have cell walls made of cellulose. They may live as solitary organisms or in great colonies where there is noted division of work by various members of the group. They can be free-floating, sessile (meaning they are attached to a structure or surface) or able to move on their own. Members of this group include the Gonium, Volvox, Spirogyra and the Ulva.

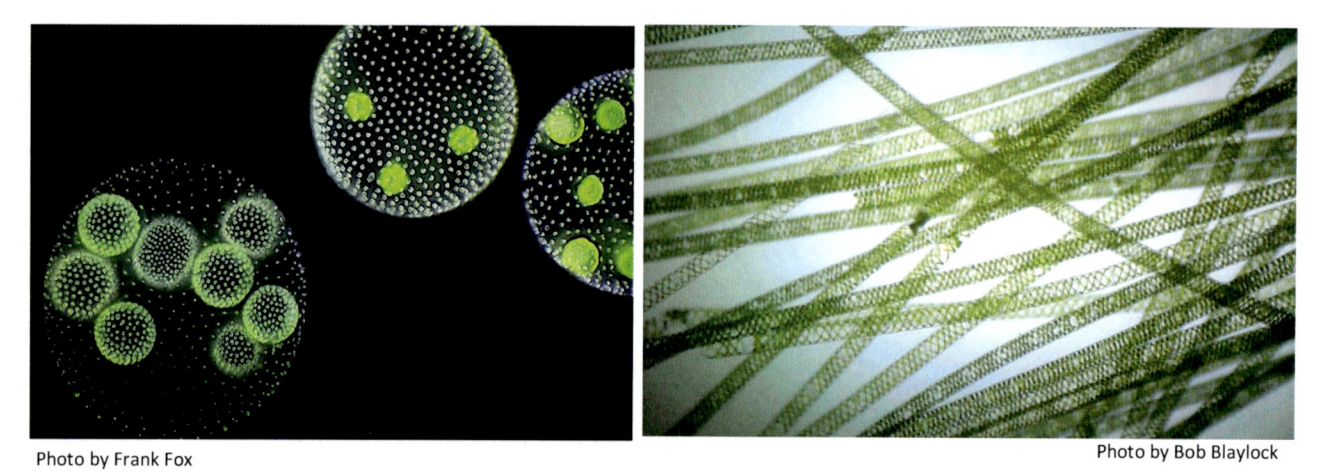

Colonies of *Volvox spp.* (left) and *Spirogyra spp.* (right) are members of Phylum Chlorophyta. Note the green chlorophyll present in this division of algae.

The second division of algae is Phylum Phaeophyta which are the brown alga (plural for algae). The prefix phaeo- refers to the color brown. Like Phylum Chlorophyta, the members of this phylum contain chlorophyll a, yet also contain chlorophyll c. Energy is stored in the form of oils and complex carbohydrates. Most of these algae are sessile, meaning they are fixed in place to a surface and can be quite large. Examples include Fucus and Sargassum commonly known as seaweed.

Sargassum spp. (left) and *Fucus spp.* (right) are members of the Phylum Phaophyta, the brown algae.

The third division of algae is Phylum Rhodophyta. The prefix rhodo- refers to red and so these are the red algae. The members of this group contain chlorophyll a and d and store their food as starches. These algae also contain another set of pigments known as phycobilins which have the capability of absorbing colors of light found at great depths of the sea hence their noted ability to produce energy deep in the ocean.

The fourth division of algae is Phylum Chrysophyta. These algae are known as the golden-brown algae. These algae contain chlorophyll pigments a and c. They store their food as oil and complex carbohydrates. A unique characteristic of this phylum is that members contain silicon in their cell walls which, when they die, leave behind tiny little abrasive "skeletal" remains. The algae are commonly referred to as diatoms and a collection of their "skeletons" is known as diatomaceous earth. Because of the abrasiveness of these remains, diatomaceous earth can be used as a pesticide for external and internal parasites in pets.

The fifth division of algae is Phylum Pyrrophyta also known as Phylum Dinoflagellata. The prefix pyrro- refers to fire which relates to a very unique ability of these algae known as bioluminescence. The prefix bio-, as you are aware, refers to living things and the root word luminesce means to create light. These algae have the ability to make light! You might think of them as being like mi-

Red algae (above) store glucose as starches..

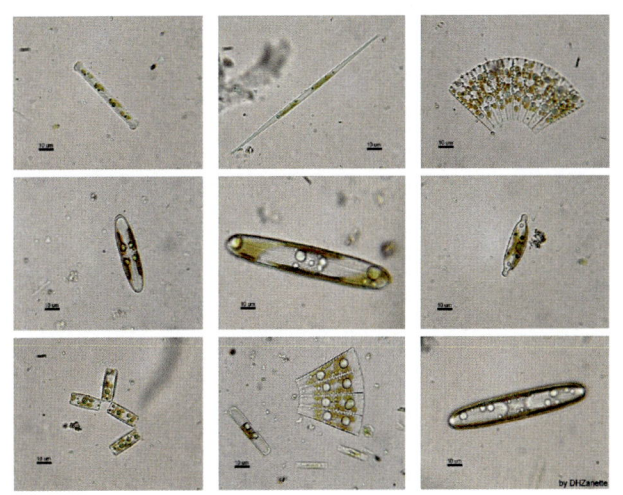

Diatoms (above) are golden-brown algae.

Dinoflagellates can create light like fireflies.

croscopic fireflies found in the ocean. These algae contain chlorophyll pigments a and c and store their food as starches. Members of this group have the ability to swim due to the presence of two flagella. Because of their position on the individual algae cell, these flagella allow the algae to spin as they move through the water. They are also responsible for the phenomenon known as red tide. When enormous populations of these algae develop in the ocean, the water appears to be a reddish color.

The sixth and final division of algae is Phylum Euglenophyta. These algae share many characteristics of Phylum Chlorophyta in that they contain pigments a and c. However, while they can make their own food (stored as a starch), they also have the ability to consume other organisms. Unlike the green algae, members of this group do not have a cell wall. This allows them to be more flexible. They have two flagella, one long and one short, which enable them to freely move about in water. They are found in freshwater lakes and ponds.

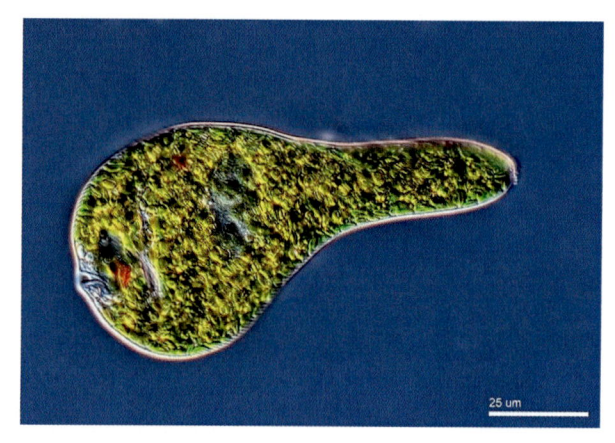

Photo by Rogelio Moreno
Euglena spp. are frequently found in ponds.

Let's pause now and review what we've learned in this lesson. We have learned that:

- Members of the Kingdom Protista are unicellular and, unlike members of Kingdom Monera, have membrane-bound organelles (they are eukaryotic);

- Members of the Kingdom Protista can be divided into three large groups: animal-like protists, plant-like protists and the fungal-like protists;

- The animal-like protists are known as protozoans and they are categorized based upon how they move about (locomotion);

- Protozoans move with cilia (hair-like projections), a pseudopodium (foot-like appendage) or a flagellum (tail-like structure);

- The fungal-like protists include slime molds, parasites of algae, white rusts and downy mildews;

- The plant-like protists include the alga and categories are based upon color of pigments present and means of conducting photosynthesis, along with method of storing energy.

Lesson 20: Kingdom Fungi

The final kingdom of living things that we'll explore is the Kingdom Fungi. You are likely very familiar with mushrooms and molds which belong to this diverse group of creatures. However, there is a vast array of fungi which are microscopic. The causes of athlete's foot and ringworm are fungi and are obviously very tiny. Regardless of size, the main feature they all share is that fungi are considered to be saprophytes. Saprophytes live off of dead organisms and are commonly known as the decomposers.

Because of this feature they are great recyclers of materials which make up living things. In the process of breaking down once-living creatures they produce carbon, nitrogen and phosphorus containing compounds which can be readily reused by other creatures. Unlike plants and protistas which contain chlorophyll pigments, fungi do not conduct photosynthesis. Fungi store food in the form of glycogen, like animals.

But, you might be thinking, when someone has athlete's foot or ringworm they are certainly not a dead organism. The explanation here is that these species of fungi live off of dead skin or hair cells. It should also be pointed out that some fungi cause diseases much more serious than itchy toes or skin irritations. Histoplasmosis, a serious lung disease caused by a fungus which often grows in bird feces, can be dangerous to persons keeping poultry or pet birds. The fungus Aspergillus causes problems for cattle while smuts and rusts are fungal diseases of wheat and other grain-producing plants. Dutch Elm disease, the killer of millions of elm trees, is caused by a fungus.

There are four divisions within the Kingdom Fungi. Criteria for these divisions include whether or not hyphae of the fungus contain crosswalls or not. Hyphae are the primary structures of fungi and can be thought of as being tubular root-like structures which grow throughout the dead material seeking nutrients. Certain divisions of fungi will have walls which span across these tubes and others do not. These crosswalls are also known as septa (plural form of septum, meaning wall).

Divisions are also based upon how the fungus reproduces. All fungi can reproduce sexually and asexually. In the sexual form, two mating strains of fungi form gametes which join together. However, the nuclei of those gametes may not fuse as they do when other organisms reproduce in this manner. Asexual reproduction occurs through the production of spores. Refer back to the mushroom growing lab in Lesson 12 to review how fungi reproduce. Let's look at each of the four divisions of the Kingdom Fungi now.

The first division of Kingdom Fungi is Division Zygomycota. Most of these fungi live in the soil and dung (animal feces). They are characterized by lacking crosswalls in their hyphae. A common member of this division is the common black bread mold, *Rhizopus spp.* These fungi are found on land.

Rhizopus spp. are "enjoying" an old pumpkin (above) and peach (right).

The second division of Kingdom Fungi is Division Basidiomycota also known as the club fungi. Members of this group have crosswalls within their hyphae. The mushrooms, bracket fungi, rusts and smuts are members of this division. Like the Division Zygomycota, members of this group are found on land. The oyster mushrooms you grew in Lesson 12 are members of this division.

Members of *Basidiomycota spp.* are characterized by their fruiting bodies.

The third division of Kingdom Fungi is Division Ascomycota which are known as the sac fungi. These fungi can be found in watery environments or on land and maybe multicellular or unicellular. They are characterized by having crosswalls with holes or perforations. They differ from the Division Basidiomycotes in that their sexual reproductive method involves the creation of asci or sacs which eventually rupture releasing ascospores which float down to the ground to produce new hyphae. The morel mushroom is a common multicellular example of a member of this division. Another very common member of this division which is unicellular is *Saccharomyces spp*. Look closely at its name. Can you recognize the saccharo- prefix from our discussion in Lesson 3 where we discussed saccharides? These fungi thrive on carbohydrates (saccharides) and familiar members include the yeasts used to make bread and ferment fruit juices and grains into wine and alcoholic beverages. Yeasts are also very important in genetic research.

Morels (left) and bread yeast (right) are both members of the Division Ascomycota

The final division of Kingdom Fungi is the Division Deuteromycota which are often referred to as the imperfect fungi. They are called this because the sexual form of reproduction in these organisms has yet to be fully understood. The fungus which causes athlete's foot belongs to this division. Another member which is very important is the fungus Penicillium from which the antibiotic penicillin is derived. The same fungus is also used to flavor certain cheeses. The Aspergillus fungus found in this division is used to ferment soybeans into soy sauce.

Penicillium roqueforti is used to flavor blue cheeses.

Let's pause here and review what we've learned in this lesson. We have learned that members of the Kingdom Fungi:

- Are saprophytes (meaning they live off of dead organisms) and they are known as decomposers;
- Can be categorized according to the presence or lack of crosswalls in their hyphae which can be considered to be like roots which grow through the substance being "eaten" by the fungus;
- Can reproduce sexually and asexually;
- Which belong to the Division Zygomycota have no crosswalls in their hyphae (common black bread mold is a member of this division);
- Which belong to the Division Basidiomycota have crosswalls in their hyphae (familiar members include mushrooms, rusts and smuts);
- Which belong to the Division Ascomycota have crosswalls with perforations and reproduce through the use of asci or sacs which produce ascospores (morels and yeasts are members of this division);
- Which belong to the Division Deuteromycota have reproductive cycles which are not yet fully understood (athletes's foot fungus and the fungus used to create penicillin and flavor cheeses belong in this division.

Lesson 21: Body Systems of Movement

We have two more sections to explore in this course: the first is a detailed look at human biology and the second, and final topic, is ecology, which is the study of how living things interact with each other and the environment. In this lesson, we'll begin our look at human biology. When we say human biology, we'll be looking at how we are made and how we function. How living things are made is known as anatomy while how living things function, is known as physiology. So, in these lessons you'll be studying the anatomy and physiology of the human. Courses which specialize in these topics are known as "A and P" courses. Let's begin.

We, like all living things, are beautifully created and are quite complex. To reduce this complexity, let's think back to our very first lesson in this course where we learned the five characteristics of living things. Can you name those five characteristics?

We said that living things:

1.

2.

3.

4.

5.

We are going to take these five characteristics of living things and make them our five themes for studying the anatomy and physiology of the human. We'll first look at the organ systems of the body which allow us to move. Then, we'll explore the organ systems which allow us to acquire and distribute energy. Next, we'll look at the organ systems which allow us to grow and develop. We've already discussed how living things (including humans) reproduce on the cellular level, so here we'll look at the organ systems of reproduction on a larger scale. And, finally, we'll look at the organ systems which allow us to respond to our environment which are the systems which make up our senses (sight, smell, taste, hearing and touch). Let's begin now with the organ systems which allow us to move.

When one thinks about our ability to move, the systems of the body which likely come to mind first are the muscular and skeletal systems. In order for these body systems to work day-in, day-out in a coordinated fashion, there must be continual control of our muscles through our nervous system as well as continual nourishment of our muscles and bones through our blood and circulatory system.

So, in this lesson we'll begin with the skeletal and muscular systems. Then in the next lesson we'll take a close look at how our nervous system plays an essential role in the control of our muscles and bones which allows for movement. The need for nourishment of these organ systems will provide our transition to the next characteristic of living things which is how we acquire and distribute energy resources throughout our body via the circulatory and digestive systems. These systems will be presented in Lesson 23. Let's begin our focus on movement now with a look at the skeletal system.

The skeletal system in the human consists of approximately 206 bones. We say approximately because at birth, a baby has over 300 bones of which several fuse together to reduce the number down to about 206 in an adult. The bones range in size from the large thigh bone which is the femur, to the tiny bones of the middle ear known as the malleus, incus and stapes. Anatomists usually divide the bones into three main groups based upon shape. There are the long bones which consist of the bones making up the arms and legs, the flat bones which are bones primarily of the skull and pelvis and the irregularly-shaped bones which are the bones of the spinal column. We'll take a closer look at the long bones in this lesson as they are the bones which allow for greatest movement.

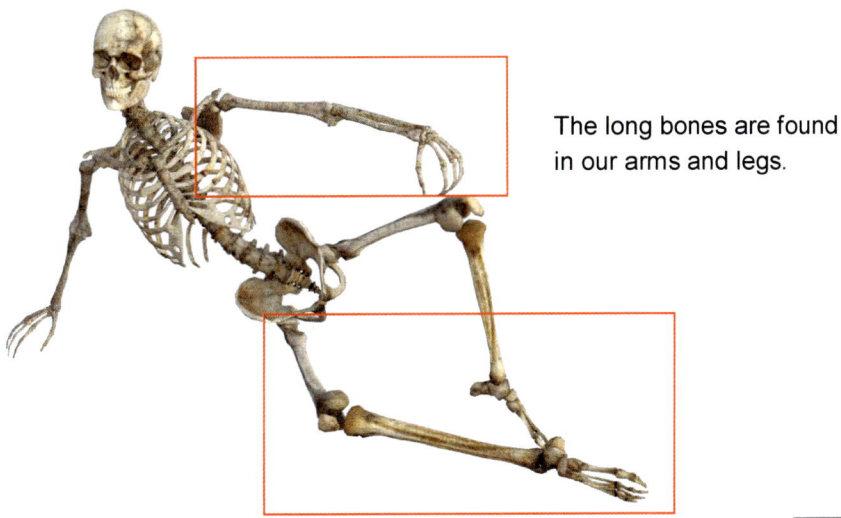

The long bones are found in our arms and legs.

To best understand the anatomy and physiology of a long bone, we need to first go back to the stage in a baby's development when its bones were first forming. Initially, the long bones were formed of cartilage which over time is replaced by calcium- and phosphorus-containing compounds which make the bone hard. This hardening process is called ossification. The prefix os- refers to bones and the cells which make-up bones are known as osteocytes. This hardening process of ossification generally begins in three or more places within a long bone. In the photo here of a thigh bone (from a calf) you can see the three main parts of a long bone.

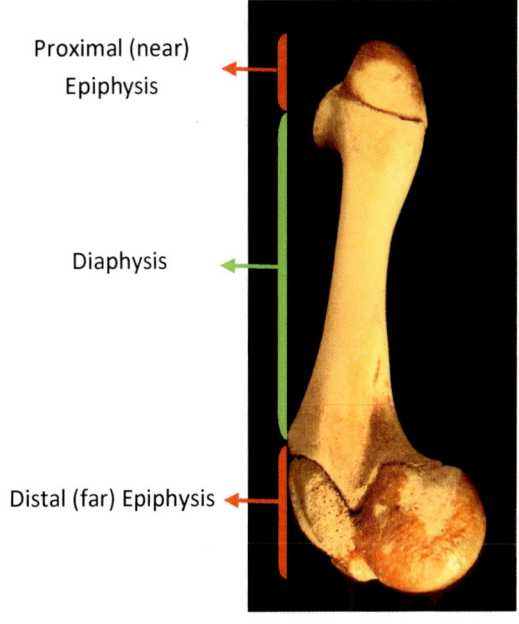

Each long bone has three main parts.

There are two epiphyses (ee-pif'-uh-sees), one found on each end, and then one diaphysis (die-af'-us-sis) which joins the two epiphyses. Ossification begins in the unborn baby in each of these locations and spreads outward. In other words, you will find what's referred to as a center of ossification in the diaphysis and then a center of ossification in each epiphysis. At birth, ossification has usually spread to fill almost all of the cartilage of each long bone, however, a "boundary" of sorts remains between the diaphysis and each epiphysis on the bone.

A center of ossification is found in each section of each long bone. The bone hardens away from these centers.

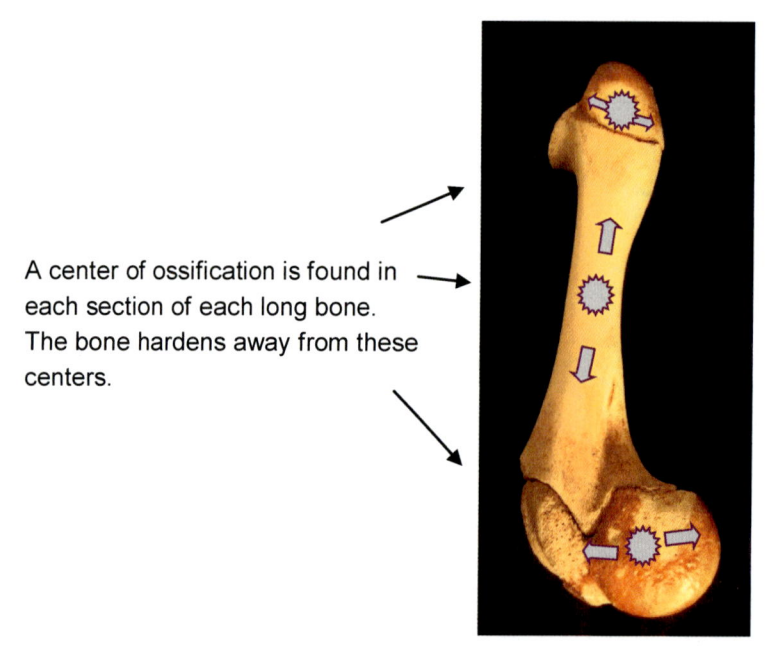

It is at this boundary where growth of the bone occurs and, consequently, this point is referred to as the growth plate or epiphyseal line of the bone. So, as a child grows, it is along this line or plate where osteocytes continue to divide to create new bone which leads to lengthening of the bone. Eventually at about age 16-20 this growth comes to an end and the growth plate "closes" meaning that it also fills with hardened bone and no longer can be readily noted on an x-ray or radiograph.

The growth plate (yellow dotted line) is the boundary between the diaphysis and epiphysis.

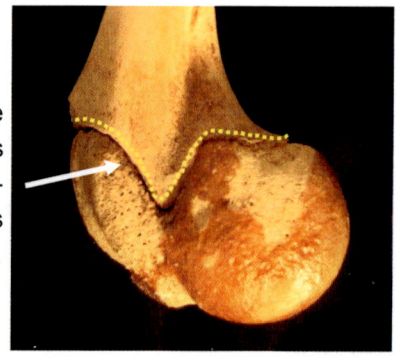

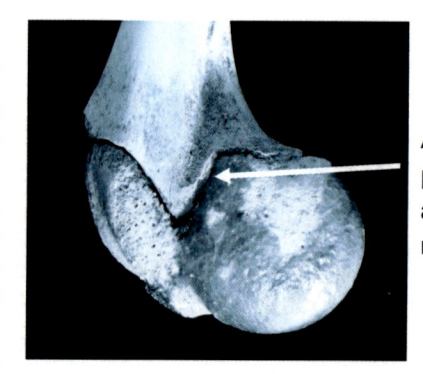

An "open" growth plate is visible on a radiograph (x-ray)

Because we are most interested in the role bones play in allowing movement, we need to now look more closely at the way bones are joined to one another. The joints between the long bones of the arms and legs are known as synovial joints. The joints between flat bones, as in the skull are referred to as sutures. We'll focus our attention in this lesson on the synovial joints of long bones.

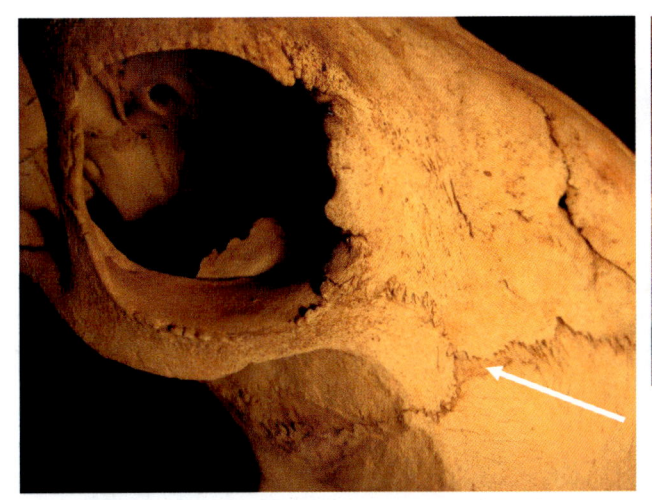

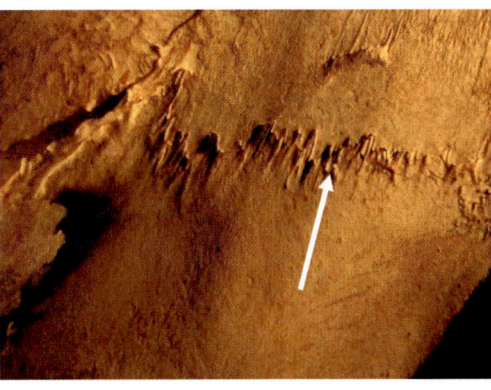

The joints of flat bones of the skull are called sutures (cow skull.)

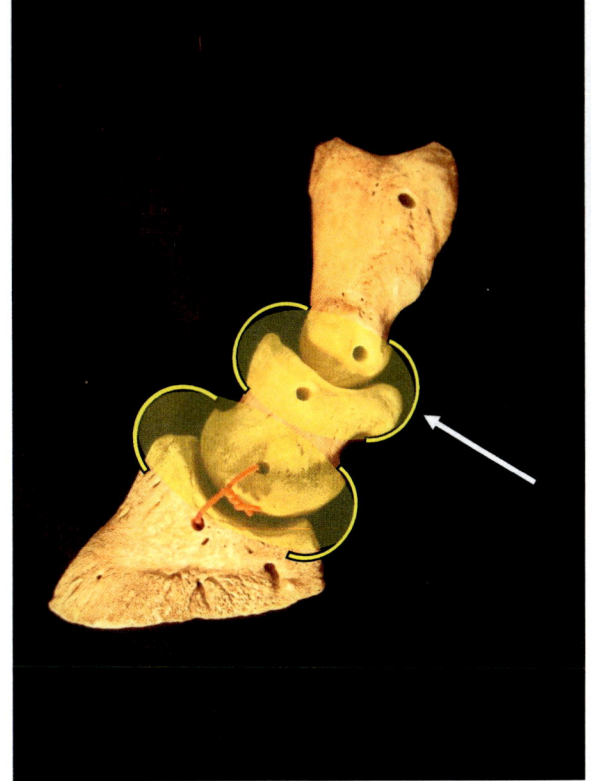

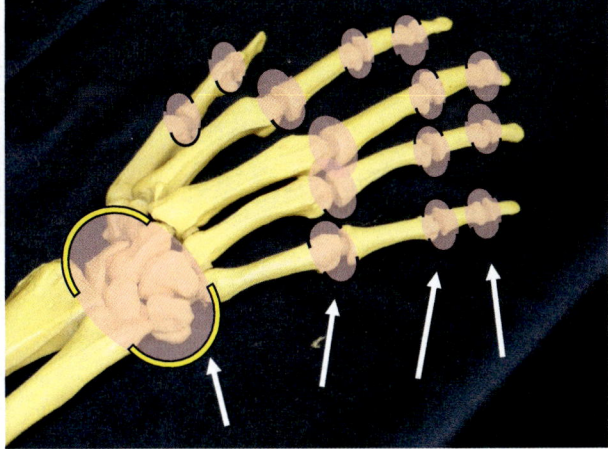

The joints of long bones are called synovial joints. The layer of tough connective tissues which encloses each joint is the synovial capsule. Within, we find synovial fluid which lubricates the surfaces of the joint for almost frictionless movement.

Because the bending of joints happens so frequently over the lifetime of a person, the joints between bones must be very smooth, durable and well-lubricated. If you look at the ends of long bones, you'll find a layer of what is known as hyaline cartilage. This cartilage is exceptionally smooth and durable and allows for easy movement. To reduce friction between the ends of two adjoining bones, a fluid known as synovial fluid keeps the bones gently sliding over each other. This fluid, which consists of complex proteins, is produced by the tough membranes which bridge the joint space between the two bones. This covering is known as the synovial capsule. You might think of this capsule as like being a portion of sock slipped over the ends of both bones. Inside the sock you'll find the hyaline cartilage covering each end of the bone as well as a continual supply of synovial fluid to keep the joint working smoothly. Over time, and with age, this slippery joint surface may become worn and result in pain and inflammation. This is known as arthritis.

There are different classifications of synovial joints depending upon the range of motion required at the joint. For example, the joint between the femur and the pelvis at the hip is classified as a ball-and-socket joint as is the shoulder joint. This kind of engineering allows for a wide range of motion in all directions. The joint at the knee and elbow are classified as hinge joints as they mainly allow for bending in one direction. The ankle and wrist joints, due to the many bones found in these joints, work like both hinge joints and ball-and-socket joints and allow for movement in many directions.

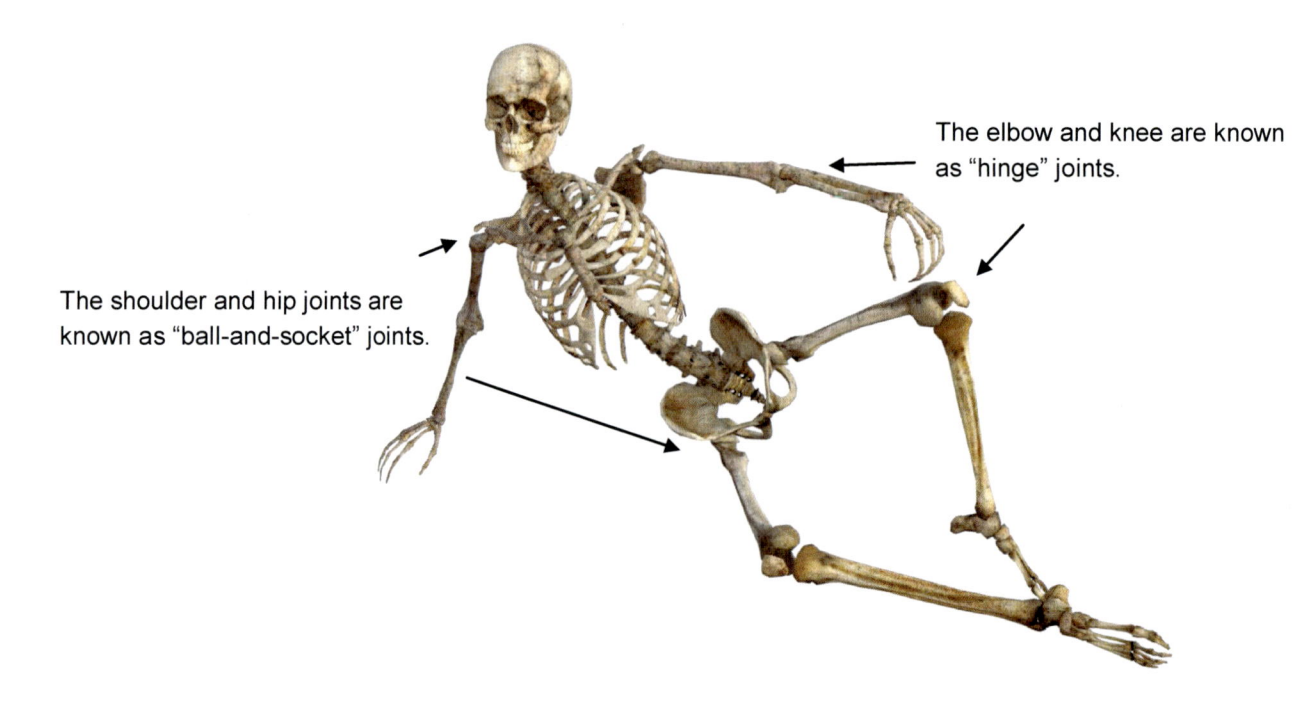

The elbow and knee are known as "hinge" joints.

The shoulder and hip joints are known as "ball-and-socket" joints.

It's important to know that while bones may appear to be hard and lifeless, they are indeed very much living organs. Earlier we mentioned that the cells which make up bone are known as osteocytes. If we examine osteocytes closer, we find that there are actually two kinds of osteocytes: the osteoblasts and the osteoclasts. The osteoblasts are bone cells which have the responsibility of building new bone, such as along the growth plate, while the osteoclasts have the job of taking apart or tearing down bone. There is a constant building up and tearing down of our bones. You might think of this as a continual remodeling process going in our bones.

When there is a break or fracture in a bone, the activity of the osteoclasts and osteoblasts is greatly accelerated. The damaged bone fragments and cells are quickly removed by the osteoclasts. The osteoblasts then begin to build new bone structure which eventually hardens into the new bone or callous. Over time, this callous, which is usually more extensive than necessary, gets remodeled further by the osteoclasts to make a solidly repaired site.

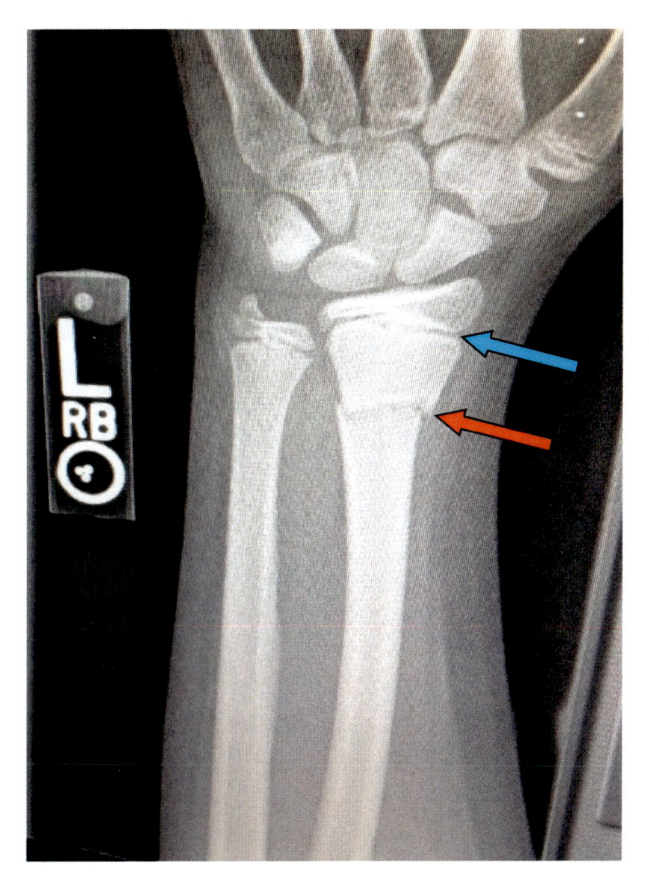

A normal "open" growth plate can mimic a fracture. In this radiograph (x-ray) of the wrist, we can see both. The blue arrow is pointing to the growth plate on the radius while the red arrow is pointing to the fracture site. Note in the ulna (just to the left of the radius) another "open" growth plate.

Before we continue with our discussion of how movement takes place with bones, we need to point out that, in addition to allowing for movement, bones are also very important when it comes to protection of vital organs in the body. The flat bones of the skull protect the brain, while the irregularly-shaped bones of the spinal column protect the spinal cord. The ribs are also very important for protecting the lungs, heart and upper abdominal organs.

In addition to protection, our bones also play a major role in the production of red blood cells. In the hollow space, known as the marrow cavity, within bones we find tissues which produce new red blood cells. This tissue is known a hemopoietic (hee-mo-poy-et-ic) tissue which literally means blood-making tissue. The marrow cavity is also a location for fat storage in adults. Recall Lesson 4 where we discussed how lipids and fats were considered to function like stores of energy. The marrow cavities are a location for these stores. Finally, our bones also function as a storage site or depot for the element calcium. Not only is calcium important for the strength of our bones, calcium plays a major role in our nervous system.

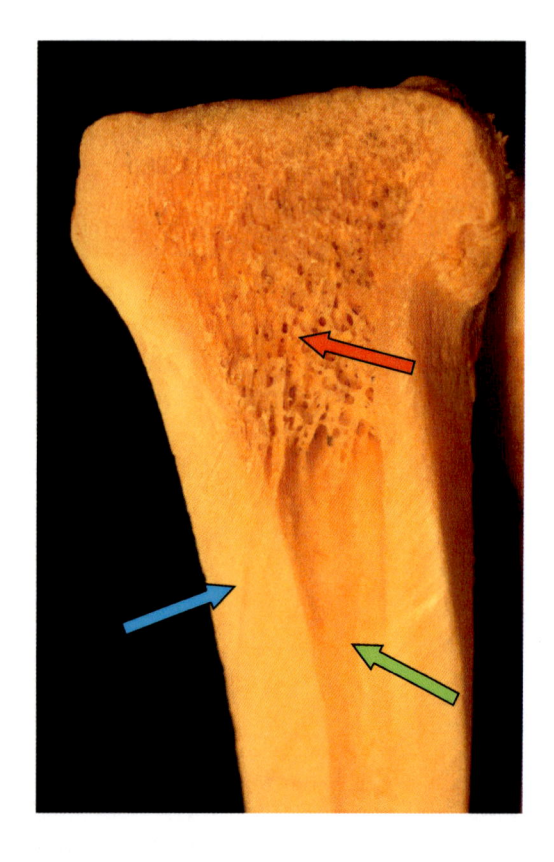

In this photograph we see a long bone that has been cut in half lengthwise (longitudinally). This enables us to view compact bone (blue arrow), spongy or cancellous bone (red arrow) and the marrow space (green arrow) where fat is stored and new red blood cells are produced.

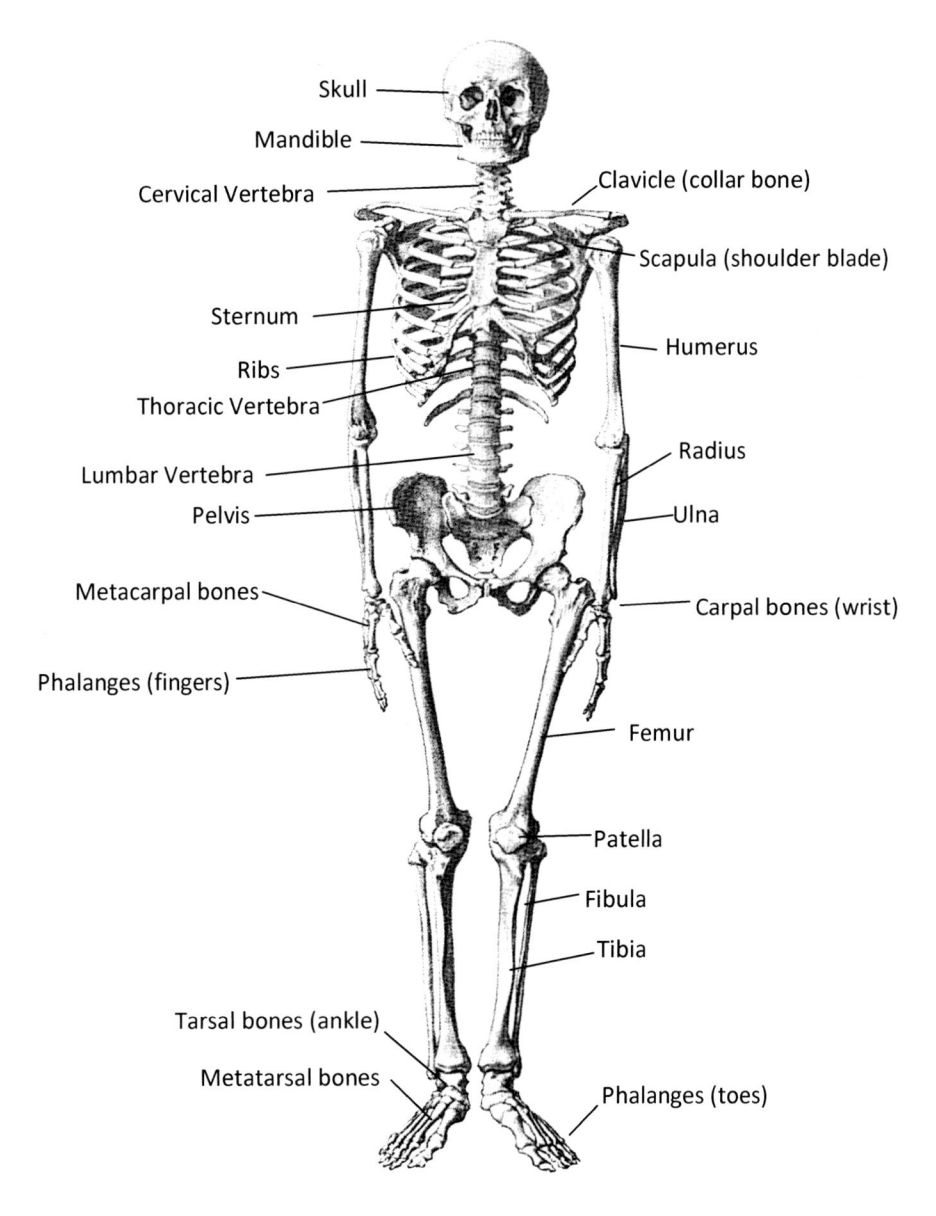

Now that we have presented the "bone side" of the idea of how humans move, let's take a look at the muscle side of movement. Before we jump (pun intended) into movement, let's look a little more broadly at muscles in general. In humans (and most animals for that matter) we find three types of muscles. These types are based primarily on the type of cells known as myocytes (myo = muscles) which make up that type of muscle. There is smooth muscle which is found in our internal organs like our stomach, intestines, bladder and respiratory tract, for example. Then there is cardiac muscle which is specifically made for our heart. Finally, there is skeletal muscle which is the muscle we are interested in here as these are the muscles which attach to our bones and enable us to move. Skeletal muscle is also referred to as striated muscle. If something is said to be striated or have striations, it has stripes. When viewed under a microscope, striated or skeletal muscle does, indeed, have stripes which run perpendicular to the long dimension of the muscle. In other words, you can see stripes which run around and around the individual fibers of a muscle.

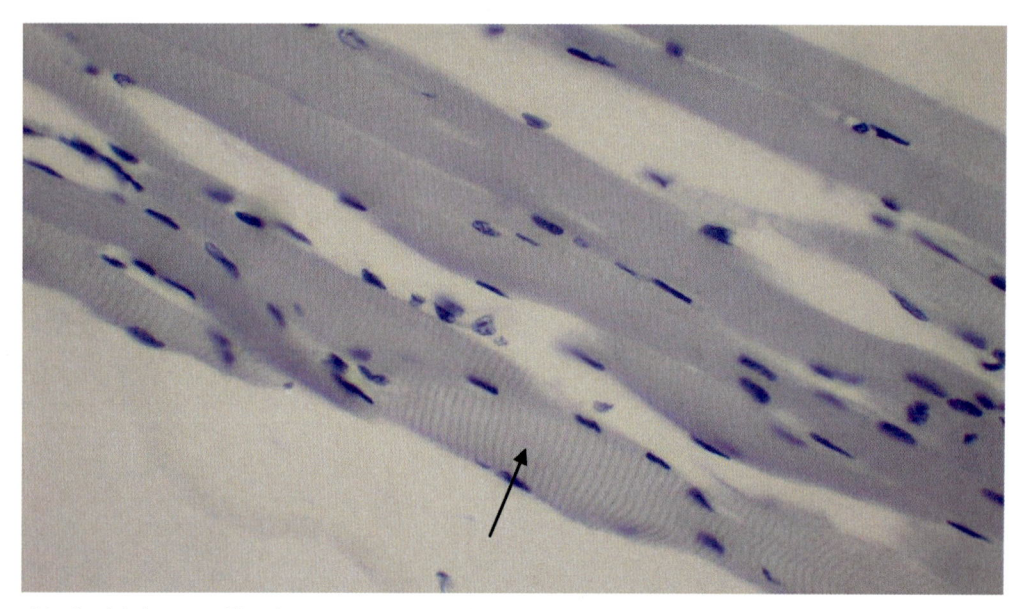

Under high magnification, you can see the striations (stripes) of skeletal muscle.

These stripes are directly related to how skeletal muscles contract. On the molecular level there are components of skeletal muscle cells made up of what are known as contractile proteins or filaments. There are two types of contractile proteins: actin and myosin. When a nerve stimulus arrives at a muscle fiber, it is the actin and myosin which work like a miniature ratchet system to make the muscle shorten or contract. When this nerve stimulus is applied to the entire muscle, the length of the muscle shortens which results in movement.

However, for this movement to have purpose, the ends of each muscle must be attached to bones. This attachment of muscles to bones is made through tough connective tissue material known as tendons. Know that tendons connect muscles to bones. Connective tissue structures which connect bones to bones are known as ligaments. And in order for a muscle to result in action, these connecting points must be on opposite sides of a joint. It would do little good for a skeletal muscle to have both of its ends attached to the same bone as no action would result! So, by attaching each end of a muscle to different, adjacent bones, we can find movement occurring.

It's also important to realize that muscles can only actively contract. They extend only through relaxation. In other words, you can make your muscle shorten (contract) but you can't make your muscle actively get longer (extend). It only gets longer by being relaxed and being stretched through the action of another muscle working to move the joint in the opposite direction. Let's look at an example of how this works.

First, we need to learn the names of the two ends of a muscle which attach to bones. The end of the muscle which is nearer to the main trunk (chest and abdomen) of the body is known as the origin of the muscle. The opposite end of the muscle which is farther from the main trunk of the body is known as the insertion of the muscle. So, all skeletal muscles have an origin and insertion. The location of each of these two points of attachment determines the action made upon the joint.

Now, we'll get back to our discussion of the concept of contraction and relaxation. Let's consider two well-known muscles which work oppositely of each other: the biceps and the triceps. Look at the photo below which illustrates the locations of these two muscles and then examine the diagram which illustrates the origins and insertions of each set of muscles. Note that both muscles cross the elbow joint and will therefore have action upon that joint.

In order for you to bend your elbow, the biceps contract. By shortening the biceps, the radius and ulna bones are moved upward or closer to the humerus. To extend your elbow, it's the shortening

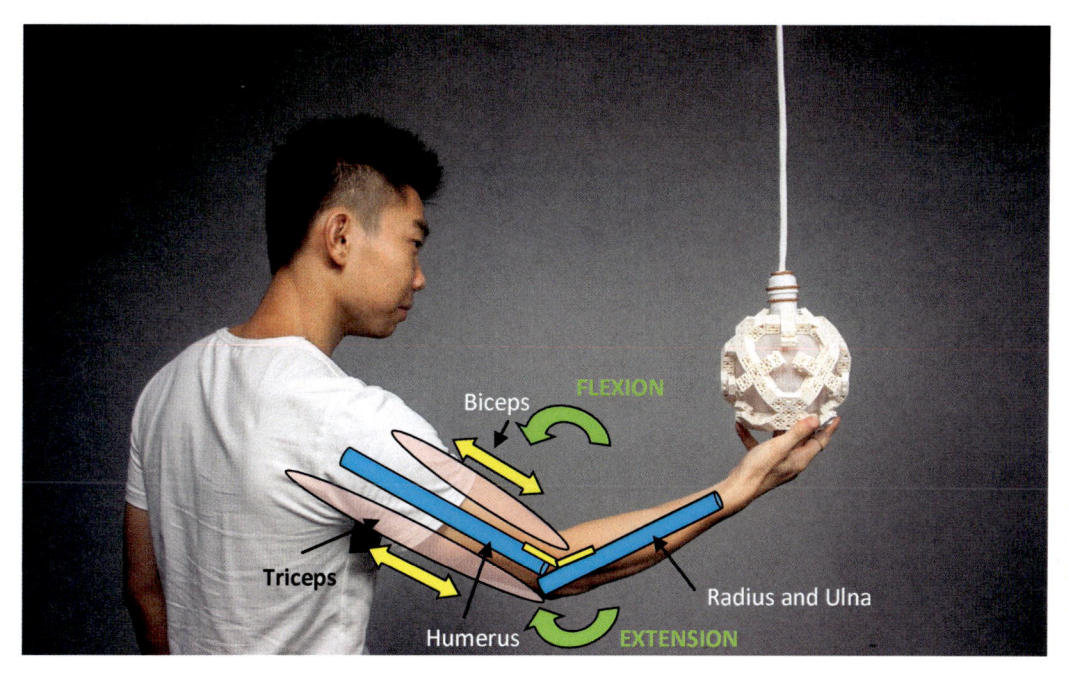

The biceps and triceps muscles work as a "team" to flex and extend the elbow.

action of the triceps which straightens the elbow. In other words, the biceps don't actively "push" the elbow straight. Instead, the biceps can only relax and allow the triceps to "pull" the elbow into an extended position. This paired-action allows you to bend and then straighten all of the joints between long bones in your body. Bending of a joint is referred to as flexing while straightening a joint is known as extending the joint. Muscles are then identified and ultimately grouped as being flexors or extensors of joints in the body. Look at the photo below to see the main groups of flexors and extensors of the arms and legs.

In addition to flexion and extension of joints, depending upon where the origin and insertion points are on bones, muscles can also move bones inward or outward and even rotate bones in a circu-

Here you can see the locations of the flexors and extensors of the arms and legs.

356

lar direction. Inward movement is referred to as adduction and outward motion is referred to as abduction. Rotation of a joint is known as pronation or supination depending upon the direction being turned. Look at the photo below to see the muscles which allow for adduction and abduction. Note again how the movement only results through coordinated action and relaxation of the two sets of muscles. Muscles don't push; they only pull.

Movement towards the body is known as adduction while movement away from the body is abduction.

Let's stop here and review what we've learned in this lesson regarding how human beings move. We said that:
- The study of structure and function of a living thing is known as anatomy and physiology;
- Bones begin as structures made of cartilage and then harden through ossification;
- There are long bones, flat bones and irregularly-shaped bones;
- The ends of a long bone are the epiphyses. the central portion is the diaphysis and the border between is the epiphyseal line or growth plate where growth occurs;
- Cells of bones are osteocytes of which there are two types: osteoblasts which work to build bone and osteoclasts which work to break down bone;
- Joints between flat bones are called sutures and joints of long bones are called synovial joints;
- Within the synovial capsule we find the lubricant synovial fluid;
- Muscles must span joints in order for movement to occur;

- There are three types of muscles: smooth (found in internal organs), cardiac (found in the heart) and skeletal or striated (which cause movement);
- Actin and myosin are the contractile proteins of striated muscle;
- Muscles can only contract (pull), they cannot push and therefore, muscles work in pairs to result in movement; and
- Actions of muscles result in flexion or extension, adduction or abduction or rotation of a joint.

Now that you have an idea of how muscles and bones work together to allow for movement, let's explore in greater detail how the nervous system controls this action. Recall from our discussion that actin and myosin proteins on the molecular level, actively work as tiny ratchet systems to shorten a muscle. We said that the "triggering" or initiation of this action came as a result of a nerve stimulating these proteins. To better understand this process, let's look first at how nerve cells, known as neurons, are made.

Neurons, like almost all other cells of the body, except for red blood cells, have a nucleus which works to control the activities of the cell. (Red blood cells have a nucleus early in their lives but lose it just before entering the blood stream.) The nucleus of a neuron is found in the region known as the soma. Extending in opposite directions from the soma are two types of projections, one being relatively short and the other being much longer. The shorter projections are known as dendrites and the longer end is known as the axon. There may be multiple dendrites extending from a single neuron but only one axon. These projections work as the contact points between connecting neurons. The dendrite of one neuron is aligned to the axon of the "next" neuron along the pathway leading from or to the spinal cord and brain. So, if you were to examine a "whole" nerve, you'd find a series of neurons, each lined-up dendrite to axon, dendrite to axon and so on. You can liken this to standing hand-in-hand with all of your family members or friends. Your body represents the soma of the neuron. Your left arm would be held close to your body representing the dendrite while your right arm would be fully extended and represent the axon. By linking hands, you and your colleagues can become a "whole" nerve.

It's important to know this anatomy of a neuron because it helps explain how the flow or transmission of a signal to or from the brain travels. The dendrite end of a neuron works only to accept signals. Once a signal is accepted by the dendrite, it travels toward the soma of the neuron and on to the axon. At the axon, the signal then gets passed on to the dendrite of the subsequent neuron which, in

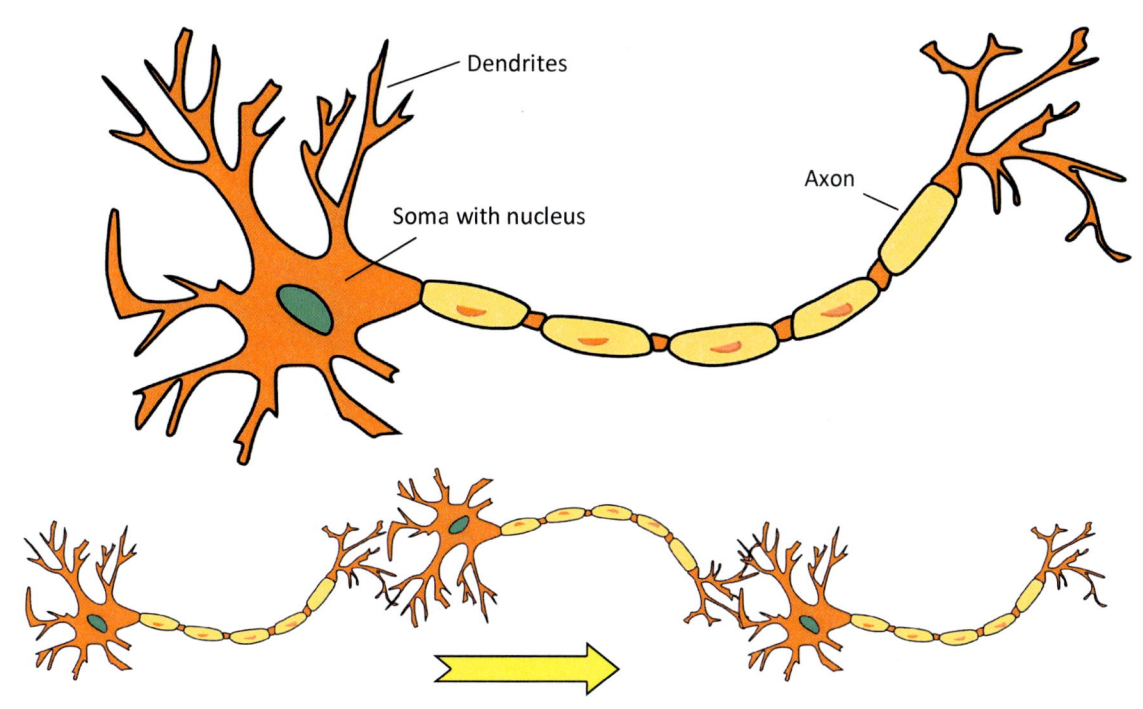

The axon of one neuron joins with the dendrite of a subsequent neuron. Flow of information is in one direction.

turn, transmits the signal onward. So, transmission of a nerve signal, known as a stimulus, is in one direction only, like a one-way street. This means that in order for information to move back and forth between, say, your fingertip and your brain, there has to be two sets of nerves: one going from the fingertip to the brain and a separate nerve going from the brain back to the fingertip. Each will carry stimuli (plural of stimulus) in one direction only. Together, they work to enable information to go both directions. So in essence, we have two one-way streets which work together to get information transmitted between the brain and various parts of our bodies.

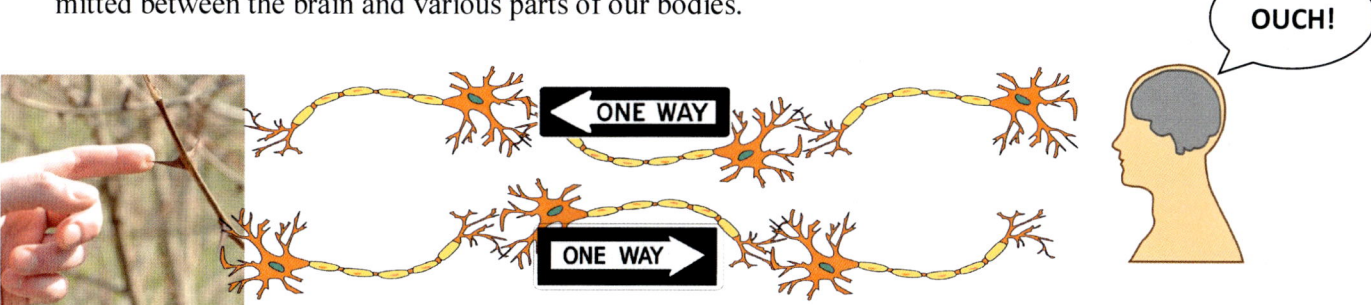

Flow of stimuli from the periphery of the body to the brain and then back, happens on two different sets of neurons.

Since we're focusing our attention on how nerve action results in movement, let's look first at the neurons which bring the stimulus from the brain out to the muscle. These neurons are known as efferent neurons or motor neurons indicating they are carrying a signal outward and the signal is intended to result in motion (motor = action or motion). So, we have a series of efferent or motor neurons lined up end-to-end which carry a signal from the brain down through the spinal cord exiting at an appropriate location to continue down to the intended muscle. Keep in mind the direction of signal movement is away from the brain and spinal cord.

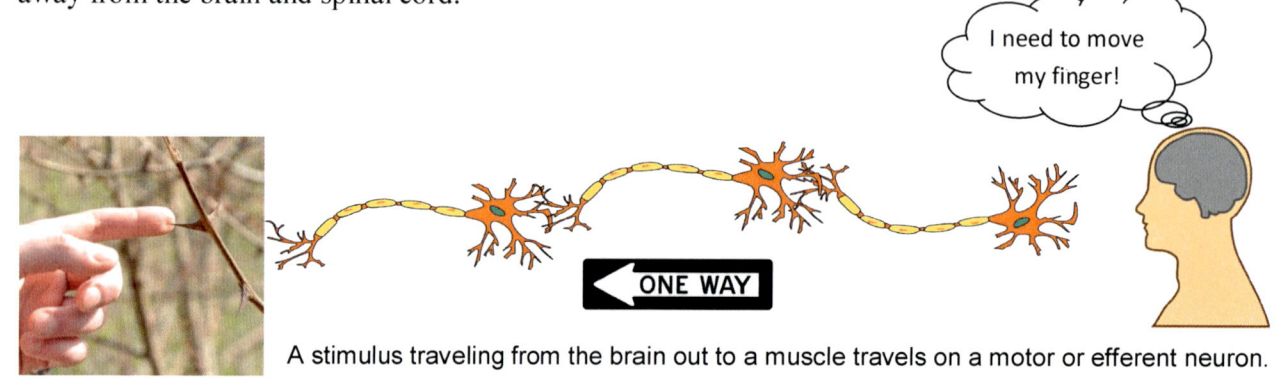

A stimulus traveling from the brain out to a muscle travels on a motor or efferent neuron.

Before we discuss the neurons which carry signals back to the spinal cord and brain, let's examine in greater detail what actually happens at the final neuron which has the "privileged" duty of stimulating the muscle. At the axon-end of this final neuron, as with junctions all along the nerve between dendrites and the axons, there are special cell structures which are known as vesicles. Vesicles contain chemicals known as neurotransmitters. You can think of these vesicles as being like little packets or blisters filled with the neurotransmitter. When the stimulus arrives at the final ending of the nerve, these vesicles are stimulated to rupture and release the neurotransmitter out into the space between the end of the nerve and onto the surface of the myocyte. This results in a change within the myocyte which results in the triggering of the actin and myosin filaments which results in a muscle contraction. This connecting point between neuron and myocyte or between neuron and neuron is called a synapse.

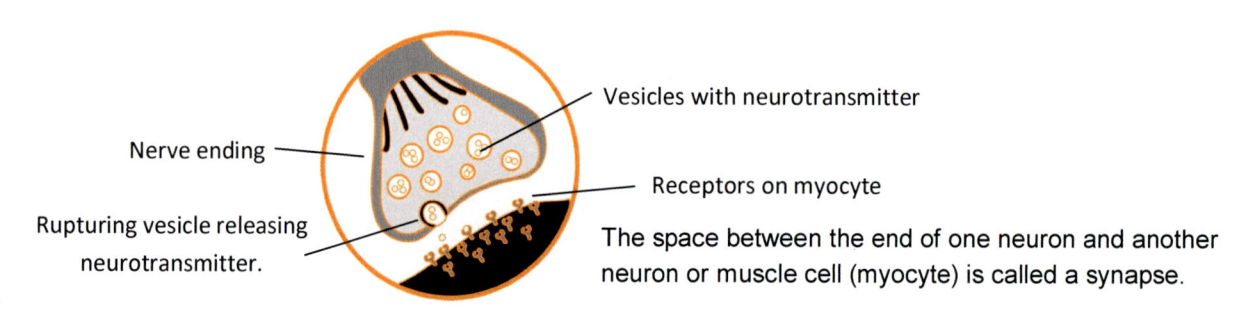

Nerve ending

Rupturing vesicle releasing neurotransmitter.

Vesicles with neurotransmitter

Receptors on myocyte

The space between the end of one neuron and another neuron or muscle cell (myocyte) is called a synapse.

The most common of all neurotransmitters is one known as acetylcholine or ACH for short. When the stimulus arrives at the terminal end of the final neuron, the ACH is "spilled" out onto the intended muscle to cause the desired contraction. As long as there is ACH present in this space, for the most part, the muscle will continue to be stimulated to contract.

So you may be wondering, how does one "get rid" of this ACH when you're ready to allow this muscle to relax so that the joint action can go in another direction. What do you think might also be available nearby that could cut or chop up these ACH molecules rendering them ineffective? If you said enzymes, you're exactly right! And, can you think of the particular name for this enzyme? Think way back to Lesson 3 where you first learned about enzymes and how they are named. Yes, this enzyme which cuts or lyses ACH is known as acetylcholinesterase or ACHase for short. So, as quickly as the ACH is being released to stimulate the muscle to contract, acetylcholinesterase is ready nearby to "gobble it up" and allow the muscle to relax and be ready for the next call to action when more ACH arrives. And to think that this is happening continuously in our bodies as we go about our daily tasks each and every day and that we never have to think about any of it in order for it to happen is amazing. It's truly amazing!

Let's now turn our attention back to the "one-way street" which carries information from the "outskirts" of the body back to the spinal cord and brain. Because these nerves are carrying information toward the brain, they are known as afferent or sensory neurons indicating that they are carrying information obtained through our senses (touch, taste, smell, hearing and sight). Like the efferent or motor neurons, the afferent neurons also work through the action of various neurotransmitters. We'll take a closer look at the organ systems which comprise our sensory system in Lesson 27.

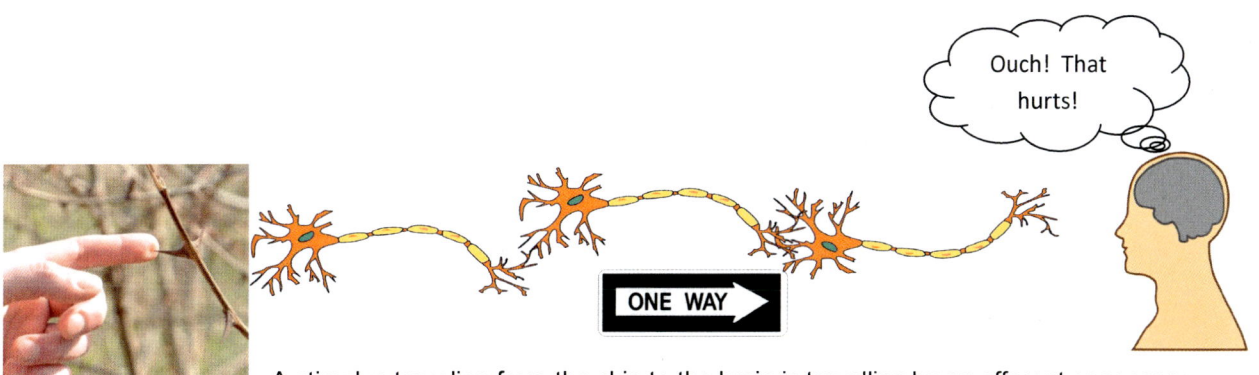

A stimulus traveling from the skin to the brain is travelling by an afferent or sensory neuron. Again, it is one-way travel.

At this point, let's take a closer look at the stimulus that we've been referring to the "entity" which is moving back and forth via the one-way nerves of the body. The stimulus is actually like an electrical impulse which begins at one point and moves along the length of the neuron. Scientifically known as the action potential, it works through the action of sodium and potassium atoms getting pumped into and out of the neuron. This pumping action is controlled by the presence of calcium atoms in the area surrounding the neuron. Recall from our discussion regarding the functions of the skeletal system that we learned how bones function as a reserve or depot for calcium. This is one location where this calcium reserve is utilized.

An interesting side note to our discussion here is the problem known as milk fever which is familiar to persons who raise cattle. While there is no actual fever or elevation in body temperature, the unwelcome situation does have to do with milk. Milk fever, also known as eclampsia, occurs in the first few days after a cow has given birth to a calf. It's in these first few days when milk production goes into "high" gear and lots of calcium is being mobilized throughout the cow's body and being diverted into the milk for the new calf. If too much calcium is moved too quickly, the normal levels of calcium along the outer surfaces of the neurons becomes depleted.

Recall how we said earlier that calcium atoms work to control the pumping of the sodium and potassium atoms into and out of the nerve cell. Without adequate levels of calcium, things go haywire and nerve impulse conduction becomes erratic. The cow usually "goes down" meaning she is unable to stand. Her muscles twitch and she becomes anxious, irritable and nervous. Eventually, she may actually become paralyzed and can die if not treated. Treatment involves giving calcium intravenously which often results in quick recovery. This condition can also happen in other mammals including humans.

Typical presentation of a cow with milk fever: recumbent (lying down) and head tucked into flank.

Photo courtesy of L. Mahin

Before we leave our discussion of nerve function, lets visit the very interesting concept of anesthesia. The root word -esthesia refers to pain and when coupled with the prefix an- we see that anesthesia means without or the removal of pain. Now that you know how neurons and neurotransmitters work and how substances move into and out of neurons, you can readily understand how certain drugs can have their actions at various points along the "streets" of nerves going to and from locations where pain or movement needs to be controlled. The anesthetic commonly used to "deaden" the skin or numb a tooth prior to being worked upon by a dentist is lidocaine (also known as xylocaine or novocaine). The mechanism of action with lidocaine is that it prevents sodium ions from moving into a neuron, thereby blocking the transmission of an action potential along a neuron. Because the neuron cannot transmit the sensation of pain, the patient does not experience discomfort from the procedure.

Another chemical agent used historically in poison arrows made by South American natives is curare. This chemical, derived from a vine, has the ability to block the ACH receptor sites on muscle cells. Consequently, when ACH is released from the nerve ending, its expected action does not occur and the muscle does not contract. This causes the injected animal to lose voluntary control of its muscles (it goes limp) including its diaphragm which causes it to suffocate and die. This agent was also used in early anesthetic practice, however, because it only blocks muscle action (i.e. the patient will lie still) it has little to no effect upon the sensation of pain being experienced by the patient. Other agents were then added to relieve pain. Curare is no longer used in modern medicine as other safer, faster-acting drugs are available.

Let's pause here and review what we've learned so far about the nervous system. We said that:
- Cells of the nervous systems are called neurons and consist of a central nucleus-containing area known as the soma which has a single axon is present with multiple dendrites and an action potential (stimulus) travels into the dendrite "end" and out through the axon "end;"
- Axons join with dendrites and terminate on myocytes (muscle cells) where neurotransmitters are released to activate muscle contraction, the most common neurotransmitter being acetylcholine (ACH);
- Neurons which carry messages out to muscles are known as motor or efferent neurons;
- Neurons which carry messages from sensory organs such as the skin or eyes are called sensory or afferent neurons;
- Conduction of a stimulus is the result of sodium and potassium being pumped in and out of the

neuron and calcium mediates this process;
- Anesthesia interferes with transmission of nerve impulses or neurotransmitter function.

Lesson 22: Body Systems of Nutrient Delivery

In the last lesson, we discussed how the nervous system works in conjunction with our muscles and bones to result in movement. In order for these systems to work, another key system, the circulatory system, must also be involved. Recall from Lesson 7 that we learned about how the mitochondrion functions to convert glucose molecules into energy-containing ATPs. We learned that this process was known as cellular respiration. We learned that in order for respiration to happen in an efficient manner, oxygen had to be present. We said that for each glucose molecule, 38 ATPs could be produced with oxygen present whereas only 2 could be produced in anaerobic conditions. The means of delivery of oxygen-as well as glucose-to our body's cells is the circulatory system. Oxygen (from the air we inhale) and glucose (derived from the foods we eat) are the sources of these vital components. It's the circulatory system which works like a delivery system to get them where they are needed.

Let's begin our discussion of this amazing delivery system down at the delivery sites of the bones, muscles and nervous tissue. Because blood is carrying oxygen and glucose molecules to the tissues, these components must find a way to leave the blood and move out among the tissue cells. The blood vessels are extremely small at this cellular level, so small, in fact, that the walls of the blood vessels are only one cell layer thick. In other words, only the thickness of one cell separates the blood from the tissue cells. At this level, the blood vessels are called capillaries.

Between the cells that make up the walls of the capillaries are openings known as fenestrations. Fenestrations can be thought of as "windows" which allow small items like oxygen, glucose and water to readily pass through. Larger blood components-like proteins-find it difficult to pass and are moved through special mechanisms of transfer.

Recall from our discussion of cellular respiration that, in addition to the production of ATPs, carbon dioxide was a waste product of cellular respiration and would need to be carried away from the cells. Like oxygen, glucose and water, carbon dioxide is also a small molecule and readily moves through the "windows" of the capillaries from the cells back into the blood to be carried to the lungs to be removed.

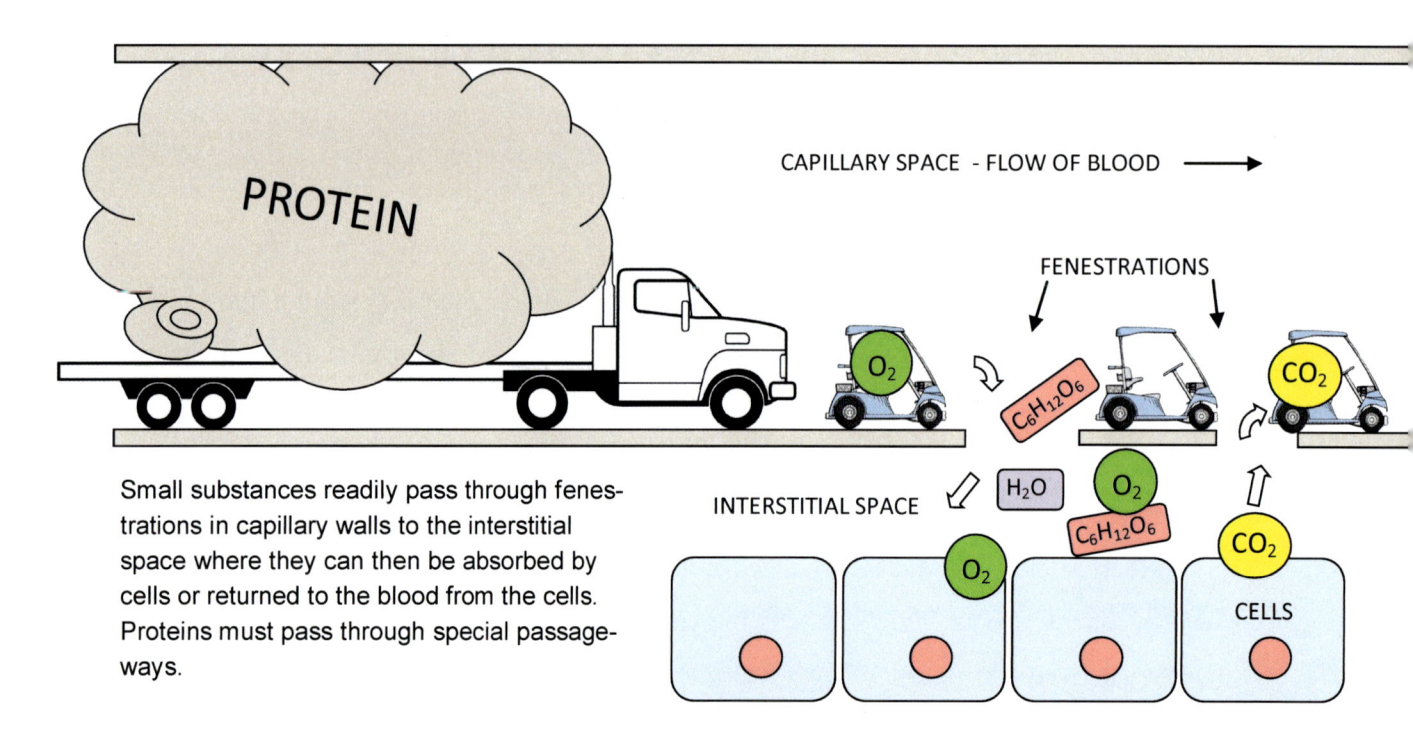

Small substances readily pass through fenestrations in capillary walls to the interstitial space where they can then be absorbed by cells or returned to the blood from the cells. Proteins must pass through special passageways.

The size of these capillary fenestrations vary from place to place through the body. In locations like the muscles, where lots of oxygen and glucose are needed, as well as other structural components, the "windows" are quite large. In the spleen the fenestrations are so large that proteins and even whole blood cells can leave and re-enter the circulatory system. This is because one job of the spleen is to remove "old" or damaged red blood cells (erythrocytes).

We find that fenestrations of the capillaries are the smallest or tightest at the brain and spinal cord. The need for oxygen and glucose and the need for carbon dioxide removal are the greatest in these locations so passage of these small molecules happens readily. However, security against any sort of invading organism or chemical substance is very high-especially at the brain. This is known as the blood-brain barrier. On a daily basis, this high security system works extremely well at preventing unwanted invaders. However, it can be challenging when a medication would work best if it actually could cross into brain tissue. Drugs which work as general anesthetics (the kind which make you fall asleep) can cross the barrier and are capable of having their desired effect.

Another substance which can also readily cross the blood-brain barrier is alcohol. Unlike lipids, which we learned about back in Lesson 4, alcohols are very water soluble and are a relatively small molecule. Because of this, they readily cross the blood-brain barrier and, therefore, have their intoxicating and damaging effects upon cells of the brain.

Let's return to our discussion of the capillaries now. It's important to realize that this level of the circulatory system, where the transfer of oxygen, glucose, carbon dioxide and other body substances takes place, is only one "stop" on the circulatory delivery system. The name of this system, circulatory, refers to a circle. The blood moving through the blood vessels is on a continuous loop picking up and releasing certain components along the way. Note that the blood *arriving* at the capillary level is higher in oxygen and glucose than the blood leaving the capillary level. As we mentioned earlier, the blood *leaving* the capillary level will be higher in carbon dioxide and other waste components.

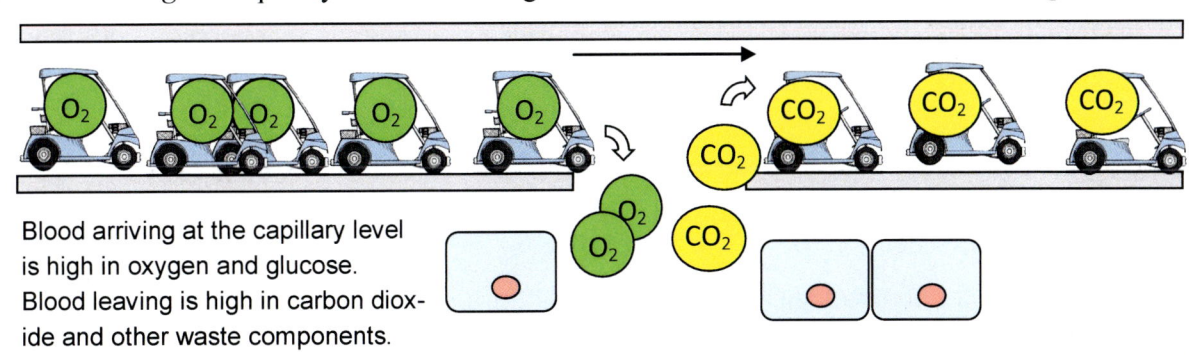

Blood arriving at the capillary level is high in oxygen and glucose. Blood leaving is high in carbon dioxide and other waste components.

Let's continue our exploration of the circulatory system by moving "upstream" in this endless circle. Recall that we said the walls of the capillaries were only one cell thick. The blood vessels which supply the capillaries consist of tubes which are many cell layers thick. These vessels are known as arterioles. If we continue "upstream," blood vessels which supply the arterioles are known as arteries. Layers of cells within these vessels include a well-defined inner lining known as the endothelium, a strong layer of muscles and then a tough outer covering.

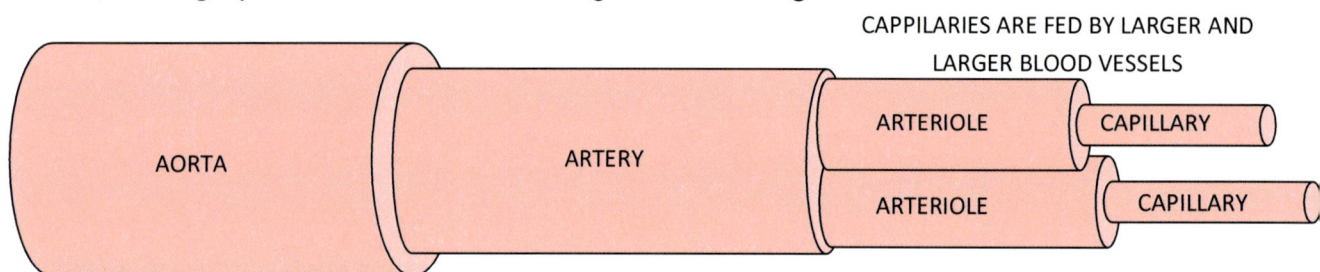

CAPPILARIES ARE FED BY LARGER AND LARGER BLOOD VESSELS

AORTA — ARTERY — ARTERIOLE — CAPILLARY

The muscles of arteries are different from the striated muscles which attach to our bones. These muscles are known as smooth muscles and are common throughout many internal organs in our bodies. In arteries and arterioles, these smooth muscles encircle the artery and work to control the diameter of the blood vessel. By contracting or relaxing, the degree of blood flow can be regulated. For example, in times when you are active, like running or chasing something, your muscles have a great need for oxygen and glucose. The arteries supplying them relax and allow ample blood to flow. Think about when you're hot and sweating. Again, lots of blood flows to your skin to be cooled and then returnS to your body to cool all parts. Also, these arterial smooth muscles allow your body to maintain an adequate blood pressure. Muscles constricting (making smaller) or dilating (making larger) our arteries allow blood to flow adequately and at an appropriate pressure throughout our body.

If we continue to travel "upstream," eventually we'll arrive at the very largest of arteries known as the aorta. The aorta is the main blood vessel to leave the heart carrying blood out to the body. Keep in mind that this blood leaving the heart through the aorta is the highest in oxygen so it makes sense that the location where oxygen is added to the blood must be close! Let's continue into the heart now.

To understand the anatomy of the heart, let's pretend the heart is like house with four rooms, two on ground level and two in the basement. The two on ground level each have an entry door. The two in the basement have no doors to the exterior, but each have a chimney. Entering the house can only be done through one of the doors on the main floor. Exiting the house can only be done through one of the

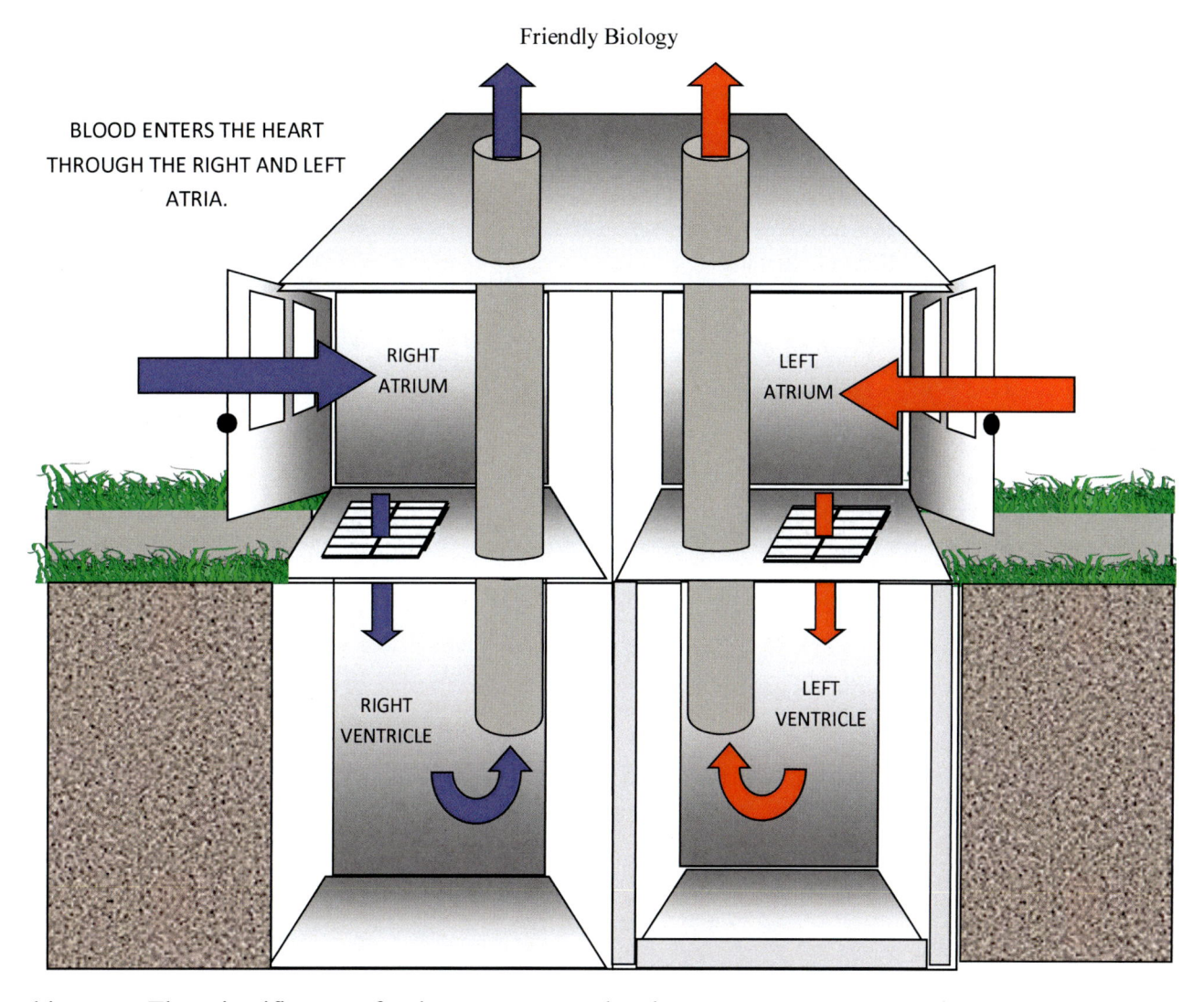

BLOOD ENTERS THE HEART THROUGH THE RIGHT AND LEFT ATRIA.

chimneys. The scientific name for these rooms are chambers.

The two rooms on ground level are known as the atria (atrium is singular). In architecture, an atrium is a reception area and, likewise, the atria of the heart are where blood is received from the body. Compared to the two rooms in the basement, the atria are much smaller in size and the walls are thinner. Recall that we said these two rooms receive blood from the body. Let's take a closer look now at exactly where this blood is coming from as it enters the atria.

The atria are identified as being the left atrium and right atrium. The left atrium is on the left side of your body while the right atrium is on your right side. Let's look at the right atrium first. The right atrium receives blood returning from the body. Because this blood has already been out to body organs, it is low in oxygen and high in carbon dioxide. This should make sense as the body cells will have "taken" oxygen from the blood while the body cells will have "given" carbon dioxide, the waste product of cellular respiration, back to the blood.

369

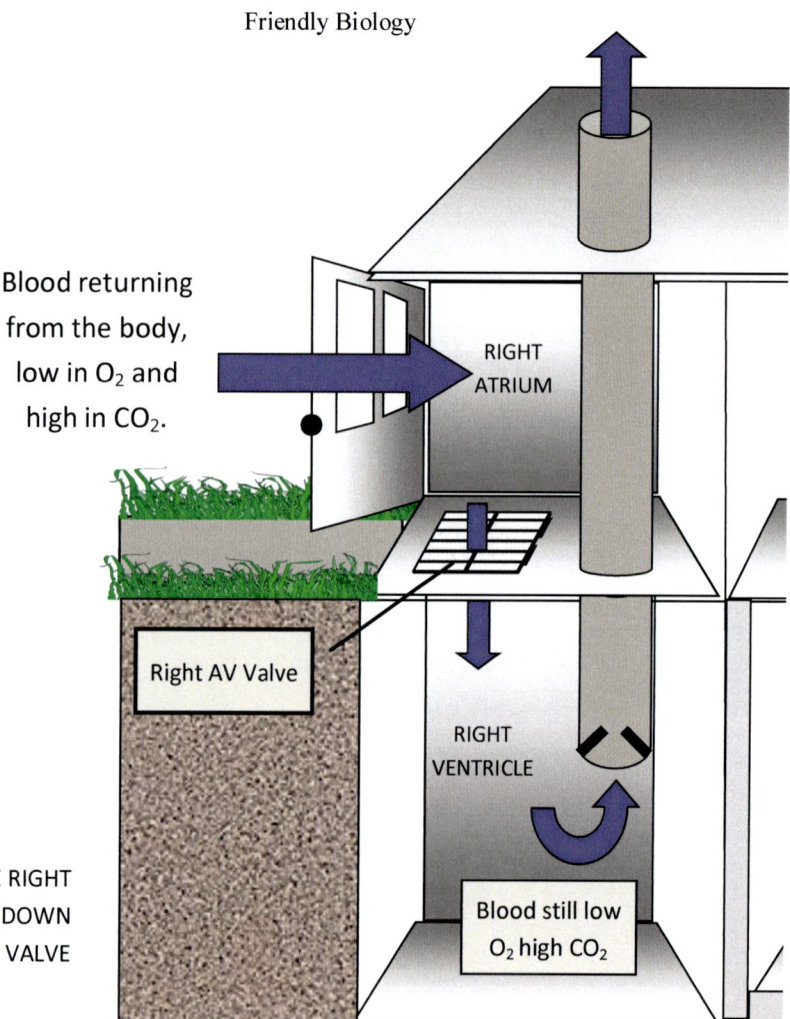

Blood returning from the body, low in O_2 and high in CO_2.

RIGHT ATRIUM

Right AV Valve

RIGHT VENTRICLE

Blood still low O_2 high CO_2

UPON CONTRACTION OF THE RIGHT ATRIUM, BLOOD MOVES DOWN THROUGH THE RIGHT AV VALVE INTO THE RIGHT VENTRICLE.

This low-oxygen blood arrives from the body at the right atrium and upon the atrium's squeezing or contraction, this blood gets dumped down through a "trap door" in its floor into the basement room (the ventricle) below. This "trap door" is the first of four one-way valves that the blood will pass through. The blood moves downward from the right atrium into the right ventricle. The valve it passes through, therefore, has the name "right atrioventricular valve" or right AV valve, for short. Note that nothing has changed yet regarding the level of oxygen or carbon dioxide in the blood.

We mentioned earlier that the walls of the atrium were much thinner than those of the ventricles. Like the arteries, the wall of the heart is made of an inner lining known as the endocardium and then a layer of specialized muscle. Outside this layer of muscle is the epicardium. The muscle which makes up the heart is known as cardiac muscle or myocardium. Cardiac muscle is only found in the heart.

Contraction of the right ventricle is the next phase of action in the heart. Recall that we said in our house analogy that the only exit from a basement room (ventricle) is up through a chimney. So after passing through a second valve positioned at the entrance to the chimney, blood will spurt up through this vessel and actually leave the heart. Blood vessels which leave the heart are called arteries while blood vessels arriving at the heart are called veins. This particular vessel leaving the right ventricle is known as the pulmonary artery. The name pulmonary refers to air or the lungs and this is exactly where this blood is destined: the lungs.

Blood now heads to the lungs to get oxygen and leave carbon dioxide.

- Pulmonary artery
- Pulmonary trunk
- Right AV Valve
- RIGHT ATRIUM
- RIGHT VENTRICLE
- Blood still low O_2 high CO_2

UPON CONTRACTION OF THE RIGHT VENTRTICLE, BLOOD MOVES UP THROUGH THE PULMONARY TRUNK INTO THE PULMONARY ARTERY AND ON TO THE LUNGS..

After leaving the heart, the pulmonary artery will branch into smaller arteries which enter the right and left lungs. Each lung is divided into sections known as lobes. The pulmonary artery divides into yet smaller arteries and eventually these divide into tiny capillaries. As we discussed earlier, capillaries are blood vessels whose walls are only one cell layer thick. This is very important because it is here in the lungs at this capillary level where oxygen is absorbed from the lungs into the blood and it is where carbon dioxide will be removed from the blood to re-enter the air sacs of the lungs.

This exchange of the gases which are dissolved in the blood takes place with each inhalation (air moving into the lungs) and exhalation (air moving out of the lungs) we take-all day long, every day of our lives. The air we inhale moves through our nose, down through the pharynx and into the trachea (windpipe). As the trachea nears the lungs, it divides into smaller branches known as bronchi which, in turn, branch into bronchioles. Eventually, these bronchioles end in tiny grape-like clusters known as the air sacs or alveoli (al-vee'-o-li). It is around the alveoli that the capillaries of the circulatory system wrap themselves to allow for gas exchange to take place.

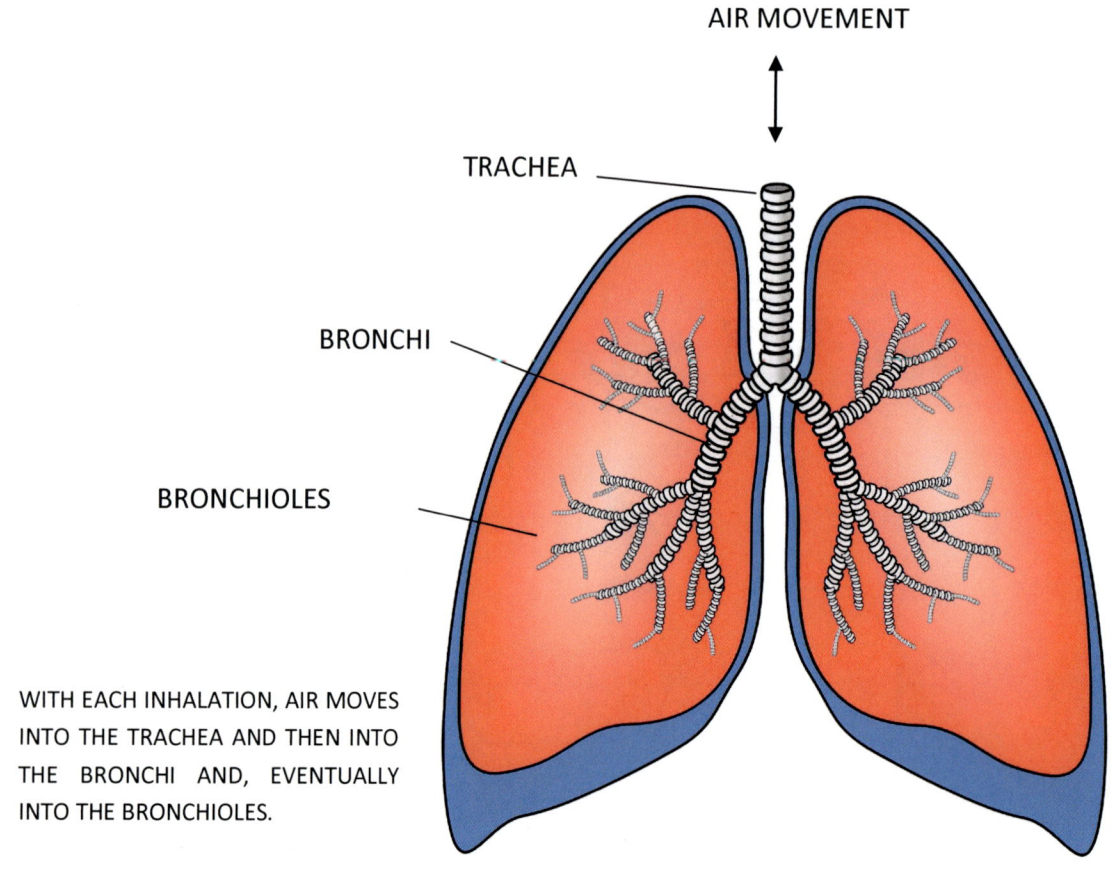

WITH EACH INHALATION, AIR MOVES INTO THE TRACHEA AND THEN INTO THE BRONCHI AND, EVENTUALLY INTO THE BRONCHIOLES.

CAPILLARIES

LUNG TISSUE

ALVEOLI

ALVEOLAR DUCT

BRONCHIOLE

BRANCHES OF PULMONARY ARTERY (LOW O_2) AND PULMONARY VEIN (HIGH O_2)

EXCHANGE OF GASES

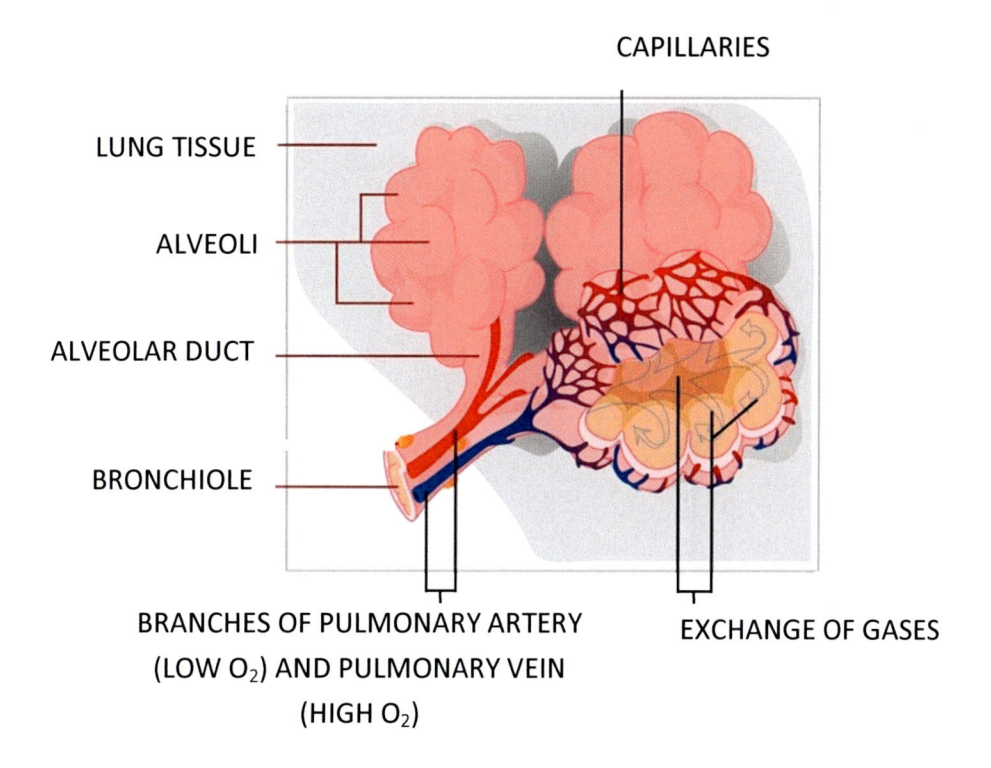

So, at this capillary level within the lungs, oxygen levels in the blood rapidly increase while carbon dioxide levels drastically fall. This newly "revitalized" blood is now ready to return to the heart. Where do you think it will enter? If you answered left atrium, you're correct. Remember, blood only enters the heart through an atrium. This blood, now *high* in oxygen and *low* in carbon dioxide enters the heart through the left atrium. The blood vessel which returns this blood from the lungs to the heart is known as the pulmonary vein. Normally, we think of veins as having low levels of oxygen, however this vein is the sole exception; it is a vein carrying blood *high* in oxygen. In fact, it is the best oxygenated blood in the whole body.

So, we now have the blood in the left atrium. As the left atrium contracts, the blood, once again, falls through a "trap door" in the floor (left atrioventricular or left AV valve) and finds itself in the left ventricle. Before we learn about the function of the left ventricle, let's take a closer look briefly at how these valves work.

Valves in the heart allow blood to flow only in one direction. They consist of flaps of tough tissue known as cusps which function like the door of a trap door. Depending upon location in the

LEFT ATRIUM

Blood returning from the lungs via the pulmonary vein, high in O_2 and low in CO_2

Left AV Valve

LEFT VENTRICLE

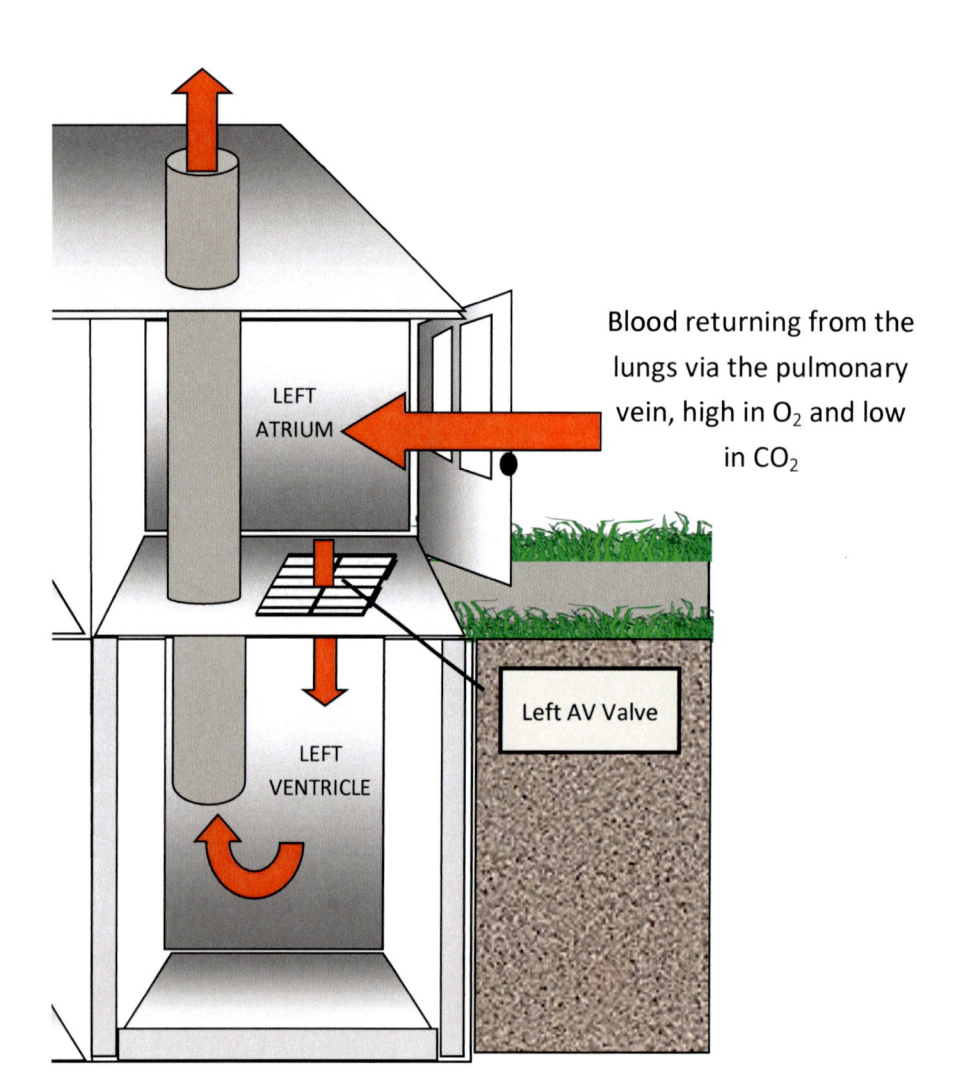

BLOOD RETURNS FROM THE LUNGS INTO THE LEFT ATRIUM. UPON CONTRACTION, BLOOD MOVES DOWN THROUGH THE LEFT AV VALVE INTO THE LEFT VENTRICLE.

heart, a valve may have three cusps and be called a tricuspid valve or have two cusps and be called a bicuspid or, sometimes, semilunar valve.

In addition to these cusps, the valves which create the openings between the atria and ventricles (the AV valves) also have special string-like structures which attach the edge of a cusp to the adjacent wall of the heart. These string-like structures are called chordae tendonae (cords of tendons) and limit how far the cusp will "swing" when opened and then closed. Look at the following diagram.

Note that when the atrium is contracting the cusps swing downward allowing blood to pass. When the ventricle below begins its contraction and the pressure upon the blood increases, the

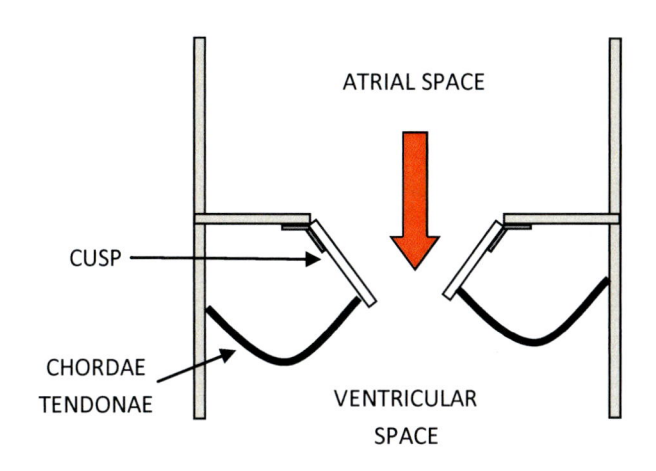

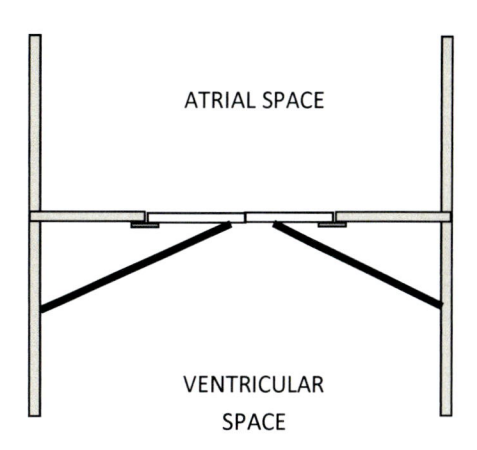

Open A-V valve.

Contraction of the atrium increases pressure, forcing the cusps of the valve to swing open. Blood flows down through the valve into the resting ventricle. The chordae tendonae are relaxed.

Closed A-V valve.

The ventricle is now filled and it contracts resulting in increased pressure. The cusps of the valve swing upward to close the valve. The chordae tendonae are stretched and prevent the valve from prolapsing (swinging upward too far).

cusps "swing shut" as far as the chordae tendonae will allow and prevent blood from flowing back "up" through the valve opening. The only exit for the blood is up through a chimney (which also has a valve to allow passage in one direction). The chordae tendonae, when functioning properly, prevent the "doors" from swinging back upward (prolapsing) and allowing blood to flow back up into the atrium above.

For different reasons, the valves in our hearts may not close exactly as intended and blood may actually squirt or seep back against the normal flow. When blood does this it can create various sounds that are not normal. Normal heart sounds that a doctor listens for are the lub dup, lub dup, lub dup sounds that are made when the valves normally shut. When there is leakage in a valve or even if a valve seals well but maybe provide too small of a passageway for blood to flow (stenosis), one hears whirring, buzzing or swishing sounds. These abnormal sounds are known as heart murmurs. Depending upon their location and degree of severity, some murmurs may be of little consequence. However, others may be of major concern and require treatment.

Let's go back now to where we were with our highly oxygenated blood that had just arrived in the left ventricle. Because we are in a ventricle, recall that the only way out is up through a chimney. On the right side, the blood that left the heart went to the lungs to get a fresh supply of oxygen and leave off waste carbon dioxide. Now on the left side, this blood, high in oxygen and low in CO_2, is destined to head back out to the cells of the body. Because this is quite a task to push blood out all

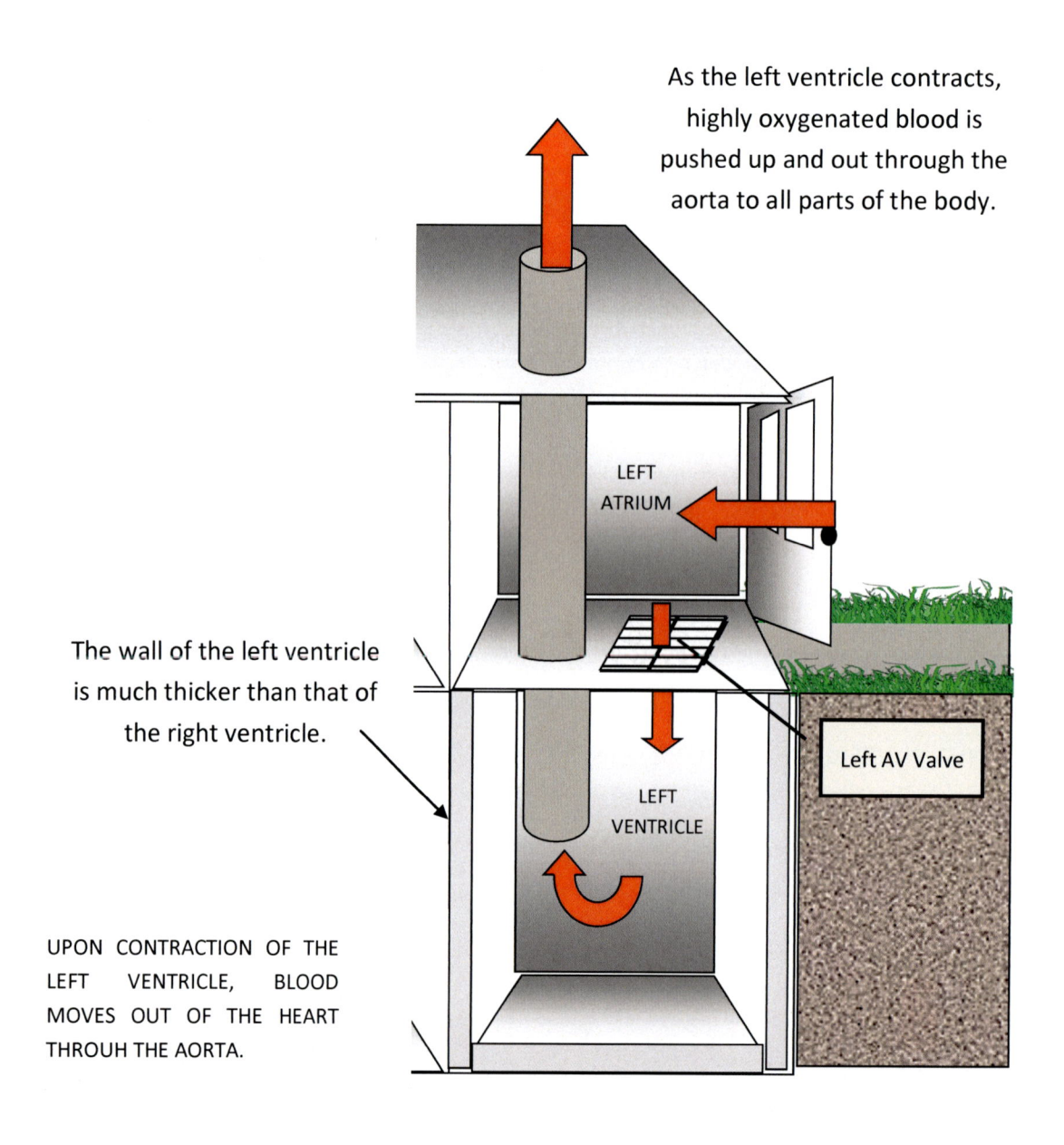

As the left ventricle contracts, highly oxygenated blood is pushed up and out through the aorta to all parts of the body.

LEFT ATRIUM

The wall of the left ventricle is much thicker than that of the right ventricle.

Left AV Valve

LEFT VENTRICLE

UPON CONTRACTION OF THE LEFT VENTRICLE, BLOOD MOVES OUT OF THE HEART THROUH THE AORTA.

across the body, the myocardium of the left ventricle is quite thick. You might think of the left side as being like a body builder: the increased work requirement results in greater muscle development. In fact, it is muscle wall thickness which readily allows a student dissecting the heart to quickly distinguish the left ventricle from the right ventricle. The wall of the left ventricle is thick and the wall of the right ventricle is thin.

As the left ventricle makes its massive contraction, blood quickly moves through a fourth valve at the entrance to the chimney, then up through the chimney and exits the heart through the aorta. The aorta branches into the many arteries supplying all organs of the body as we discussed earlier.

At this point we need to examine a very special set of arteries which immediately branch off the aorta just as it leaves the heart. These special vessels are known as the coronary arteries and it's their job to supply blood to the heart itself. Just like the skeletal muscles and organs all over the body need a supply of oxygen and glucose to adequately function, the myocardial cells also need a constant supply. The coronary arteries are that vital pipeline and we find many of them coursing across the surface of the heart.

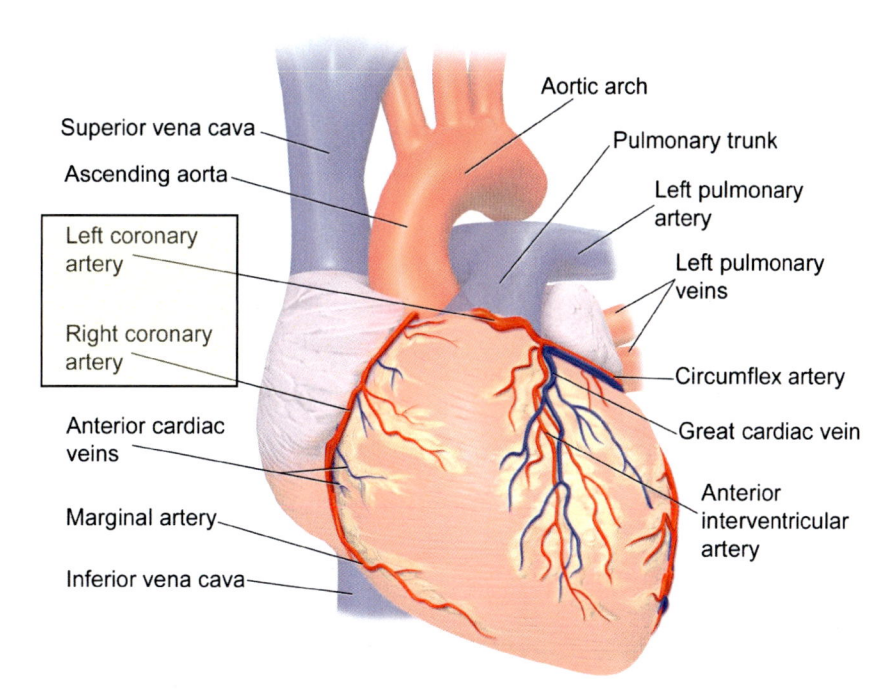

Coronary Circulation (Anterior)

Diagram courtesy Blausen Medical Communications, Inc.

Understanding the significance of these vessels is extremely important because if these vessels get clogged, as you might expect, major life-threatening problems can result. Just like when we discussed how strokes can happen in the brain, a clog in a coronary artery results in heart cells being starved for oxygen and glucose. Without fuel, these cells lose their ability to contract in a normal rhythmic pattern and eventually fail. The result is commonly known as a heart attack or scientifically, a myocardial infarction.

If the clog is severe and treatment is delayed, death can quickly result. If the clog is not yet complete, yet a person has symptoms of a heart attack such as chest pain, dizziness or shortness of breath, a cardiologist (a doctor who specializes with the heart) can perform treatments to reduce the clog or make a detour around the affected site. This is commonly known as a bypass surgery or stent. Making good choices regarding our diets has been shown over and over again to help our coronary arteries remain capable of delivering adequate supplies of blood to our heart cells.

At this point in our journey through the circulatory system, we have just exited the heart via the aorta and have continued down various arteries to all parts of the body. Recall that the large arteries branch into smaller arteries which in turn branch into arterioles which supply the capillaries. As the blood completes its deliveries and pick-ups at the capillary level, it begins its trek back to the heart through veins.

The first tiny veins that attach to the capillaries are known as venules. The venules then drain blood into small veins which in turn drain blood into larger veins which eventually drain into the largest vein known as the vena cava. The vena cava then drains blood into the right atrium and the process begins again. Unlike arteries, veins have a much thinner wall and actions of skeletal muscles near veins contribute to flow of blood through the veins back to the heart. Also, unlike arteries, veins have small valves within them to prevent the flow of blood back upstream.

Now that we've explored the structures of the circulatory system, let's look at the blood, itself, which moves through these "pipelines." It's important to realize that our blood is not just a red-colored liquid but, instead, it's actually a clear, watery liquid which has millions of red-colored cells mixed within. The watery portion of blood is known as plasma or serum. The cells which give blood its red color are known as red blood cells or erythrocytes (ee-rith-ro-sites). Not only are there red blood cells but there are also white blood cells, known collectively as leucocytes (loo-ko-sites). Let's look first at the erythrocytes.

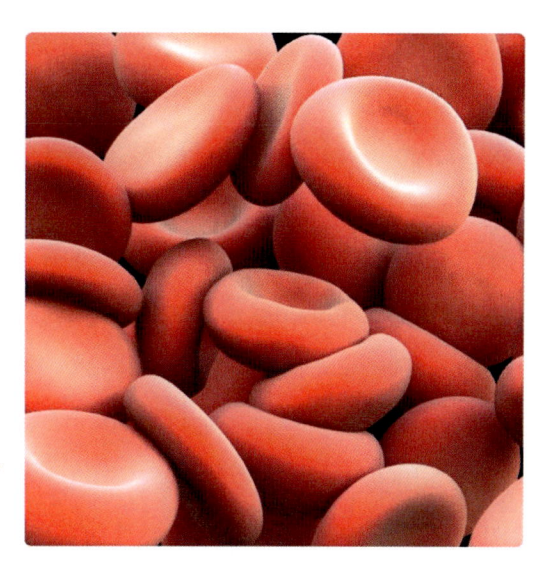

Red blood cells (erythrocytes) are bi-concave disks without a nucleus.

In one drop of blood there are estimated to be over 300 million red blood cells and the average human has about eight liters of blood. You can see that we have a tremendous number of erythrocytes. We have learned that erythrocytes are made in bone marrow tissue within our bones. While immature erythrocytes have a nucleus, fully matured red blood cells do not have a nucleus and are termed anucleated cells.

The primary job of the red blood cells is to carry oxygen to cells in the body and this capability is due to a special protein found within erythrocytes known as hemoglobin. Each hemoglobin molecule has the capability of carrying four molecules of oxygen. As red blood cells arrive at the lungs, each cell "loads up" its hemoglobin with oxygen molecules with the intent of delivering that oxygen to a cell somewhere in the body. The element iron plays a major role in the ability of hemoglobin to gather oxygen molecules. As you might imagine, if the number of red blood cells is decreased, due, for example, to blood loss, a decrease in red blood cell production or possibly infection with a parasite like malaria, delivery of oxygen to the body's cells is directly affected. A deficiency in iron also directly affects the delivery of oxygen to cells of the body. This decrease in the functioning number of red blood cells is known as anemia. All cells in the body suffer due to decreased oxygen availability in conditions of anemia.

As we mentioned earlier, in addition to the erythrocytes, we also have the leucocytes commonly referred to as white blood cells. While there are different types of leucocytes in our blood, each works as part of our body's defense system against invading organisms or substances. We can liken them to soldiers on constant patrol ready to attack an invader.

The most abundant type of leucocyte is the neutrophil. These are distinguished by having a polymorphic nucleus (meaning its nucleus can be in many different shapes). These white blood cells are known as the attackers as they literally attack and destroy invading microorganisms.

The next most abundant type of leucocyte is the lymphocytes. These white blood cells are the ones which create antibodies or immunoglobulins which we discussed in Lesson 5 and again in Lesson 18 where we learned about vaccines.

Antibodies can be thought of as "weapons with a memory" in that they are created through exposure to specific invaders known as antigens. During your first exposure to an antigen, your lymphocytes create antibodies which continue to be present in your circulation for many years. Should you get exposed to the antigen again, these "weapons with a memory" immediately attack the antigen, hopefully preventing you

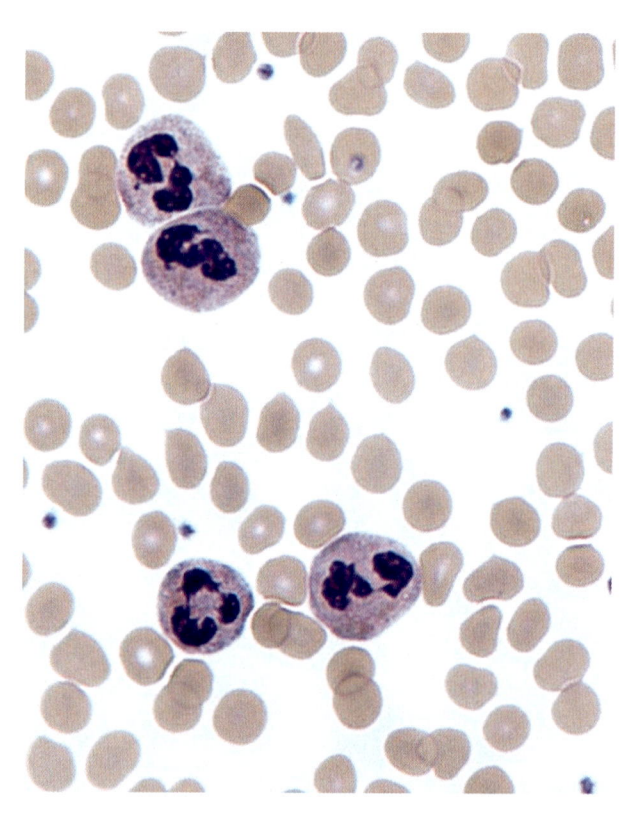

After staining, you can readily distinguish neutrophils which stain purple and have "lumpy" nuclei. The pink-staining cells are erythrocytes.

from becoming ill. As we learned earlier, the mechanism of vaccination works in this way in that the vaccine is a purposeful, mild exposure to the antigen. The intent is for your body to begin building its "arsenal" of "weapons with a memory" in the case you get exposed to that antigen later in life.

Eosinophils (ee-oh-sin-oh-fils), basophils and mast cells round out the remaining leucocytes found in the blood. These white blood cells contain little vesicles containing histamine as well as other chemicals. Histamine is the chemical which, when released by these cells, triggers your body's response to an invading organism or substance known as an antigen. A good example of histamine release is when you are bitten by a mosquito. The saliva that the mosquito injects into your skin triggers these white blood cells to release histamine. The histamine causes capillaries in the area to "open wide" resulting in extra flow of plasma (the watery portion of the blood) to the area which results in

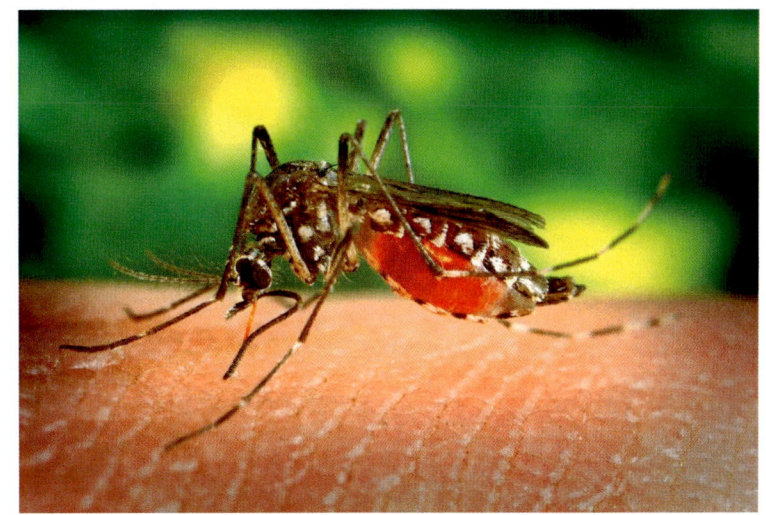

The saliva of the mosquito triggers eosinophils to release histamine. Histamine results in dilation of nearby capillaries which results in swelling and redness about the site. The sensation of itchiness is also a result of histamine.

swelling. Consequently, you see a welt develop around the bite. Histamine also triggers the sensation of itchiness so you respond by scratching the site of the bite. Increased blood flow to the area results in the redness you see around the bite. This response of pain/itchiness, redness and swelling is known as inflammation.

An overreaction to an invasion by an antigen is known as an allergy. In these situations, these white blood cells release a massive dose of histamine which can result in common allergy symptoms like runny, itchy eyes, a runny nose or a rash. Treatment with medications known as antihistamines can alleviate these symptoms.

In some cases, however, in persons who are highly allergic to antigens, such as bee stings or peanuts, for example, a super massive "dumping" of histamine from eosinophils takes place. This can cause the tongue to swell and air passageways to constrict resulting in extreme difficulty in breathing which may result in suffocation. Rapid treatment with epinephrine (Epi-pen) is required to reduce the constriction of the airways to restore adequate oxygen supplies to the body. Asthma attacks are often the result of these types of reactions as well.

Often grouped with white blood cells of the blood are much smaller cells known as platelets. Platelets, along with the fibrin proteins we discussed in Lesson 5, are responsible for allowing blood to clot.

Before we leave our discussion of blood, let's take a brief look at the unfortunate situation of carbon monoxide poisoning. You are likely familiar with the term carbon monoxide which is a by-product found in the burning of fossil fuels such as gasoline that we use in our cars as well as methane or propane used to heat our homes. Note that this chemical compound is carbon MONoxide and not carbon DIoxide (waste compound from cellular respiration).

The reason carbon monoxide is dangerous is because it has the ability to readily "jump on board" hemoglobin molecules arriving on red blood cells within capillaries within the lungs. In fact, carbon monoxide can be absorbed even more quickly than oxygen atoms can. By "filling these slots" intended for oxygen on hemoglobin molecules, carbon monoxide decreases the overall level of oxygen in the blood and cells within the body are affected. The cells of the brain suffer most quickly and sleepiness ensues. This results in less coherence of the need to get to "fresh" air which leads to unconsciousness and, eventually, death.

Because carbon monoxide is odorless and invisible, it can collect in closed spaces without our knowing. Carbon monoxide detectors are an inexpensive means of detecting the presence of this toxic gas in our homes. Cars running in a closed garage or having exhaust pipes covered in a snow drift can quickly produce enough carbon monoxide to be deadly. Opening the garage door and ensuring the exhaust pipe is open to the air outside a running vehicle can save your life.

We've covered quite a bit in this lesson on the circulatory system. Let's pause here and review the major points we've discussed. We learned that:

- The means of delivery of oxygen as well as glucose to our body's cells is the circulatory system;
- Capillaries are the smallest blood vessels;
- Capillaries have openings known as fenestrations which can be thought of like "windows" which allow small items like oxygen, glucose and water to readily pass across;
- Carbon dioxide is a waste product of cells and readily moves through the fenestrations of the capillaries from the cells back into the blood to be carried to the lungs to be removed from the body;
- The size of these capillary fenestrations vary from place to place through the body;
- Capillaries are fed by arterioles which are fed by arteries which are fed by the aorta;
- In arteries and arterioles, smooth muscles encircle the artery and work to control the diameter of the blood vessel;
- By contracting or relaxing smooth muslces, the degree of blood flow to parts of the body can readily be regulated;

- The aorta is the main blood vessel to leave the heart carrying blood out to the body;

- The scientific name for "rooms" in the heart are chambers;

- The atria of the heart are where blood is received from the body;

- The right atrium receives blood returning from the body while the left atrium receives blood returning from the lungs;

- From the right atrium, blood moves down into the right ventricle and the valve it passes through is the right atrioventricular valve or right AV valve;

- The muscle which makes up the heart is known as cardiac muscle or myocardium;

- Blood leaves the right atrium via the pulmonary artery on its way to the lungs and blood returns from the lungs via the pulmonary vein;

- Returning blood, now high in oxygen and low in carbon dioxide, enters the heart through the left atrium;

- Blood moves down into the left ventricle through the left AV valve and through strong muscular contraction the left ventricle forces blood up through the aorta to all parts of the body;

- As the trachea nears the lungs it divides into smaller branches known as bronchi which in turn branch into bronchioles which eventually end in tiny grape-like clusters known as the air sacs or alveoli (al-vee'-o-li) and it is at this level that transfer of oxygen and carbon dioxide takes place;

- The heart muscle itself is nourished via the coronary arteries and veins;

- Compromise of the coronary arteries and veins may result in a heart attack or, scientifically, a myocardial infarction;

- On the venous side, capillaries drain into venules which drain blood into small veins which in turn drain blood into larger veins which eventually drain into the largest vein known as the vena cava which drains into the right atrium of the heart;

- Veins have a much thinner wall than arteries and actions of skeletal muscles near veins contribute to flow of blood through the veins back to the heart (also, unlike arteries, veins have small valves to maintain correct direction of blood flow);

- The watery portion of blood is known as plasma or serum;

- The cells which give blood its red color are known as red blood cells or erythrocytes (ee-rith-ro-sites);

- The primary job of the red blood cells is to carry oxygen to cells in the body and this capability is due to a special protein found within erythrocytes known as hemoglobin;

- The decrease in the functioning number of red blood cells is known as anemia;

- Leucocytes function as part of our body's defense system against invading organisms;

- The most abundant type of leucocyte is the neutrophil;

- Lymphocytes are white blood cells which create antibodies or immunoglobulins which work as weapons for later attack of invaders;

- Eosinophils, basophils and mast cells work to mediate inflammation in the body;

- And, finally, carbon monoxide causes problems in our body because it replaces oxygen being carried by hemoglobin within red blood cells causing the body's brain cells to starve for oxygen which results in loss of consciousness and eventual death.

Lesson 23: Acquisition of Energy Sources

In the past two lessons we've explored the systems of our bodies which allow for movement (skeletal, muscular and nervous systems) as well the systems which deliver oxygen, glucose and other nutrients to these systems (the respiratory and circulatory system). We've learned that all living things require a source of energy and we know that in humans, this energy ultimately comes in the form of glucose. In this lesson we'll focus on the digestive system whose function it is to take the food we eat and convert it into glucose to be delivered to cells by the circulatory system.

If we consider the origin of the word "digest," we find that it means to divide or break down into smaller parts. This breaking down into smaller parts begins immediately when a forkful of food lands in your mouth. Skeletal muscles in your lips, cheeks and tongue allow you to take bites of food into your mouth to initiate the process of digestion. By moving the bite (scientifically known as a bolus of food) left to right and forward and backward you are able to utilize your teeth to mechanically break the food into smaller pieces. While this chewing is taking place, saliva (spit) from salivary glands found beneath your tongue and just below your ears is mixed with the bolus to make it easier to swallow. The enzyme amylase is also present in saliva. Can you recall the carbohydrate that gets lysed (chopped-up) by the action of amylase? Hopefully, amylose comes quickly to mind. Amylose, as you recall as the scientific name for starch, is immediately "attacked" by amylase in your saliva.

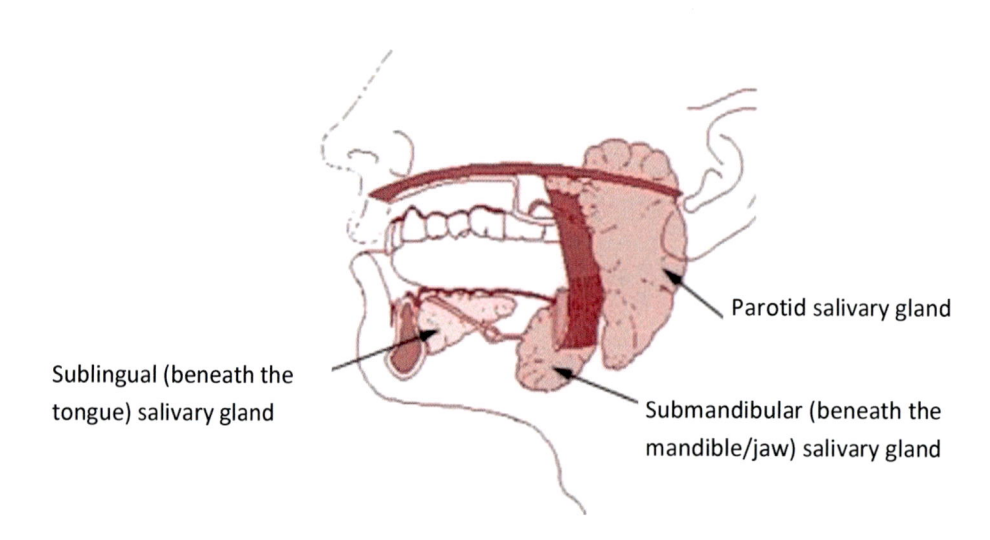

DIGESTION BEGINS IN THE MOUTH WHERE SALIVA IS MIXED WITH THE BITE OF FOOD.

Once the bolus of food is well-chewed and mixed with saliva, the action of swallowing (also the result of work of skeletal muscles) takes place. The bolus then finds itself traveling down through a tube known as the esophagus. Movement down through the esophagus happens due to the action of smooth muscles in the wall of the esophagus. Up to this point, digestion was taking place due to the action of skeletal muscle action. Beginning at the esophagus and continuing almost to the very end of the digestive tract, movement of the bolus will occur due to smooth muscle action. The rhythmic movement through the digestive tract is known as peristalsis.

When the bolus arrives at the lower end of the esophagus, it encounters the door-like opening of the stomach. This "front door" of the stomach is called the cardiac sphincter. The name cardiac refers to the heart indicating this is the end of the stomach nearer the heart. A sphincter is a set of muscles which encircle an opening. When contracted, these muscles can effectively open and close the stomach's opening. When the bolus of food arrives, the sphincter muscles relax and the food moves into the stomach.

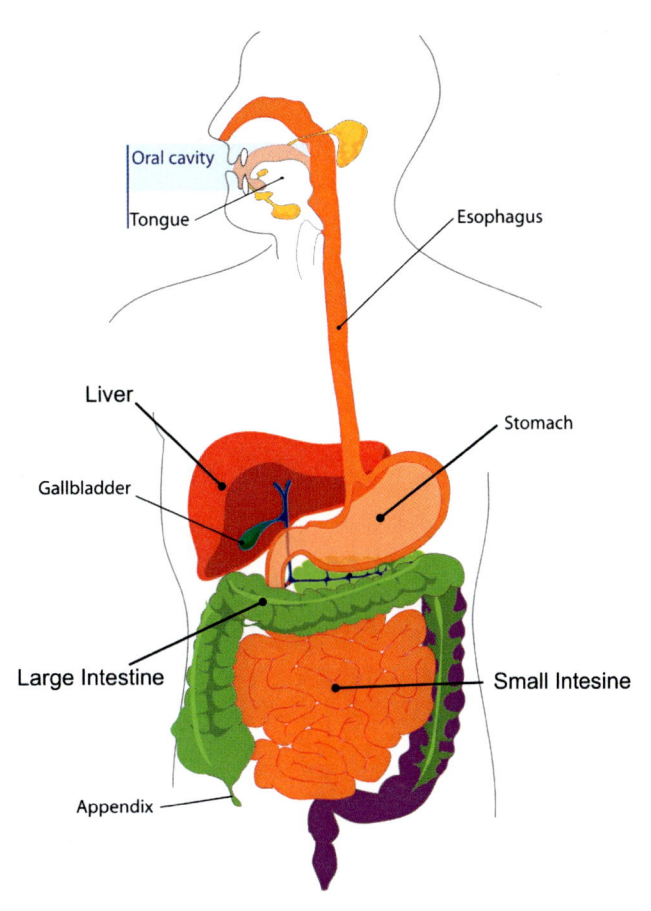

AFTER SWALLOWING, FOOD MOVES DOWN THE ESOPHAGUS TO THE STOMACH.

Like the esophagus, the stomach consists of smooth muscles which work to mix the food with the secretions produced by cells in its inner lining. There are primarily two types of cells on the lining: the parietal (puh-rye-a-tul) cells and the chief cells. The chief cells may also be called the peptic cells. Recall from Lesson 6 that we learned about pH. We discussed that the very strong hydrochloric acid (also known as gastric acid) is produced in the stomach. The cells which produce this acid are the parietal cells. The presence of the gastric acid works to break apart bonds held between the carbohydrates, lipids and proteins of the food bolus.

The chief cells produce pepsinogen which, when it comes into contact with the hydrochloric acid made by the parietal cells, converts into pepsin. Pepsin is a proteolytic (protein cutting) enzyme which breaks peptide bonds between amino acids. This results in proteins getting "chopped-up" into individual amino acids. Another secretion of the chief cells is the enzyme gastric lipase. Do you recall upon which set of compounds a lipase has it action upon? If you said lipids, you are correct. So, together, the secretions of the parietal cells and chief cells continue the digestive process upon the food we eat.

An interesting use for pepsin is in the making of cheese. The pepsin is gathered from the lining of hog (or sometimes calf stomachs) and is given the name rennet. When placed into warmed milk, the rennet acts upon the milk proteins causing them to coagulate or curdle. These curds are then separated from the watery portion (the whey) and are pressed and dried to make cheese.

Recall, also, from our discussion on pH, how the lining of the stomach is protected from the harshness of these enzymes and acid through the secretion of mucus. After all, the stomach itself is made up of proteins and would readily digest itself had it not this layer of mucus cover it. The mucus is produced by epithelial cells of the stomach lining known as faveolar cells.

When the bolus of food that has been mixing about in the stomach is ready to continue down the digestive pathway it collects at the "back door" of the stomach known as the pyloric sphincter. Like the cardiac sphincter, the pyloric sphincter controls the emptying of the stomach contents into the next organ of the digestive tract which is the small intestine.

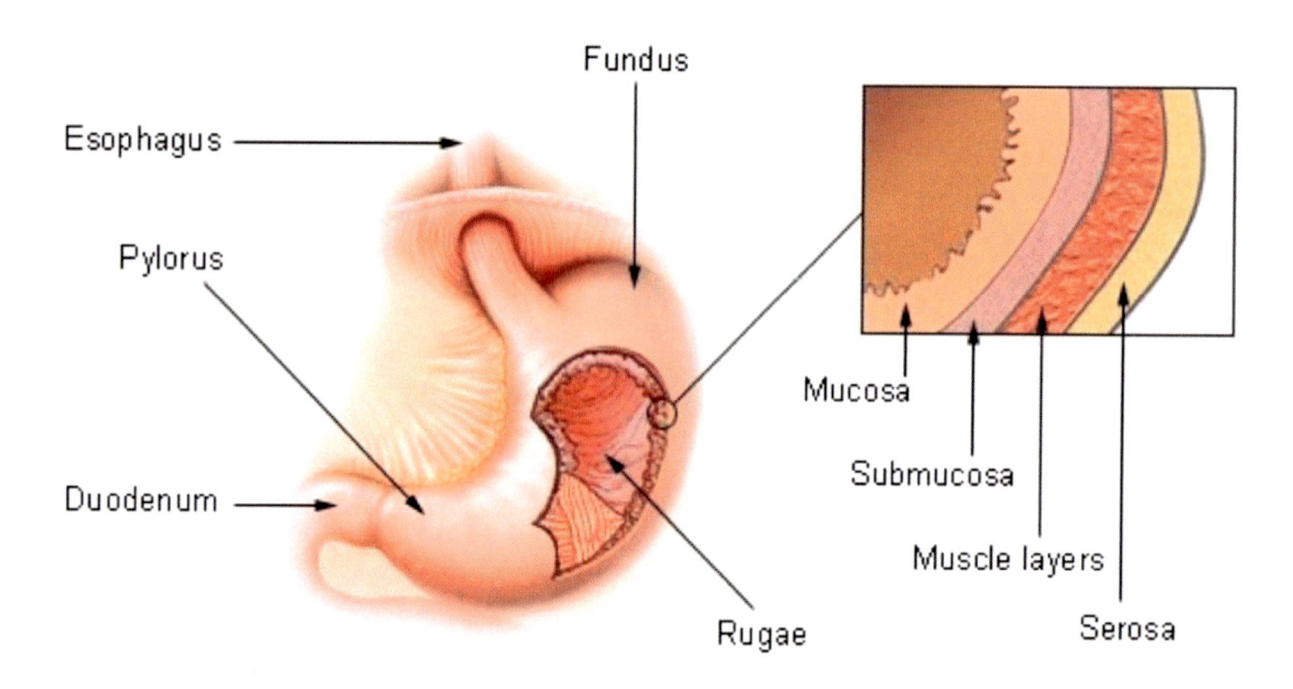

HYDROCHLORIC ACID AND ENZYMES ARE PRODUCED BY CELLS LINING THE STOMACH TO CONTINUE THE DIGESTION PROCESS

Up to this point, work done by the digestive tract has been to break down the food material that we have eaten. In the mouth, it was mechanically broken apart and starches were being acted upon by amylase. In the stomach, it was primarily chemical action which continued the dividing process. As we move into the small intestine, we'll begin to find that the food is now in small enough molecules that it can be absorbed across into the blood supply. Let's look more closely now at the structure and function of the small intestine.

The small intestine can be divided into three segments. The first segment immediately following the pyloric sphincter (and is about 20-15 cm long) is the duodenum (doo-odd-duh-num). Positioned right alongside the duodenum is another organ which has direct influences upon the digestive process. This organ is the pancreas. The pancreas has two main functions with cells designed specifically for these two jobs. The first is the secretion of the hormones insulin and glucagon (which we'll discuss is greater detail in Lesson 24) and the second function is the secretion of additional digestive enzymes. These enzymes are delivered through small tubes leading from the pancreas directly through the wall of the duodenum. As food, now scientifically called chyme (kime, as in lime) passes by, these enzymes have their actions upon proteins, lipids and carbohydrates.

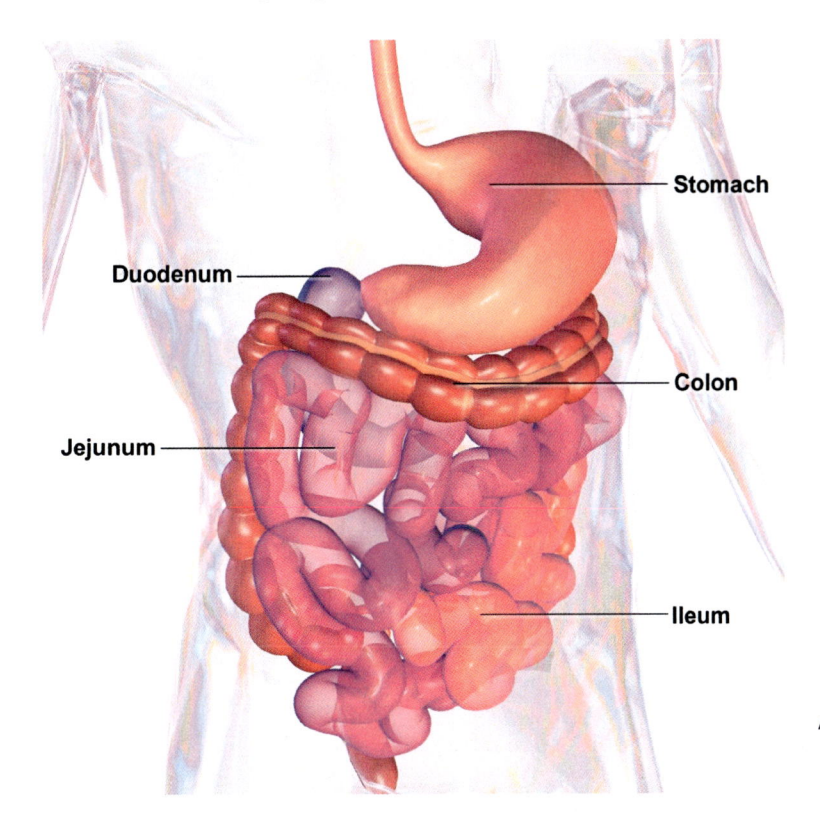

FOOD LEAVES THE STOMACH AND BEGINS ITS TREK THROUGH THE SMALL INTESTINE.

Artwork courtesy Blausen Gallery 2014

Also, in the duodenum we find additional tubes delivering yet another set of substances into the chyme. These substances, known as bile, originate in the liver but are stored in the gall bladder until fatty foods in the chyme move by. The bile contains what are known as bile salts which have the capability of attaching themselves to molecules of fats causing them to break down into separate fatty acids and monoglycerides which are readily acted upon by enzymes coming from the pancreas. This enables the fat to be more readily absorbed into the blood stream. Because the bile is basic in nature with a pH of 7-8 it also works to neutralize the highly acidic chyme which has just exited the stomach. Bile also contains remnants of red blood cells which were being recycled by the liver. This substance is known as bilirubin and is considered a waste product which continues on its path to be removed from the body through the digestive tract.

The second portion of the small intestine is known as the jejunum. The jejunum is the longest section of the small intestine and extends to a length of approximately 2500 cm. Because of its length it coils over itself many times in the abdominal cavity. If you examine the internal lining of the jejunum you will find it is covered with little finger-like projections known as villi (vill-eye). Upon closer, microscopic examination, you would find that each villi itself is covered with even tinier projections known as microvilli. The presence of these millions of microvilli greatly enhances the surface area within the jejunum to allow for maximum absorption of nutrients across the intestinal wall.

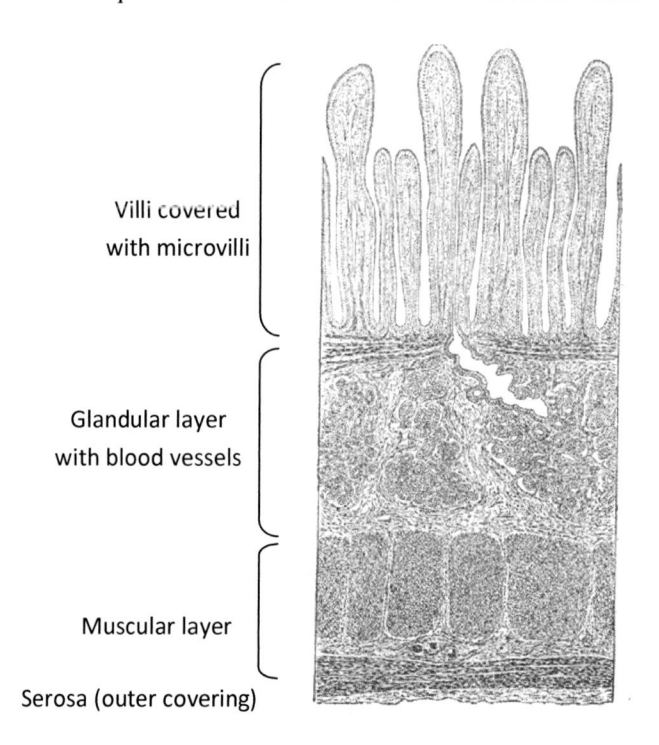

THIS IS A CROSS SECTION OF THE WALL OF THE SMALL INTESTINE AS SEEN USING A MICROSCOPE.

- Villi covered with microvilli
- Glandular layer with blood vessels
- Muscular layer
- Serosa (outer covering)

Within each villi we find capillaries "eager" to pick up newly arriving nutrients. These capillaries carry these nutrients into larger veins which eventually drain into a very large vein known as the hepatic portal vein. The hepatic portal vein then drains directly into the liver. The liver then works to make sure nothing that has been absorbed from the digestive tract is harmful to the body before allowing the food nutrients to enter the main circulatory system. You might think of this job of the liver as being like a very important filter which prevents unwanted items from entering circulation.

In addition to the production of bile and filtering of incoming substances from the digestive tract, the liver is known to have close to 500 specific, vital functions in the body. The cells which make up the liver are known as hepatocytes (where the prefix hepat- refers to the liver). You may have heard of hepatitis which refers to inflammation of the liver. Recall from Lesson 5 that we learned that albumin was a major protein found in our blood which helps regulate the water content of the blood. Albumin is made in the liver!

Another very important function of the liver is the production of glycogen. Recall from Lesson 3 that you learned that plants have the capability of storing glucose as starches or amylose. In humans, including animals, glucose is stored as glycogen in the liver. In times of need, this stored glycogen is converted through the use of enzymes into glucose which can then be delivered out to cells in the body. With all of these vital functions a healthy liver is extremely important to living a long life.

Let's move back now to our trek through the digestive tract. The third and final segment of the small intestine is known as the ileum. The ileum is the shortest segment of the small intestine and serves as the connecting point with the next portion of the digestive tract which is the large intestine. At this junction there is another portion of the digestive tract present which has yet to be fully understood. This organ is the cecum, commonly known as the appendix. The appendix in humans is usually only 9 cm long and has a blind end, meaning it goes nowhere. The food material that enters the appendix must come back out through the same opening. For various reasons, the appendix may become inflamed which is known as an appendicitis. If the appedix gets to the point of breaking open (rupturing), the contents of the digestive tract (which is normally filled with billions of bacterial organisms) can spill out among other organs in the abdomen. This can result in a massive, life-threatening infection. For this reason, an inflamed appendix is very routinely removed before this serious situation develops.

The chyme now enters into the large intestine, which, as its name implies is much larger in diameter (but not length) than the small intestine. The large intestine, also known as the colon, is approximately five feet long (150 cm) in the adult human. If we examine the lining of the large intestine, we find that it is not covered by villi and microvilli since the primary job of the colon is no longer the absorption of nutrients. The primary role of the large intestine is water reabsorption.

When the chyme first entered the small intestine upon leaving the stomach, a tremendous amount of water was added to dissolve the food particles making them more readily absorbed as the food moved through the small intestine. The consistency of the chyme all the way through the small intestine is much like that of diarrhea. The large intestine reabsorbs about a quart and a half of water every day.

Besides water absorption, another important function of the large intestine is the production and absorption of vitamin K. Normally-present bacteria in the colon digest what remains of the food material and, as a byproduct, produce vitamin K. Recall from Lesson 5 that we learned about the blood protein fibrin. We learned that fibrin was essential for allowing blood to clot. The steps involved in the clotting of blood are directly dependent upon the presence of vitamin K and the large intestine is the source of this vital nutrient.

The remaining unwanted foodstuff is now called feces and continues onward to the rectum. It collects there until nerves are stimulated to initiate a bowel movement. After food entered the esophagus and up to this point, the muscle action moving the food through the digestive tract was all done by smooth muscles. However, a circle of skeletal muscles known as the anal sphincter creates the final door to the digestive tract. When the anal sphincter relaxes, feces leave the digestive tract.

Our journey through the digestive system is now complete. Let's review what we've learned in this lesson. We learned that:

- The main function of the digestive system is to take the food we eat and convert it into glucose to be delivered to cells by the circulatory system;
- A bite of food is known as a bolus;
- Saliva, which contains enzymes, is mixed with the bolus in the mouth to begin digestion;
- Food moves from the mouth to the stomach through the esophagus by the action of smooth muscles;

- The stomach mixes the food with substances produced by the stomach lining's chief and parietal cells (the chief cells create pepsinogen which reacts with hydrochloric acid from the parietal cells to produce pepsin, a proteolytic enzyme);

- The "front door" of the stomach is the cardiac sphincter while the "back door" of the stomach is known as the pyloric sphincter;

- The small intestine is divided into three segments: the duodenum (which is closely associated with the pancreas), the jejunum (where microvilli allow for great surface area for nutrient absorption) and the ileum;

- The pancreas secretes the hormone insulin which regulates the ability of glucose to enter cells of the body and other enzymes which continue the digestive process;

- The liver is known to have close to 500 specific jobs in the body which include the filtration of nutrients arriving from the small intestine as well as the manufacture of albumin and glycogen (the storage form of glucose);

- At the junction of the ileum and the large intestine (colon) is the appendix or cecum;

- The primary role of the large intestine is water reabsorption and the production and absorption of vitamin K which is vital for blood clotting.

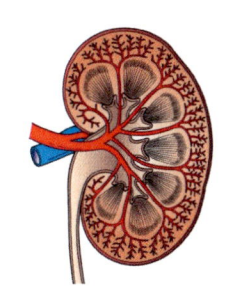

Lesson 24: Systems of Waste Management

In the last lesson, we explored the many organs of the digestive tract and we discussed how nutrients are absorbed from the intestine into circulation for the use of cells all over the body. As cells are constantly utilizing these nutrients, there are "waste" portions of the nutrients which are not needed. There are also unneeded waste byproducts of the many bodily reactions. Additionally, because cells in our bodies are constantly dying and being replaced in many organs, parts of cells will find themselves destined to be removed as waste from the body. In this lesson, we will look at the organ systems which serve the purpose of removing wastes from the body. These systems are the renal and urinary systems. We'll begin first with the renal system.

The renal system is made up of our two kidneys along with associated blood vessels which carry blood to and from them. Because the main job of the kidneys is to filter out unneeded waste products from all parts of the body, it makes sense that blood would be the best substance for getting those wastes to the kidneys. Branching from the lower aorta after it enters the abdomen are two large arteries known as the renal arteries which deliver blood to the kidneys. For the sake of our discussion, we'll refer to this blood being delivered to the kidneys as "dirty" blood and the blood leaving the kidneys via the renal veins as "clean" blood. Let's follow the pathway of the flow of the "dirty" blood into the kidney and see how it becomes "clean" blood.

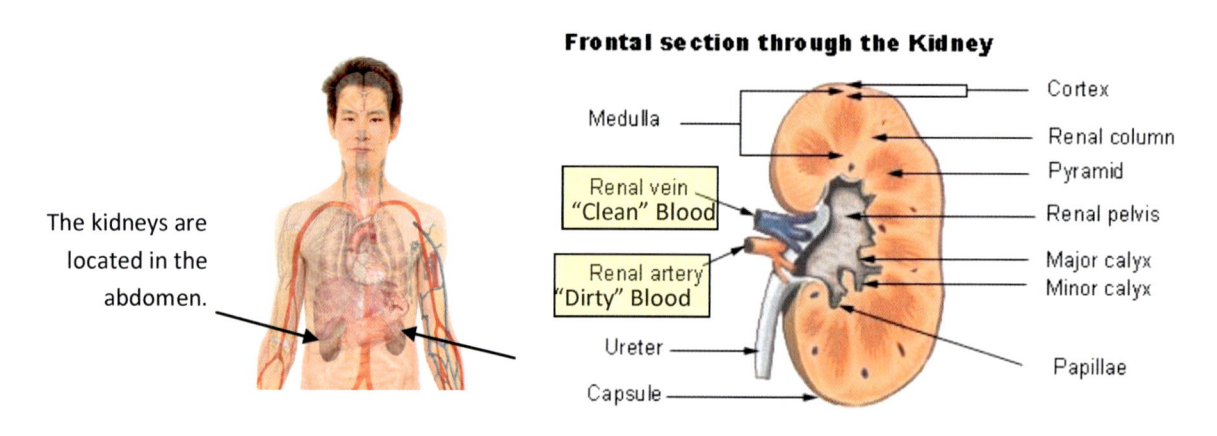

Frontal section through the Kidney

The kidneys are located in the abdomen.

Because the wastes being produced by the cells and dead cell parts are relatively small substances, the filtering which takes place in the kidneys happens at the cellular level. This means that the renal arteries bringing in the dirty blood quickly branch into smaller blood vessels known as the interlobar arteries which branch into the smaller vessels known as the arcuate arteries. The arcuate arteries then branch into afferent arterioles which bring blood to the filtering unit. Efferent arterioles carry blood away from the filtering unit. Look at the diagram here to see this vessel arrangement.

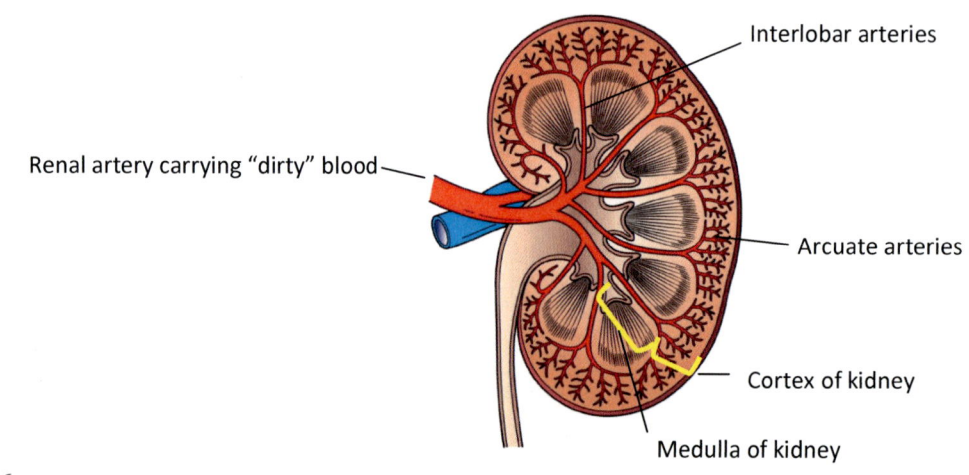

Note also in the diagram the two main layers of the kidney: the outer cortex and the inner medulla. Within the cortex we find the filtering units of the kidneys known as the glomerulae (glow-mare-you-lay, plural). This is the location where filtration actually takes place. There are small openings in the walls of the afferent arterioles which allow wastes to leave the blood. These wastes are collected in the glomerular capsule (also known as Bowman's capsule) and will eventually become urine. The blood then leaves the glomerulus by the efferent arterioles as "clean" blood and continues back to the renal vein where it eventually enters the vena cava and heart for continued circulation.

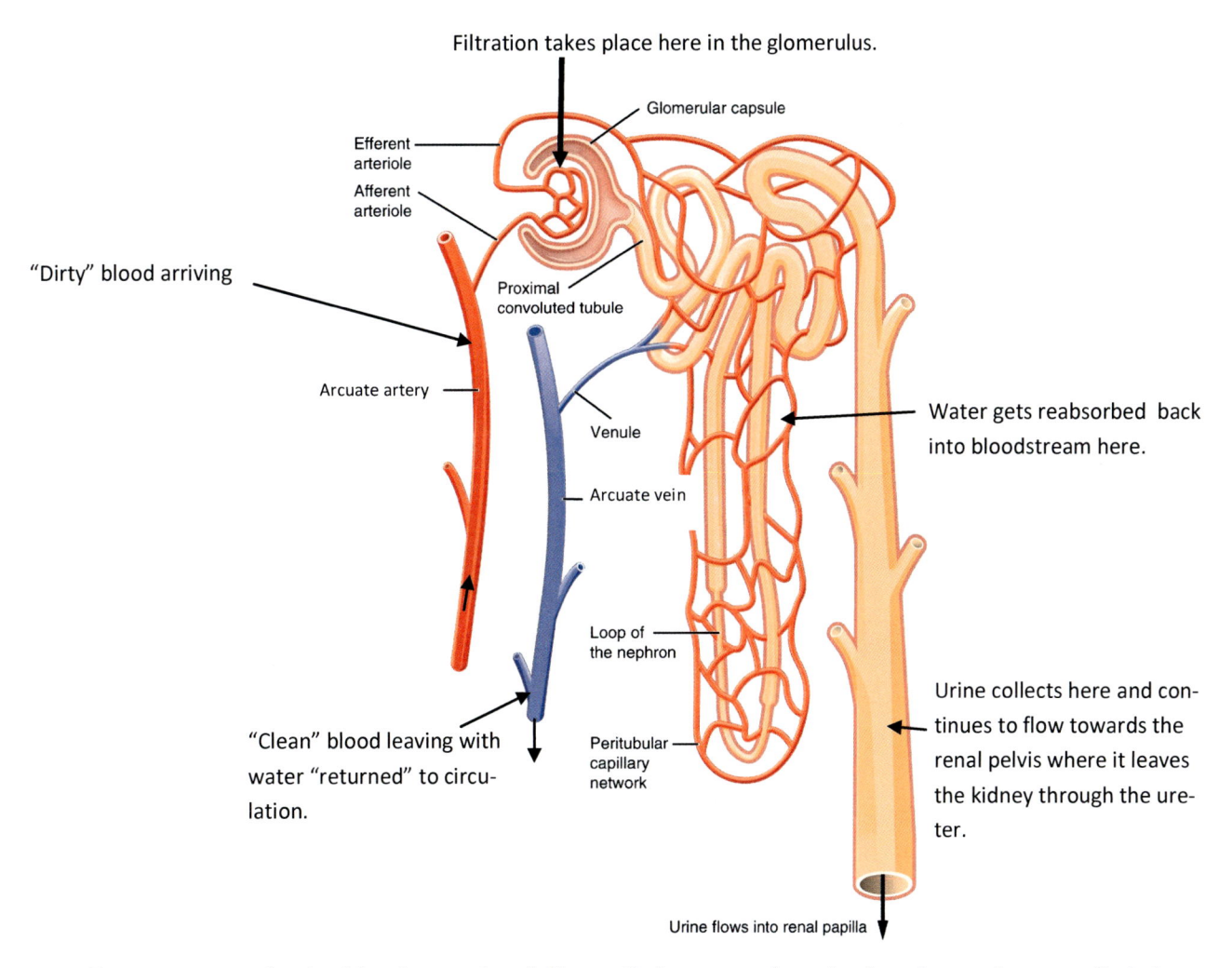

Because water in the blood can also "slip out" these openings in the glomerulus, we find that the wastes that are being collected are in quite a diluted stage initially. By moving through a series of tiny tubules in a loop down towards the center of the kidney (Loop of Henle), this water is reabsorbed back into circulation and not lost into the urine. Several gallons of blood are filtered everyday through our kidneys and if the kidneys did not have the capability of conserving water, we would rapidly dehydrate. This more centrally located area of the kidney where water conservation takes place is known as the medulla. When the bulk of the water is removed from the wastes, the resulting liquid is our urine. While our urine may appear to be "watery" it is actually highly concentrated compared to the volume of liquid that has been filtered.

In addition to filtering blood, the kidneys also work to maintain the appropriate pH of the blood. This is accomplished through the movement of various chemical compounds into and out of the blood as it moves through the kidney tissue. Two other important functions of the kidney are that it produces two hormones which have their effects in other parts of the body. The first hormone is renin which combines forces with angiotensinogen produced by the liver to create angiotensin. Angiotensin has the ability to control blood pressure.

The second hormone produced by the kidney is erythropoietin (ee-rith-ro-poy-ee-tin). The prefix for the name of this hormone, erythro-, refers to erythrocytes or red blood cells. While this hormone is produced by the kidney, its effect is upon blood marrow to stimulate the production of red blood cells. So in cases of anemia where erythrocyte levels have been reduced, the kidney responds by putting erythropoietin into the blood stream which is delivered to the bone marrow to stimulate the production of more red blood cells.

Let's take a closer look now at the urine that has been produced by the kidneys. Normally, the kidney conserves water as well as the very small chemical elements of sodium and potassium which were initially lost through filtration. In addition, we should not find any glucose or proteins in the urine. Recall from Lessons 5 and 22 that we learned that proteins are very large molecules and find it difficult to pass through any fenestrations in the circulatory system which includes the glomerulae of

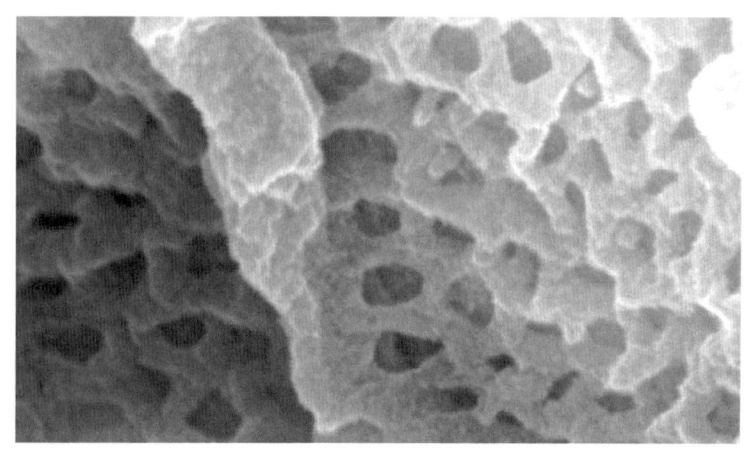

Fenestrations function like "windows" in the walls of capillaries are found within the glomerulae of the cortex of the kidney. In this photo they are magnified 100,000 times.

the kidney. If proteins are found in a urine test (urinalysis) this could mean there is damage to the kidney which has caused it to "leak" proteins into the urine or there is damage to some other component of the urinary tract such as the bladder or urethra.

The presence of glucose in the urine is also another cause for concern. As you are well aware, there is always a relatively constant level of glucose in our blood. Should it become too high, it is said to "spill-over" a filtration threshold of the kidney and be found in the urine. You might think of the glucose level of the blood as being like the lake behind a dam. If the level of glucose in the blood gets too high, the glucose spills over and is found in the urine. This situation is a primary symptom of diabetes. Historically, diabetes was referred to as sugar diabetes as the urine had a sweet taste due to the presence of glucose (sugar) in the urine.

In this photo, you can think of the lake as being the blood supply moving through the kidney. As long as glucose levels in the blood are within a normal range (70-80 mg/dL blood) all glucose continues through the kidney and back into circulation (stays in the lake). If levels get too high, this "overwhelms" the capabilities of the kidney and glucose "spills" over the dam in the spillway and into the river (urine) below. This glucose will then be evident on a test of the urine (urinalysis).

Recall from our earlier discussion of the pancreas that we said that in addition to the digestive enzymes the pancreas produces, it also has cells which produce insulin. Insulin is a hormone which has its effects upon all of the cells in the body. Insulin can be described as being like a key which opens the door of a cell membrane to allow glucose to enter. With insulin present, glucose can get into cells where it used as fuel.

Without insulin, the glucose that has been placed into circulation by the digestive tract cannot move into cells and therefore builds up in the bloodstream. In the meantime, cells all across the body begin to "starve" for glucose. The "groceries" can't get into the cells because the "key" to open the doors of the cell membranes is not present.

Effects of this lack of fuel for cells is noted all across the body. One devastating result is damage to small blood vessels everywhere in the body which leads to poor circulation, especially in the feet and legs of diabetic patients. Damage to the small vessels within the eye also leads to blind-

Insulin, produced by the pancreas, can be thought of as working like a key which opens "doors" within cell membranes to allow glucose to enter. A deficiency of insulin or "broken locks," as in the case where cells don't respond well to insulin, can result in diabetes.

ness. Routine screening of the urine for any substances not normally found can aid in the prevention or management of diabetes.

Let's continue now with the remaining organs of the urinary system. Urine that forms in the medulla of the kidney collects in a central location known as the renal pelvis. It then drains into a thin tube known as the ureter. Each kidney has its own ureter. The ureter carries the urine down to the urinary bladder for temporary storage. It's important to realize that our kidneys are constantly filtering

Frontal section through the Kidney

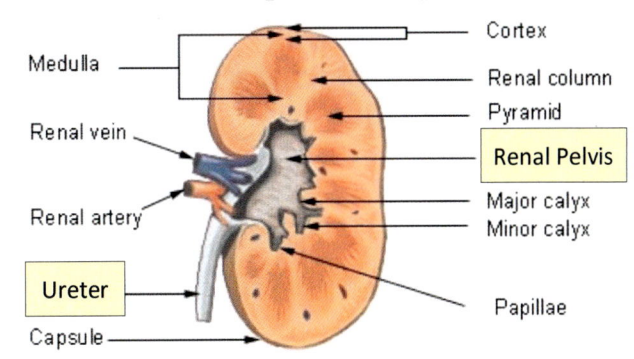

Labels: Medulla, Cortex, Renal column, Pyramid, Renal vein, Renal Pelvis, Renal artery, Major calyx, Minor calyx, Ureter, Papillae, Capsule

our blood, 24 hours a day, seven days a week and, therefore, urine is constantly being produced and delivered to the bladder.

When the bladder begins to reach its capacity, nerves in the wall of the bladder send signals to the brain and spinal cord telling you it's time to take a visit to the bathroom. Urine then is released from the bladder through a tube called the urethra. The urethra carries urine to the exterior of the body.

Urine drains constantly from the kidneys into the urinary bladder. When full, the bladder sends signals to the brain to let you know it's time to go to the bathroom.

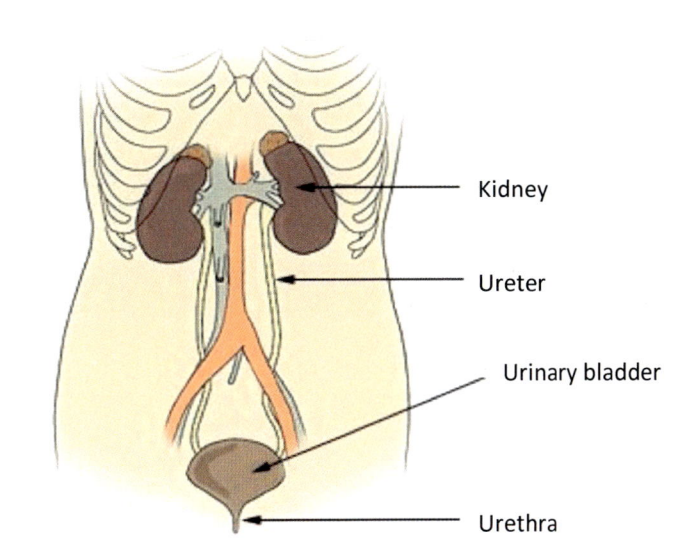

Labels: Kidney, Ureter, Urinary bladder, Urethra

Let's stop here and review what we've learned in this lesson. We learned that:
- The renal and urinary systems of our body work to remove wastes created by cells, as well as cell remnants, from the body;
- The renal system consists of our two kidneys and blood vessels which carry "dirty" blood to and "clean" blood away from the kidneys;

- The filtration unit of the kidney is the glomerulus;
- Waste products, as well as water, sodium and potassium, can leave through fenestrations in the walls of the arterioles within the glomerulus, however, water, sodium and potassium are returned back to circulation and therefore conserved;
- The kidneys also maintain pH of the blood;
- The kidneys produce hormones which control blood pressure and the creation of red blood cells;
- The presence of proteins in the urine indicates possible kidney damage or damage to other parts of the urinary system;
- Glucose in the urine indicates that glucose has exceeded the threshold capacity of the kidney;
- Insulin, produced by the pancreas, functions to open "doors" of cell membranes to allow glucose to enter;
- Lack of insulin can result in high levels of blood glucose, presence of glucose in the urine and "starving" cells all over the body;
- Urine leaves the kidneys through the ureters which drain into the urinary bladder;
- When full, the bladder is emptied to the exterior of the body through the urethra.

Lesson 25: Systems of Growth and Development

In our study of human biology, we've explored the organ systems which allow for movement (which were the skeletal, muscular and nervous systems). We then found that an energy source must be available for these systems to function and we learned about the circulatory, respiratory and digestive systems. To better understand how our bodies deal with waste products from these systems, as well as maintenance of important factors like water levels, pH and chemical element levels, we explored the function of the renal system and the urinary tract.

In this lesson, we'll focus on the organ systems which allow us to grow and develop. Obviously, all of the organ systems in our body are highly dependent upon each other for proper function. The body system which provides for growth and development is no exception as its functions are spread across all organs of the body. The organ system we will focus on here is known as the endocrine system. The endocrine system consists of various glands found in several places of the body.

Glands consist of cells which have the ability to produce substances which are exported from the cell to have effects somewhere else in the body. There are two main types of glands: endocrine glands and exocrine glands. Endocrine glands, which we'll be studying in this lesson, are those which do not have a tube or duct leading away from the gland to deliver its products. On the other hand, exocrine glands, which we'll study in Lesson 27, are those glands which do have a tube or duct leading from it to carry its product to another part of the body. The salivary glands we discussed in Lesson 23 which produce saliva are exocrine glands.

Instead of having a tube to carry away its products, endocrine glands are closely affiliated with the circulatory system. The products of the gland move directly from the gland into the blood stream. An endocrine gland is also capable of monitoring how much of the product it produces by sampling levels in the blood. When the level of its product gets low, cells of the gland are stimulated to begin producing more. This method of monitoring is known as a negative-feedback mechanism.

You might think of this system as being similar to the thermostat in your house which controls the heating system. When the temperature of the house cools down, the thermostat, which is constantly monitoring the temperature, is triggered to start up the heating system. As the temperature of the house rises again, the thermostat senses when the desired temperature is reached and eventually shuts off the heater. Many glands of the endocrine system function in this same way. Others are controlled through the nervous system. Let's begin our study of glands of the endocrine system.

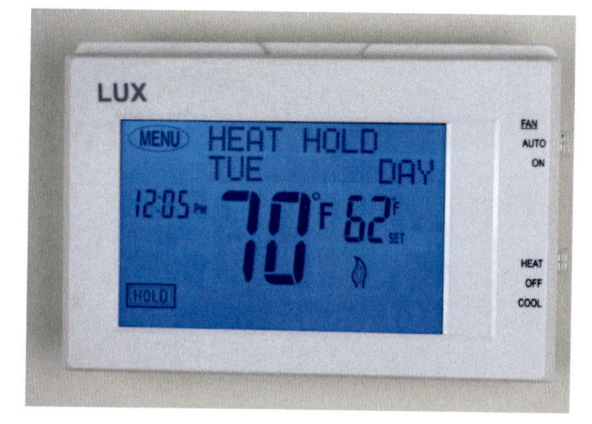

Photo by Dennis Murphy
Glands of the endocrine system respond like a thermostat.

We'll begin our study by first examining the pituitary gland. The pituitary gland is found on the underneath side of your brain and just above the roof of your mouth. It is a small teardrop-shaped gland and is often referred to as the "master gland." It is called the "master gland" because it not only produces substances which affect other organs in the body, but also actually produces substances which control other glands which, in turn, affect other organs in the body. The pituitary gland receives its "orders" primarily from the brain, particularly a segment known as the hypothalamus.

Diagram: A cross-section of the human head showing the Cerebrum, Pituitary Gland, Cerebellum, and Spinal Cord. Patrick J. Lynch, Medical Illustrator.

The pituitary gland is found beneath the brain.

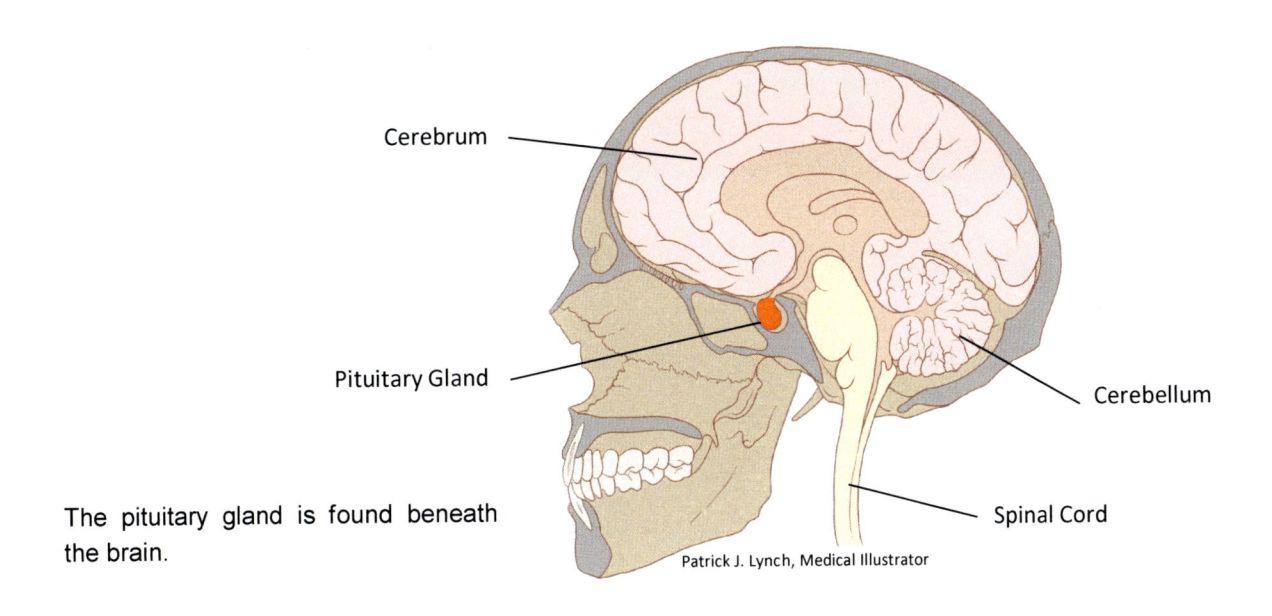

Before we look more closely at the pituitary gland, let's briefly discuss in greater detail the substances produced by endocrine glands. Products of endocrine glands are known as hormones. In our discussion of the digestive tract, we learned about the hormones insulin and glucagon produced by the pancreas. We learned that insulin worked like a "key" to open cell membranes to absorb glucose. We learned that glucagon had its effect upon the liver to release more glucose into the blood stream. Hormones are made by glands in one part of the body destined to control activity in another area of the body. Like insulin, a hormone is like a "key" which has an associated "lock" or receptor site somewhere else in the body. While most hormones are made from proteins, there are some hormones which are built from lipids. These hormones are known as steroids.

Let's take a closer look now at the pituitary gland, the "master gland." The pituitary gland can be divided into two parts: the front side known as the anterior lobe and the rear side known as the posterior lobe. Specific hormones are produced in each of these lobes. Let's look at the hormones of the anterior lobe first.

There are five main hormones or sets of hormones produced by the anterior lobe of the pituitary gland. The first we'll discuss is known as thyroid-stimulating hormone or TSH for short. As its name says, this hormone has its effect upon the thyroid gland, another endocrine gland. The thyroid gland is found on the front side of your neck. Thyroid-stimulating hormone stimulates the thyroid gland to produce two hormones known as T3 and T4 which, in turn, have their effects upon all cells of the

Thyroid Gland

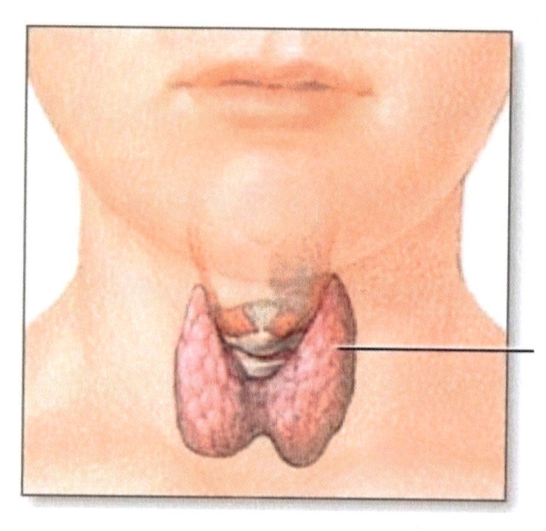

The thyroid gland is found on the front side of the neck.

body. The primary role of T3 and T4 is regulating the rate of metabolism of the cells. Metabolism is the overall process whereby cells use glucose to produce energy. You can think of T3 and T4 as being like the accelerator in your car. The secretion of more T3 and T4 results in greater pressure on the accelerator and therefore increases the rate of metabolism.

The next hormone we'll explore is one that, like thyroid-stimulating hormone, acts directly upon another endocrine gland in the body. This hormone has a complicated name: adrenocorticotropic (uh-dree-no-cor-tic-oh-tro-pic) hormone or ACTH. Let's break its name apart to reveal where ACTH has its effect. Note first the root word in the name, -tropin, which means to stimulate or activate. The prefix adreno- tells us it's the adrenal gland which receives the stimulus of ACTH and, more specifically, it's the adrenal cortex or outer layer (cortico) where the receptors of ACTH lie. So, ACTH stimulates cells in the outer layer or cortex of the adrenal glands. Humans have two adrenal glands, one atop each of our kidneys.

Adrenal gland

The adrenal glands sit on top of each kidney.

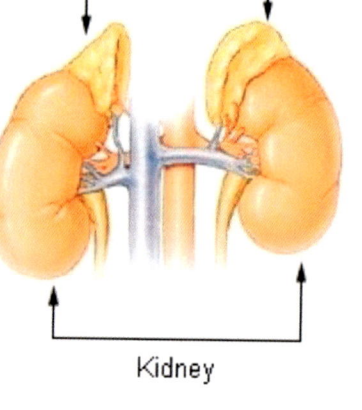

Kidney

In response to ACTH, the adrenal cortex of each adrenal gland produces two main sets of chemicals: cortisol and aldosterone. Cortisol is the body's natural anti-inflammatory agent. So, if you have inflammation, which means you have pain, swelling or heat at a specific location in your body, cortisol released by the adrenal cortex works to keep this inflammation under control. You may be familiar with cortisone creams or ointments used to reduce skin rashes or hives. Prednisone is another member of this group of antinflammatory agents which are commonly known as steroids.

The second hormone released by the adrenal cortex is aldosterone. The receptor site for this hormone is the kidney and, more specifically, the tubules within the kidney which are involved in conservation of the sodium, potassium and water molecules. Through its role of conservation of these substances, aldosterone plays a major role in regulation of blood pressure.

While we are discussing the adrenal gland, we'll momentarily move away from our discussion of hormones and take a brief look at the remaining portion of the adrenal glands. In addition to the outer adrenal cortex, the adrenal glands also have an inner layer of tissue known as the adrenal medulla. The

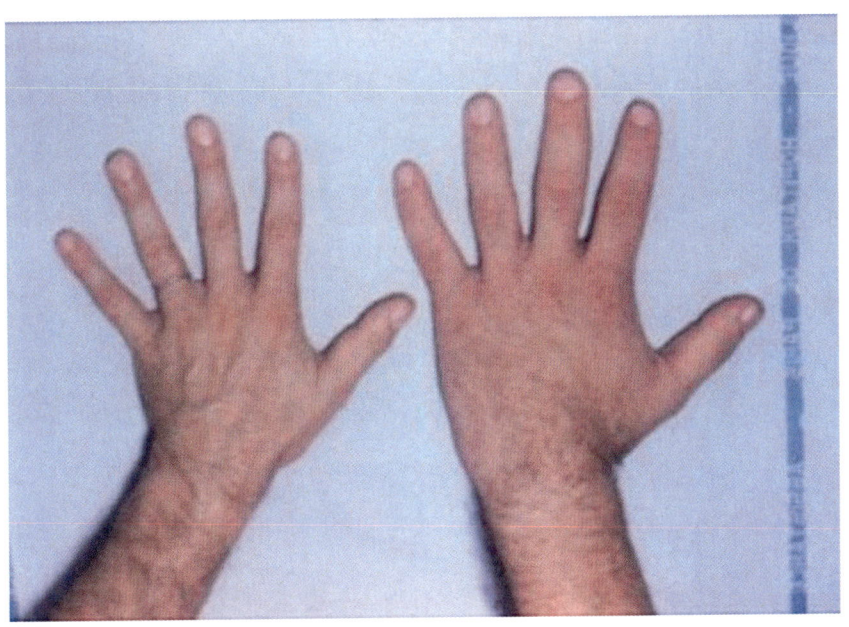

The hand on the left is that of a person with normal amounts of growth hormone. The hand on the right shows symptoms of having excessive growth hormone (acromegaly).

Philippe Chanson and Sylvie Salenave, *Orphanet Journal of Rare Diseases* 2008, 3:17.

adrenal medulla is responsible for producing epinephrine, commonly known as adrenaline. Epinephrine is the main neurotransmitter which quickly prepares the body in moments of serious crisis, known as the fight-or-flight (or sympathetic) response. In situations of extreme emergency such as an accident or encounter with something like a ferocious tiger or bear, the brain sends high-priority signals to the adrenal medulla to immediately pump out epinephrine which, in turn, stimulates the heart to beat fast with greater force, the lungs and airways to open wide, the pupils of the eye to dilate to allow for greater vision, all in an effort to either face the emergency situation "head-on" (fight) or run from the situation with great ability (flight). So, the adrenal gland with its two layers of tissue provides anti-inflammatory agents as well as neurotransmitters for dealing with difficult, emergency situations.

The next hormone we'll look at which is produced by the anterior lobe of the pituitary gland is known as growth hormone. As its name says, this hormone is responsible for growth throughout the body. The main receptor sites for growth hormone are the bones and muscle cells. A deficiency in growth hormone can result in dwarfism while excessive growth hormone results in a condition known as gigantism or acromegaly.

Prolactin is the next hormone we'll discuss that is produced by the anterior pituitary gland. If you look closely at its name you will note the root word lact-. Recall from Lesson 3 that the name given to milk carbohydrates was lactose. The receptor sites for prolactin are the mammary glands or breast tissue in females. Prolactin is responsible for development of the breasts as well as continued milk production while a baby is nursing.

The final set of hormones which arises from the anterior pituitary gland is collectively known as the gonadotropins (go-nad-oh-tro-pins). The root word here again (tropin) means to stimulate or activate. The prefix tells us the organ(s) being stimulated. In this case it is the gonads which are being stimulated. Gonad is a term used for the reproductive organs both of females and males. So, this set of hormones from the anterior pituitary gland includes those which stimulate the reproductive organs of both men and women.

Before we look at this set of hormones, let's go back one step to learn that it's a hormone from the hypothalamus (base of brain) which initially stimulates the pituitary to release the gonadotropins. This hormone from the hypothalamus is called gonadotropin-releasing hormone or GnRH. So, the hypothalamus releases GnRH which stimulates the anterior pituitary to release gonadotropins.

There are two gonadotropins from the anterior pituitary gland that we'll examine. The first gonadotropin is called follicle-stimulating hormone or FSH. In women, this hormone has its effect upon the ovary where it stimulates an ovum (egg) to develop inside a follicle. A follicle is a blister-like structure on the surface of the ovary which contains the ovum. Follicle-stimulating hormone stimulates

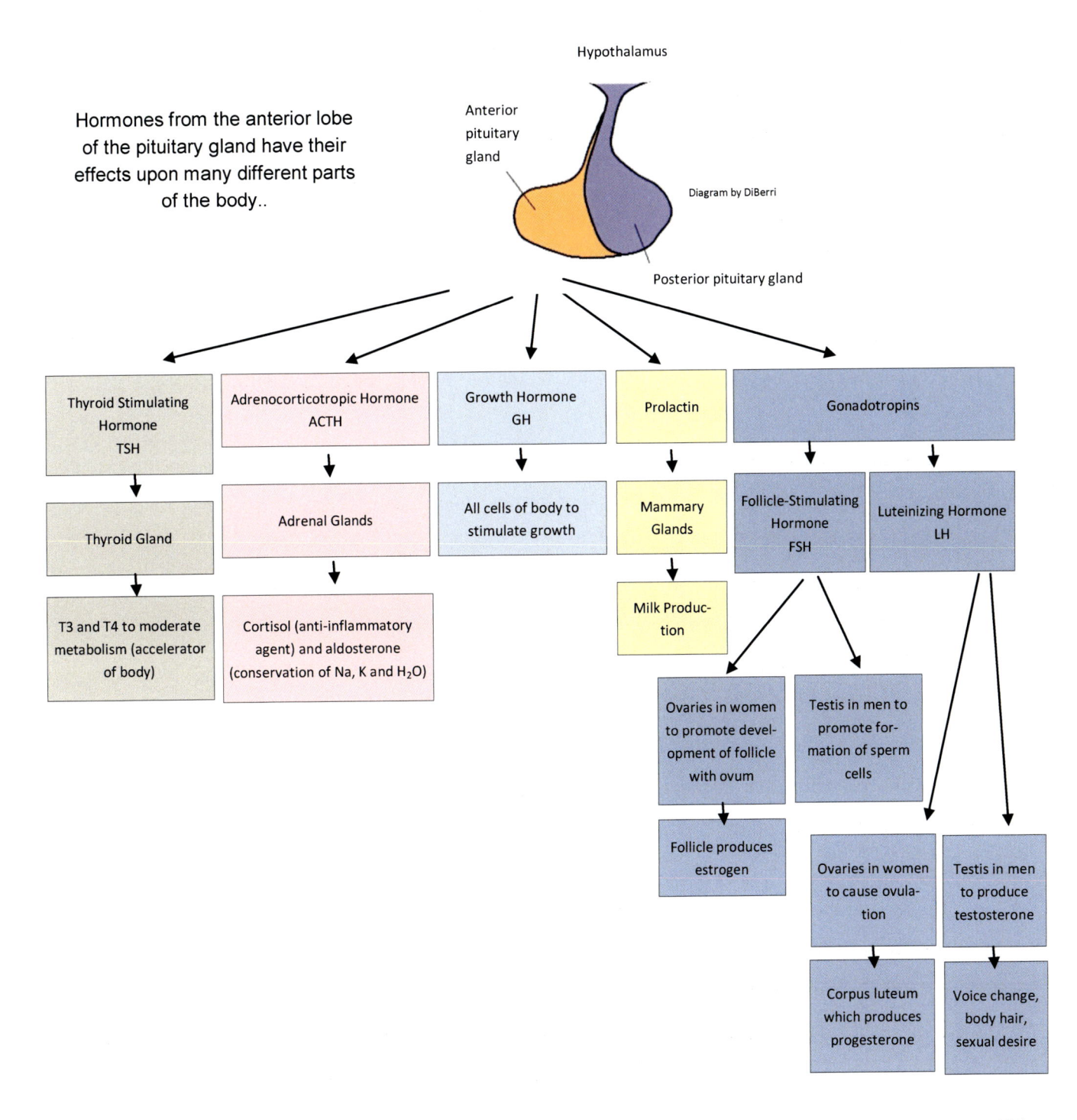

Hormones from the anterior lobe of the pituitary gland have their effects upon many different parts of the body..

Hypothalamus
Anterior pituitary gland
Posterior pituitary gland
Diagram by DiBerri

- Thyroid Stimulating Hormone TSH → Thyroid Gland → T3 and T4 to moderate metabolism (accelerator of body)
- Adrenocorticotropic Hormone ACTH → Adrenal Glands → Cortisol (anti-inflammatory agent) and aldosterone (conservation of Na, K and H₂O)
- Growth Hormone GH → All cells of body to stimulate growth
- Prolactin → Mammary Glands → Milk Production
- Gonadotropins
  - Follicle-Stimulating Hormone FSH
    - Ovaries in women to promote development of follicle with ovum → Follicle produces estrogen
    - Testis in men to promote formation of sperm cells
  - Luteinizing Hormone LH
    - Ovaries in women to cause ovulation → Corpus luteum which produces progesterone
    - Testis in men to produce testosterone → Voice change, body hair, sexual desire

the follicle to grow, usually one each month of a women's reproductive cycle. The follicle itself also produces a hormone known as estrogen. Estrogen in animals is responsible for the changes in behavior noted when a female is said to be "in heat." In men, FSH stimulates primordial germ cells in the testis to develop into sperm cells.

The second gonadotropin produced by the anterior pituitary gland is called luteinizing hormone or LH. The target organ for LH is also the ovary in women and the testis in men. In women, a surge in LH causes the developing follicle on the ovary to rupture, a process known as ovulation. Ovulation allows the ovum to be released to begin its travel down the reproductive tract with the intention of joining a sperm cell. Following ovulation, the cells that made the follicle are replaced by cells known as the corpus luteum (CL). These cells, in turn, begin producing yet another hormone known as progesterone. Progesterone is the hormone responsible for maintaining pregnancy. In other words, once the ovum is fertilized by the sperm and begins development, it's progesterone, produced by the CL, which "quiets" the reproductive system and allows the woman to remain pregnant until it's time for delivery of the baby.

In men, as we mentioned above, LH has its effects upon the testis. Between the tiny tubules which house the developing sperm cells (seminiferous tubules), there are cells known as interstitial cells (literally meaning the "in-between" cells). These cells, when stimulated by LH from the anterior pituitary gland, respond by producing testosterone. Testosterone, also a hormone, has its effect upon different areas in a man's body. Influence of testosterone includes changes in the larynx (voice box) which results in the lowering of the voice, growth of body and facial hair and the desire for sexual activity.

Up this point, all of the hormones that we have been discussing arise from the anterior lobe of the pituitary gland and have their effects upon other parts of the body. Let's complete our study of the hormones of the pituitary gland by looking at the two hormones which arise from the posterior lobe of the pituitary gland.

The first hormone is known as vasopressin. Another name for this hormone is antidiuretic hormone. The target organ of this hormone is the kidney and, more specifically, the tubules of the kidney which conserve water. This hormone works to stimulate greater conservation of water. The root word -diuretic means to cause greater urine output. The name antidiuretic hormone makes sense as its presence decreases urine output. This hormone affects both men and women in the same manner.

The second hormone which arises from the posterior lobe of the pituitary is called oxytocin. Oxytocin has its effects upon two sets of organs in women. The first organ is the uterus which is the organ which holds the developing baby (commonly called the womb). When the time has arrived for the baby to be born, oxytocin is released from the posterior lobe of the pituitary gland. As it arrives at the uterus it causes the smooth muscles of the wall of the uterus to contract resulting in the contractions of labor. In cases where a baby has gone past the expected due date or there is need to prompt labor to begin, a doctor may give injections of oxytocin (Pitocin) to initiate contractions.

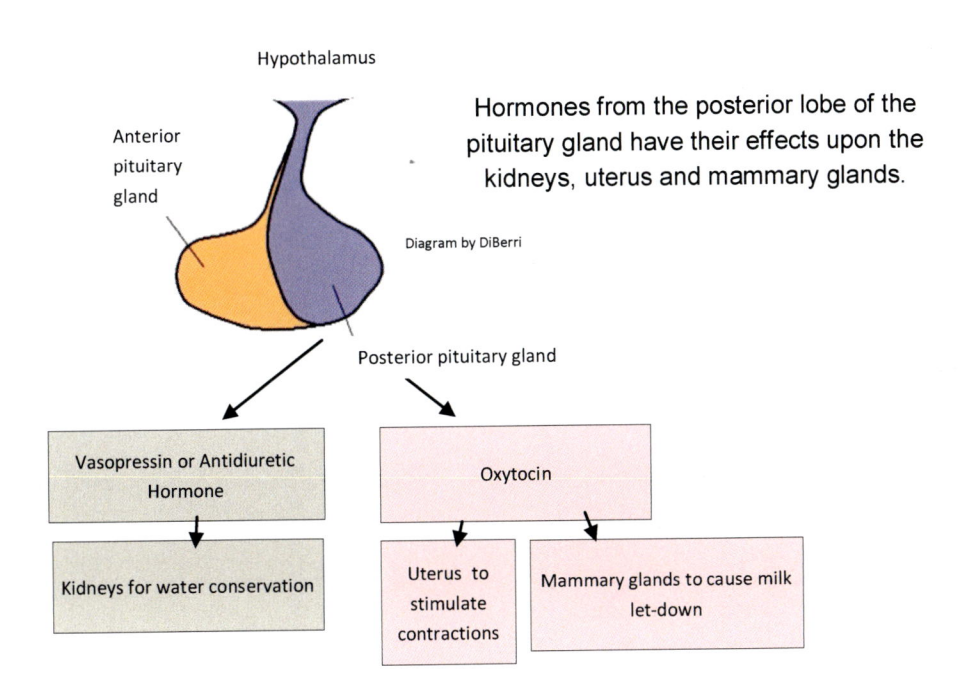

Hormones from the posterior lobe of the pituitary gland have their effects upon the kidneys, uterus and mammary glands.

Diagram by DiBerri

Vasopressin or Antidiuretic Hormone → Kidneys for water conservation

Oxytocin → Uterus to stimulate contractions / Mammary glands to cause milk let-down

The second organ which responds to the presence of oxytocin is the mammary or milk-producing tissue of women. When a baby begins to nurse, the suckling action of its mouth stimulates the breast which, in turn, sends a message to the pituitary gland to release oxytocin. The oxytocin then returns to the breast tissue to cause the milk-secreting glands to contract resulting in what is commonly known as milk let-down. The milk then collects near the nipple for the baby to consume by nursing.

In addition to the pituitary gland, there are three other endocrine glands in the human body that we need to discuss. The first is a set of glands known as the parathyroid glands. These small glands are found just behind the thyroid gland and function to maintain appropriate levels of calcium and phosphorus in the body through the hormone known as parathormone. The target organs are the bones

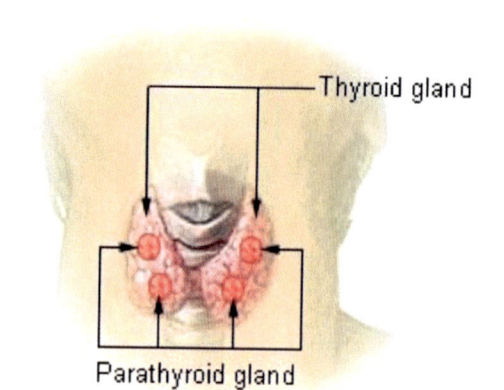

The parathyroid glands are found embedded within the thyroid gland found in the neck.

Thyroid gland

Parathyroid gland

of the body. When calcium levels fall, the parathyroid glands respond by secreting parathormone which acts upon bone cells. The bone cells respond by moving calcium out of bone tissues and back into the bloodstream where it can be delivered to cells needing calcium. A hormone produced by the thyroid gland works to reduce blood levels of calcium. This hormone is known as calcitonin. So, parathormone works to raise blood levels of calcium while calcitonin works to reduce blood levels of calcium.

The second gland is the pineal (pie-knee-uhl) gland which is found in the front portion of the brain. This small gland secretes the hormone melatonin which is known to regulate daily rhythms of life such as sleep and wakefulness.

The third and final gland we will discuss is the thymus. The thymus is found in the chest and is present in babies and children but diminishes in size as a person reaches adulthood. The thymus produces a hormone known as thymosin which stimulates the production and maturation of T-lymphocytes. These lymphocytes are known to aggressively fight organisms which invade the body. While the thymus practically disappears in the adult, the T-cells it helped to produce remain present with the individual throughout life.

This ends our discussion of the endocrine system and its major roles in controlling the development and growth in humans. In this lesson we learned that:

- The endocrine system is the body system which regulates the activities of many other body systems through the function of glands;
- There are two types of glands: exocrine which have ducts or tubes in which substances are exported and endocrine glands which deposit substances directly into the circulatory system;

- Levels of hormones in the body are maintained using a negative-feedback mechanism whereby low levels trigger the endocrine gland to begin production of the hormone;
- The pituitary gland is known as the "master gland" as it produces hormones which affect many organs in the body as well as other endocrine glands which in turn affect body organs;
- The anterior pituitary gland produces thyroid-stimulating hormone (TSH), adrenocorticotropic hormone (ACTH), human growth hormone (HGH), prolactin and the gonadotropic hormones;
- TSH affects rate of metabolism;
- ACTH signals the adrenal glands to produce anti-inflammatory substances and water conservation hormones;
- HGH regulates growth in the body;
- Prolactin regulates development of milk-producing tissues as well as milk production in women;
- The gonadotropic hormones in women stimulate the ovaries to produce ova and then maintain pregnancy;
- The gonadotropic hormones in men stimulate the testis to produce sperm cells and testosterone;
- Hormones of the posterior lobe of the pituitary gland include vasopressin, which regulates water conservation by the kidneys, and oxytocin, which stimulates the uterus to contract during birth and milk ejection while nursing.
- The parathyroid glands control calcium levels in the body;
- The pineal gland regulates daily sleep and wakefulness patterns;
- The thymus, while only active in children, functions to produce T-lymphocytes which function throughout life to defend against disease.

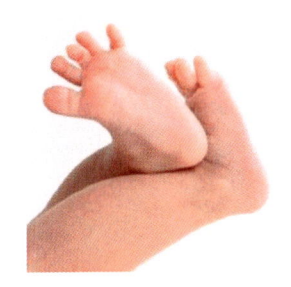

Lesson 26: Systems of Reproduction

We have been studying the organ systems of the body which enable us to achieve the characteristics of a living organism. By studying the skeletal, muscular and nervous systems, we've learned how human beings are able to move. Through study of the circulatory, respiratory and digestive systems, we've learned how humans are able to obtain and then deliver an energy source to all parts of the body. In the last lesson, we learned how the endocrine system enabled humans to grow and develop. In this lesson we will focus on how human beings can reproduce.

We have already learned quite a bit about the reproductive process of humans in earlier lessons in this course. In Lesson 12, we explored how cells which form the ovum and sperm develop from primordial germ cells found in the ovary and testis. Recall how we learned that these primordial germ cells undergo meiosis which takes them from being diploid, 2N cells to haploid, 1N cells. We also learned that the ovum and sperm cell are designed to join through the process of fertilization, again creating a 2N cell which, through a multitude of mitotic divisions, becomes the new human being. In the most recent lesson on endocrinology, we learned about specific hormones (FSH, LH, estrogen, progesterone and testosterone) which control the events of reproduction in humans. In this lesson, we'll focus on the process from a slightly broader perspective as we learn the names and functions of the organs which allow reproduction to take place.

We'll begin our discussion with the male reproductive system. We've already learned that the male gametes, the sperm, are produced in the testis, which are located in the scrotum. While still in the testis, the sperm are unable to fertilize an ovum, however, they do gain the ability to do so as they journey out of the man's body. The tiny tubes within the testis (in which the sperm cells are developing) empty the sperm cells into a central location in the testis known as the rete (ree-tee) testis. This structure then empties into the epididymis (ep-id-did-uh-miss). The epididymis is a continuous tube which lies on the surface of the testis where the sperm cells continue their maturation process. The epididymis then connects with a larger-diameter tube which exits the scrotum and continues upward into the lower abdominal cavity. This tube is called the vas deferens.

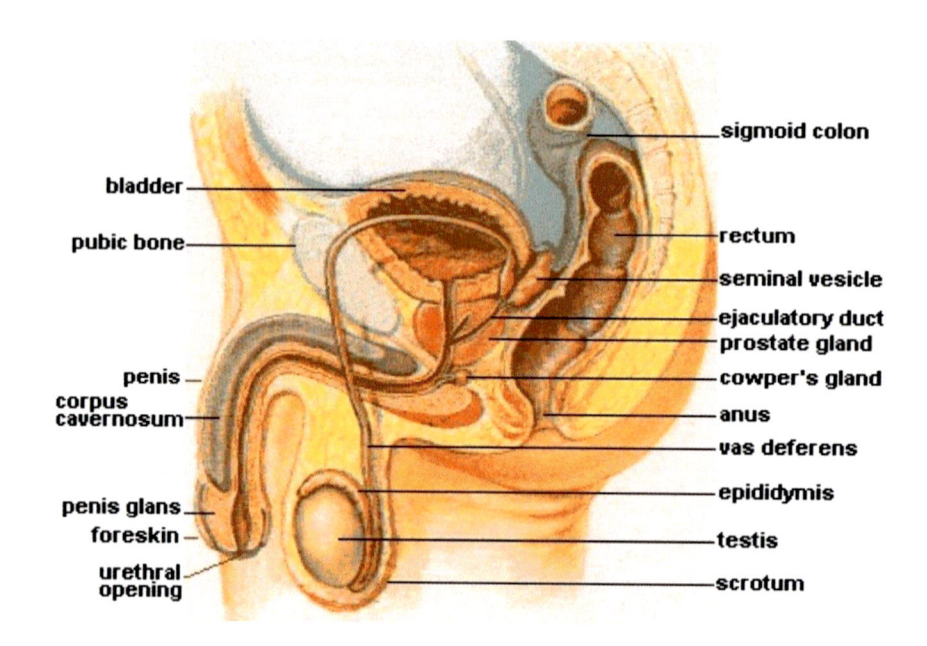

Parts of the male reproductive system.

The vas deferens continues and passes by glands which provide vital supplies to the sperm cells for their continued journey. These glands are known as the accessory sex glands and include the seminal vesicles, Cowper's gland and prostate gland. These glands secrete a mixture of energy-supplying fructose and pH-adjusting chemicals to the sperm which, all together, creates semen.

Next, the vas deferens from each testis joins the urethra to leave the body. Recall from our study of the urinary tract that the urethra is also the tube which allows urine to empty from the bladder to the exterior. The pH of urine is too acidic for the sperm cells, so the secretions from the accessory sex glands raise the pH in order for the sperm cells to survive their trip through the urethra. The urethra extends through the penis which is the organ that enables the man to deposit the semen within the reproductive system of the woman.

The part of the reproductive system of the woman that accepts the penis of the man is called the vagina. At the innermost point of the vagina is a door-like structure known as the cervix. Depending upon the influence of the hormone estrogen, the cervix opens and closes. Recall the discussion in our previous lesson where we introduced the idea of the follicle. We said the follicle was a blister-like structure on the surface of the ovary which held the ovum within and, as the follicle developed, it pro-

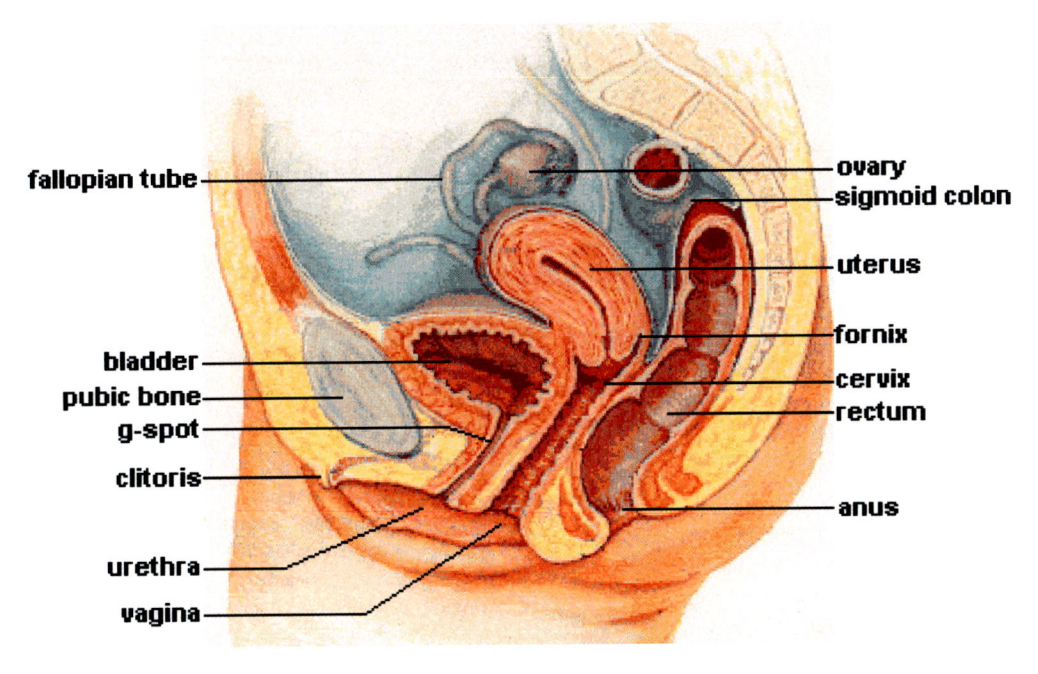

Parts of the female reproductive system.

duced estrogen. High levels of estrogen cause the cervix to open. It makes sense then, that as the follicle prepares to eventually rupture to release the ovum, the cervix would be open to allow sperm cells to enter. Once past the cervix, the sperm cells continue their upward journey by using their tail to swim. The head of the sperm contains the genetic material destined to create the new baby, while a collar of mitochondria is found just below to continually provide ATPs from the fructose supplied by the accessory sex glands.

Following ovulation (the rupturing of the follicle), the egg (under ideal circumstances) is collected by the infundibulum which works like a funnel to move the egg down though the Fallopian tube towards the uterus. Fertilization, in most cases, takes place within the Fallopian tube. The developing embryo then slowly moves down into the uterus where it affixes itself to the wall which has become thickened with a rich supply of blood vessels. This stage is known as implantation. Failure to move down to the uterus results in the condition known as an ectopic pregnancy.

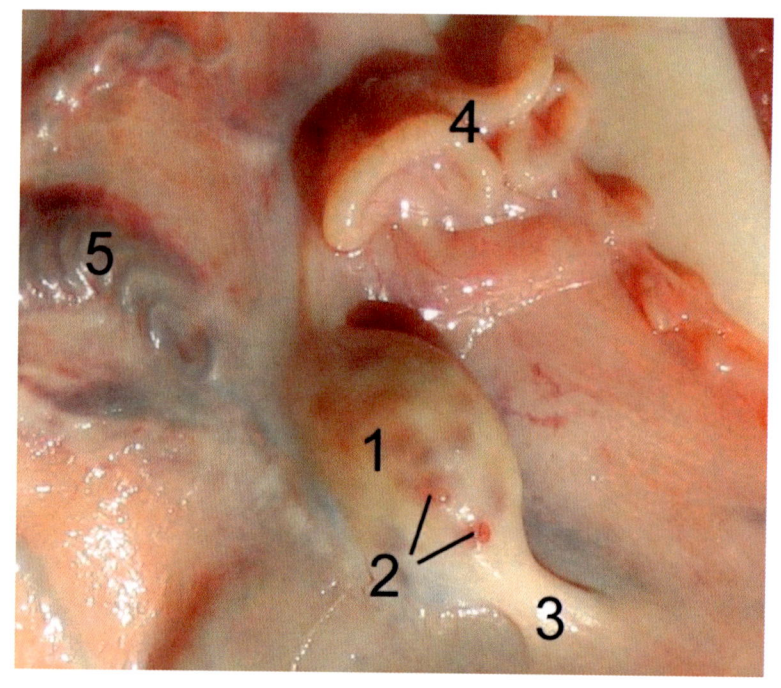

Ovary of a sheep: Point 1 is the ovary. Points 2 indicate follicles. Point 3 is the ligament which attaches the ovary to the body wall. Point 4 is the Fallopian tube which collects the ovum and provides a passage way to the uterus. Point 5 shows blood vessels which supply the ovary.

Uterus with blood vessels from mom

Placenta with blood vessels from baby

Umbilical cord

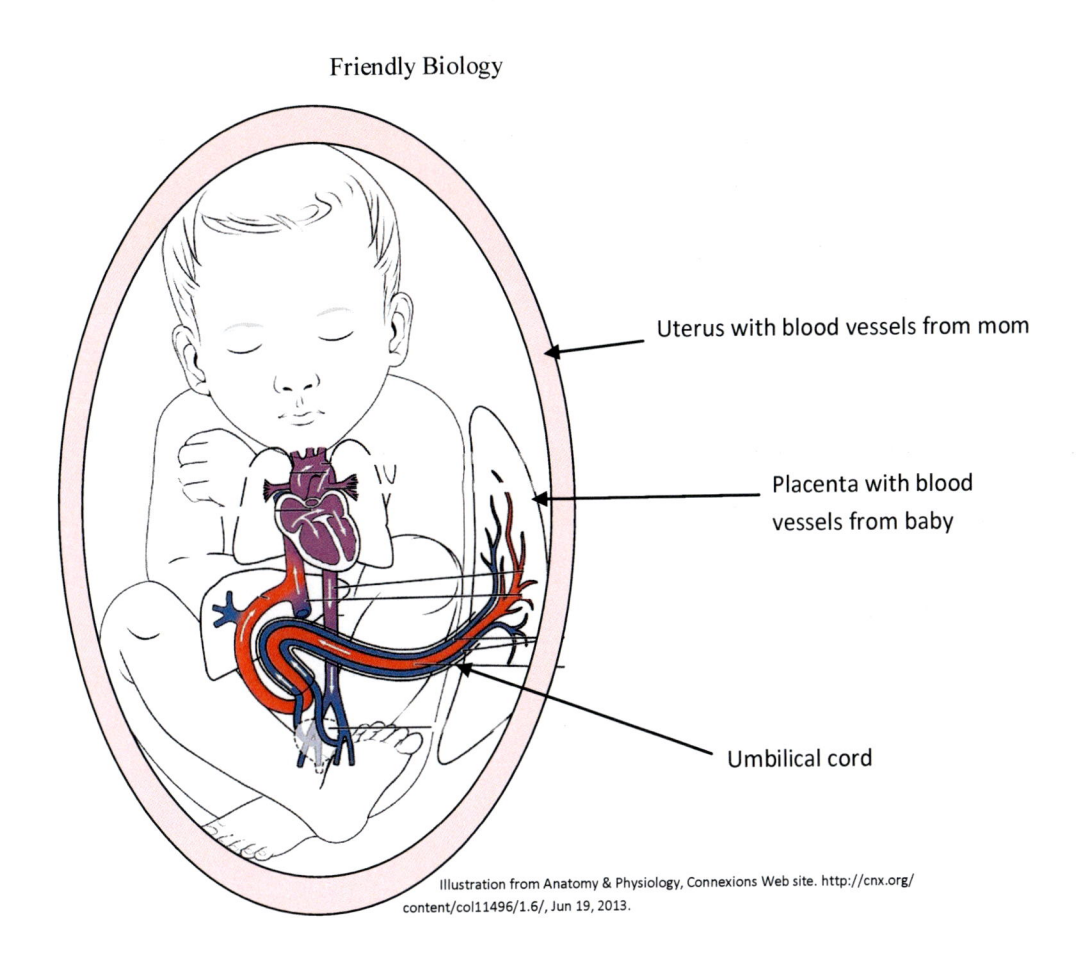

Illustration from Anatomy & Physiology, Connexions Web site. http://cnx.org/content/col11496/1.6/, Jun 19, 2013.

Blood vessels grow from the embryo (the umbilical cord) to meet with the blood supply from the uterine wall to create the placenta. It should be noted that the blood vessels in the umbilical cord never actually join with the blood vessels from the mother. The vessels come right along side each other, however the blood supplies actually never mix. Oxygen and glucose, along with all other vital nutrients, are passed through the cell membranes of the capillaries on each side of the junction. Likewise, carbon dioxide and other waste products generated by the baby are moved back into the mother's blood. The blood cells themselves never move from baby to mother or vice versa.

As we discussed earlier, following ovulation the cells that once made up the follicle on the ovary are replaced by the corpus luteum (CL) which begins secretion of progesterone. Progesterone then influences the cervix to close and remain closed throughout pregnancy. Progesterone is also produced by the membranes surrounding the developing baby to hopefully insure that the baby has enough time to fully develop inside the uterus before the time of delivery arrives.

In this discussion, we've been assuming that at least one sperm cell has arrived at the ovum to result in fertilization. If that does not happen, the ovum eventually will die and the cells which created the corpus luteum begin to decrease their secretion of progesterone. Because of this, the lining of the uterus that had been prepared to accept the potential baby begins to separate from the wall of the uterus. The cervix opens and the uterine contents are expelled from the body. This is the monthly "period" experienced by women. Once the contents are completely expelled, the process begins again with another follicle and ovum being prepared on the ovary.

If fertilization has occurred and the baby has successfully implanted on the wall of the uterus, the baby will grow and develop for approximately nine months. It is thought that it is actually a hormone created by the baby itself which initiates labor, the process of giving birth, to begin. Recall from our previous lesson that we discussed the role of oxytocin which is produced by the posterior lobe of the pituitary gland. The smooth muscles of the uterus are the receptor site of the oxytocin and begin rhythmic contractions. The levels of progesterone begin to decline and the cervix begins to open. This is called dilation. The cervix will eventually have to dilate to at least 10 centimeters before the baby's head, which is usually born first, can be delivered.

As the cervix continues to dilate and the uterine contractions increase in frequency and intensity, the membranes which have been surrounding the baby all through its development (known as the amniotic membranes), eventually rupture. The fluid surrounding the baby known as the amniotic fluid will be expelled. This event is commonly known as the "water breaking." Once this takes place, the baby will soon be born.

The uterine contractions continue to intensify and eventually the baby's head will be delivered followed soon by the rest of his or her body. The umbilical cord will remain attached to the placenta still inside the mother to be clamped-off and cut a few minutes later. The amniotic membranes and placenta will also eventually be delivered. In animals, these tissues are referred to as the afterbirth.

If problems arise with delivery of the baby, such as the baby is too large to safely pass through the birth canal or there are complications with the health of the baby or mother, the attending doctor may elect to perform a Cesarean section or C-section. The C-section involves making an incision though the skin and abdominal wall of the mother, continuing though the uterine wall and amniotic membranes to reach the baby. The baby is then gently lifted from the uterus and all layers that were passed through are sewn back into place.

Regardless of means of delivery, the baby will no longer have access to the supply of oxygen previously being provided through its umbilical cord. The mucus and amniotic fluids in its mouth and nose must be quickly sucked-out or aspirated to allow free movement of air into its lungs. Once a few breaths are taken, the lungs take over supplying oxygen to all cells of the body. The process of birth is truly a beautiful miracle.

Let's pause here and review what we've learned in this lesson. We have learned that:

- Sperm and ova are 1N cells having under gone meiosis (sperm are formed in the testis and ova form in the ovary);
- The testis are located outside the body in the scrotum and the ovaries are located in the abdomen;
- Sperm exit the testis through the epididymis and continue through the vas deferens to the urethra;
- Accessory sex glands which include the seminal vesicles, Cowper's gland and prostate gland provide nutrients and pH-adjusting substances to the sperm to create semen;
- The penis allows the man to deposit semen into the woman's vagina;
- The cervix functions like a door to open at specific times in response to estrogen to allow sperm to enter or be closed during pregnancy;
- The release of the ovum is known as ovulation and the ovum is gathered by the infundibulum and moved down the fallopian tube where fertilization usually takes place;
- The embryo moves down to the uterus to implant and continue development;
- Blood never mixes directly between the mother and developing baby (nutrients and wastes, along with oxygen and carbon dioxide, readily move across capillary membranes of the placenta);
- Pregnancy in humans is nine months;
- Oxytocin, from the posterior pituitary gland, stimulates the uterus to begin contractions while decreasing progesterone levels allow the cervix to dilate;
- The amniotic membranes break, which is the "water breaking," and the baby is soon delivered;
- Should the baby be too large to deliver, a Cesearean section may be performed to surgically remove the baby from the uterus.

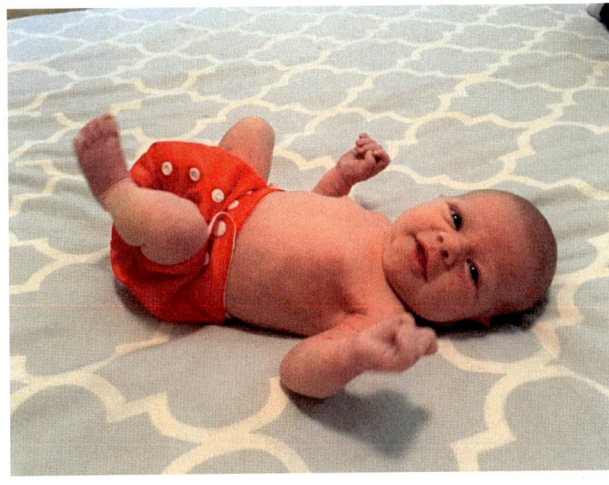

Photo courtesy Rachel and Jacob Hajda

Babies are miracles.

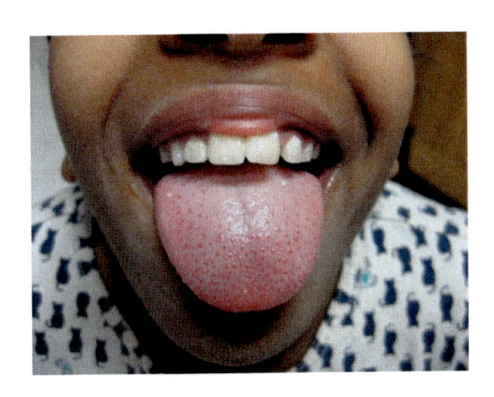

Lesson 27: Sensory Systems

We have now discussed the organ systems which allow us to move (skeletal, muscular and nervous systems), mobilize and distribute energy resources (circulatory, respiratory and digestive and renal systems), develop and grow (endocrine system) and reproduce (male and female reproductive systems.) In this final lesson of our exploration of human anatomy and physiology, we'll look at the organ systems which allow us to respond to our environment.

Think for a moment. How do you known when something is happening around you? You likely can see it, hear it or smell it. If something comes close, you may feel it and if you were to place it into your mouth, you could taste it. Seeing, hearing, smelling, touching and tasting are our five means of sensing what is happening in our environment. By utilizing our five senses, we can then respond in an appropriate way. In this lesson, we'll look at the organs in our body which allow us to see, hear, smell, touch and taste.

Let's begin with our sense of sight. Our eyes have the unique ability to take light energy and transform it into a nerve stimulus which is detectable and decodable by the occipital lobe of the cerebrum of the brain. Let's look at the structures of the eye to understand how this takes place.

The clear outer covering of our eyes is called the cornea. While it appears clear, it is actually made of many layers of cells similar to how our skin is made. The corneal cells on the outer surface are constantly dying and being replaced by newer cells below. While it's difficult to see, the cornea has blood vessels supplying nutrients, as well as many super-sensitive nerve endings present. You know how sensitive these nerve endings are when you have something very tiny "in" your eye and it feels "huge"! On its outer perimeter, the cornea transitions into a very tough, opaque covering known as the sclera (sklara, as in Clara) or the "whites" of the eye. The sclera encloses the remainder of the eye.

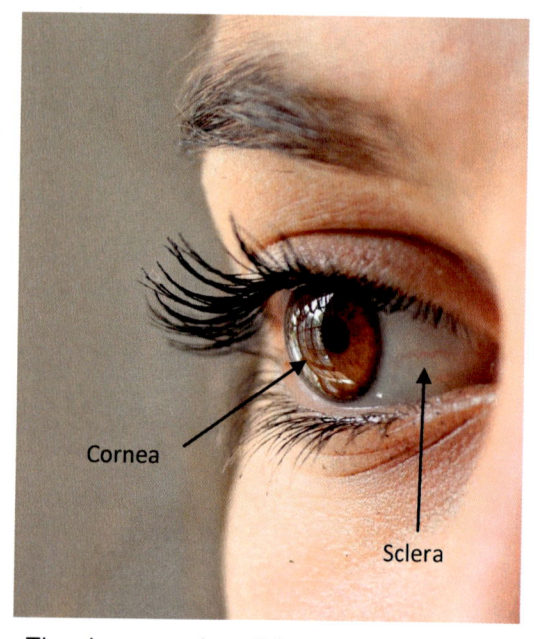

The clear covering of the eye is the cornea.

If the cornea becomes inflamed, as with an infection, it may become cloudy or even appear a bluish color. This is due to increased water which has moved into the cells from the capillaries supplying the cells in an effort to bring in disease-fighting white blood cells. In extreme cases, visible blood vessels may extend over the surface of the cornea, again in an effort to reduce infection and inflammation. With time and healing, these vessels will recede and the cloudiness of the cornea will go away.

Because the cornea consists of living cells, it must be bathed frequently with tears to maintain proper hydration and to remove dust and debris, as well as invading microorganisms. These tears are produced by tear (lacrimal) glands located on the upper and outer surface (lateral side) of the eye. With each blink of the eyelid, a new bath of tears cleanses and hydrates the surface of the cornea. The "dirty" tears drain into the central corner (medial side) of the eye into the tear duct which drains into the nasal cavity and evaporates. That's why, when you get something in your eye or you cry, your nose runs, too!

If we continue deeper into the surface of the eye, just inside the cornea, we find a clear watery liquid known as the aqueous humor. This liquid is constantly being produced and a special drainage

Anatomy of the Eye

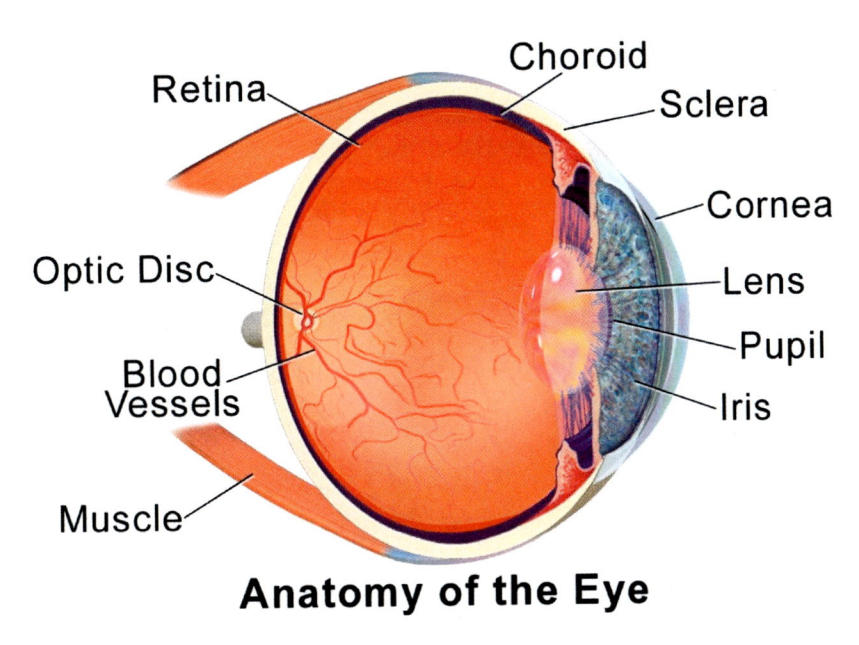

system exists to allow for the correct pressure to be maintained within the eye. Too much fluid production or a blockage in the drainage system can cause damage to blood vessels and nerves within the eye. This condition is called glaucoma.

The next structure that we encounter if we continue deeper into the eye is the iris. The iris is the circular ring of tissue which gives you your eye color. In the center of the iris is a circular hole known as the pupil which allows light that has entered through the cornea and passed through the aqueous humor to continue. The iris is made of muscles which allow it to adjust the size of the pupil which, in turn, adjusts the amount of light passing through. In locations of darkness or very low light levels, the size of the pupil increases. In bright light, the pupil constricts or becomes smaller. This quick adjustment in size is called the pupillary response (and what a doctor is observing when he or she quickly shines a bright light into your eyes).

So far we've passed through the cornea, aqueous humor and moved through the pupil made in the iris of the eye. If we continue deeper into the eye, we next find the lens. While circular in shape with convex surfaces on both its front and rear surfaces, the lens is actually much like an onion. The lens consists of many layers of cells much like the layers of an onion. The cells are alive and, like the cornea, are constantly being replaced. Because of its curved surfaces and ability to readily transmit light, the lens can effectively bend light to focus an image onto the light-sensitive cells of the retina just behind it.

Because we are able to focus on objects very close to our eyes as well as objects very far away, the curvature of the lens must be constantly adjusted. The adjustment is made by many tiny muscles which attach to the perimeter of the lens. These muscles are called ciliary muscles. They contract to "stretch-out" the lens and then relax to allow it to return to its original shape. By doing so, images can be focused on the retina.

The lens is suspended by this ring of muscles within the eye and together they form a barrier between the watery aqueous humor in front and the much thicker, jelly-like vitreous humor behind. The term vitreous refers to glass. The vitreous humor is exceptionally clear and functions to maintain the correct shape of the eye and pressure behind the lens.

The retina is the next layer we encounter as we continue deeper into the eye. The retina is the light-sensitive layer of cells which is capable of converting light energy into nerve stimuli. It consists of two main types of cell structures: the rods and the cones. The rod cells are sensitive to light and darkness while the cones allow us to visualize colors. The retina is actually 10 cell layers thick and, at the deepest extreme, the retinal cells make their connection with nerve endings of the large optic nerve. The optic nerve is the connection between the eye and the brain.

This is a high magnification slide made of the retina. It has been colored by a stain known as H and E Stain. H stands for hematoxylin and E stands for eosin. These substances stain cells the red and blue colors you see here. The top light pink and then darker pink layers are the rods and cones of the retina. The blue dots you see are the nuclei of cells.

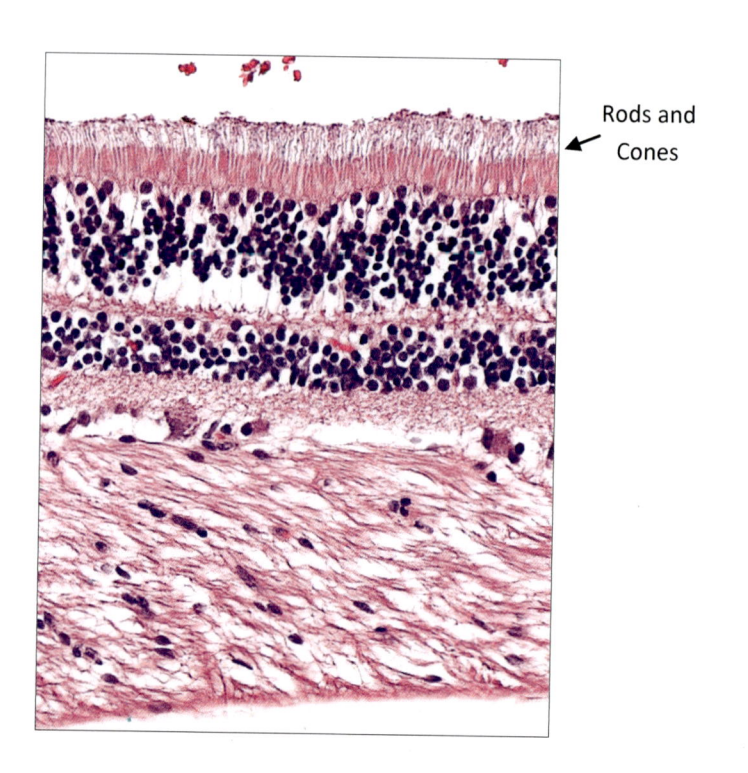

Rods and Cones

Upon leaving the back side of the eye, the optic nerve enters into the brain cavity and crosses to the opposite side of the brain. Most, but not all, of the nerves within the optic nerve from each eye make this crossover. So, for the most part, the left eye's information gets transmitted to the right side of the brain and the right eye's information gets transmitted to the left side of the brain. Ultimately, each nerve terminates in the occipital lobes of the brain. It's interesting to note that, due to the properties of light passing through the small pupil of the iris, the image projected on the retina is inverted, that is, it's upside down. The brain is capable of turning it right side up.

The optic nerves linking the retinal cells to the brain are one pair of a set of twelve pairs of large nerves which communicate directly with the brain and not the spinal cord. These twelve pairs of nerves are known as the twelve cranial nerves. They are numbered according to the position they are found as they enter and exit the underneath surface of the brain. The optic nerve is cranial nerve II. Roman numerals are used to indicate the cranial nerve number.

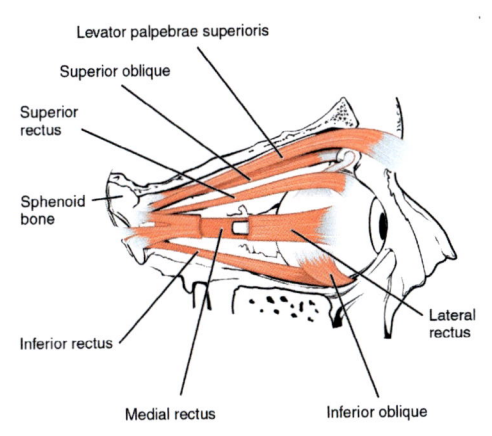

Movement of the eyes are done by striated muscles attached as specific points around the surface of the eye.

Eye movement is made possible by opposing pairs of striated muscles which attach at various points along the junction between the cornea and the sclera of the eye. Muscles which attach at the 12, 3, 6 and 9 o'clock positions allow for up and down and right and left eye movement. These muscles are given the name "rectus" muscles. Rectus means straight, and as you see in the diagram above, these muscles attach at the positions identified above and extend straight back to the center of the ocular orbit (eye socket). Contraction of these muscles allows for movement straight up or down or straight left or right. An additional pair of muscles attach at the 7 and 11 o'clock positions which allows the eye to rotate. The muscles are known as the oblique muscles since they allow the eye to be turned at various angles. By combining the action of the rectus and oblique muscles, you can move your eyes in an infinite number of positions in just the "blink of an eye." These eye muscles are controlled by cranial nerves III (3) (oculomotor nerve), IV (4) (trochlear nerve) and VI (6) (abducens nerve).

Let's move to another set of sensory organs of our body: the ears. The ears have the capability of converting sound energy into mechanical vibrations which, in turn, stimulate nerve endings to allow us to hear. Let's look at the basic parts of the ear to see how this takes place.

The outside visible part of our ear is called the pinna. Its job is to collect and funnel sound waves down into the hole of our ear known as the auditory canal. The prefix audio- refers to sound. At the inner "end" of the auditory canal we find the ear drum, scientifically known as the tympanic membrane. This thin, but tough, membrane is the structure which vibrates when sound waves strike against its surface. This vibration creates mechanical energy which is transmitted onward on its ultimate trek to the brain. The structures we have discussed so far up to the tympanic membrane make up the outer ear.

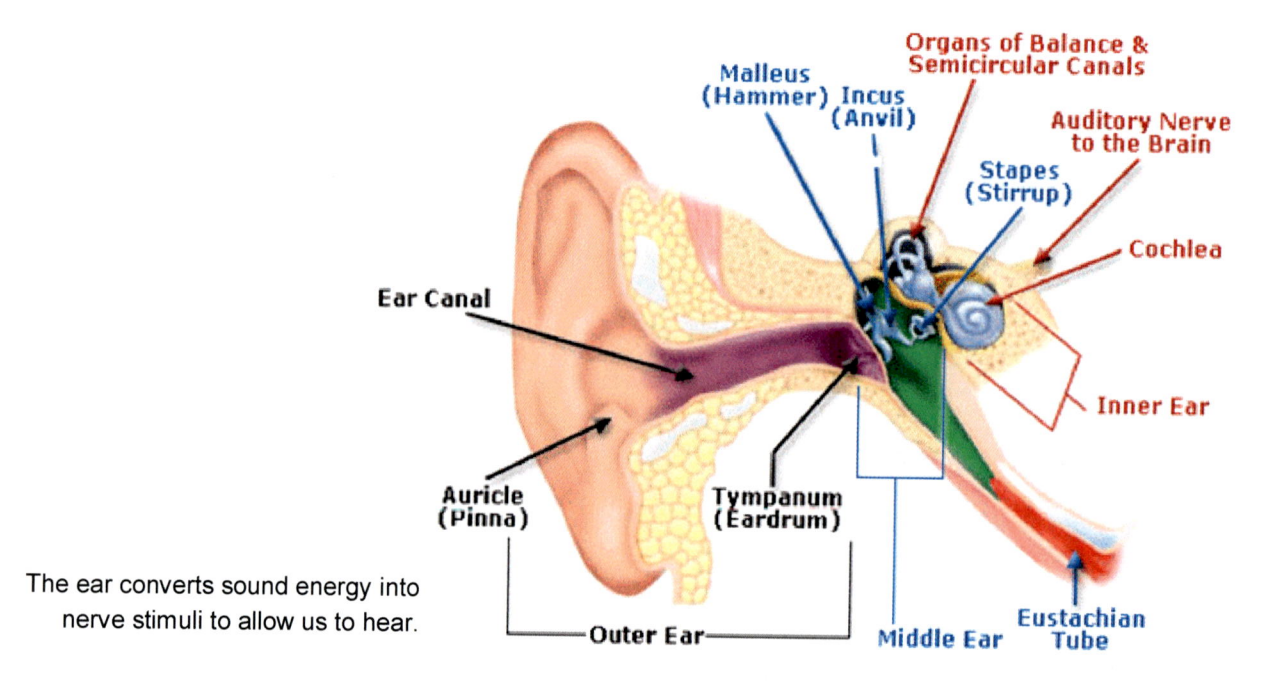

The ear converts sound energy into nerve stimuli to allow us to hear.

On the inner surface of the tympanic membrane are attached three tiny bones known as the malleus, incus and the stapes or, commonly, the hammer, anvil and the stirrup. The malleus is directly attached to the tympanic membrane and begins to vibrate right along with the tympanic membrane. The vibrations are transmitted next to the incus and, finally, the stapes. The stapes is then connected to a snail-shaped structure known as the cochlea. The structures between the tympanic membrane and the cochlea make up the middle ear.

It's important to know that there is another tube which opens into the middle ear space. This tube is the eustachian tube and it creates communication between the middle ear and the nasopharynx (the point at the back of the nose where it joins the throat). Having this connection allows you to equalize pressures between the outer and middle ear. If you've ever been traveling in the mountains or have flown in an airplane, you likely experienced this attempt to equalize pressures within your ear. You may have said, "My ears are popping."

Inside the cochlea in the inner ear is a fluid which bathes the inner surface (which is covered with many hairs which are connected to nerve endings). The nerve endings are connected to the auditory nerve which is cranial nerve VIII (8). This cranial nerve collects signals from the nerve endings within the cochlea and transmits them to the brain.

Also present in the same inner ear location are three circular tubes known as the semi-circular canals. These circular tubes exist in three different planes and are also filled with fluid and tiny hair-like projections. However, they are not part of the hearing system. Instead, these circular tubes are used to help us maintain our balance. Every time we move our head, the fluid inside these tubes gently moves across the hairs which send messages to the cerebellum and cerebrum of the brain to continuously allow us to monitor our position in an effort to maintain balance. Twirling about or spinning in the same spot often over stimulates these nerve endings which results in dizziness.

The nerve endings in the semi-circular canals also join cranial nerve VIII (mentioned above). This nerve is known as the vestibulocochlear nerve. To complete our discussion, consider the region of the ear extending from the cochlea to the location of the connection to cranial nerve VIII as being the inner ear.

We've covered the senses of sight and hearing. Let's turn now to the sense of smell. In our nose there are nerve endings which are capable of taking chemical stimuli created by contact with tiny molecules of substances floating in the air and converting them to nerve impulses. The sense of smell is much less complex than that of the eye or ear. The nerve endings in the nose connect directly to cranial nerve I (known as the olfactory nerve). The olfactory nerve enters into the brain cavity and continues to the brain where the sensation of smell is interpreted.

The fourth sense we will explore is the sense of taste. The main organ of taste sensation is the tongue, although there is evidence that there are cells that receive taste sensation on the sides and roof of the mouth and, maybe, even some in the nose. On the surface of the tongue there are visible bumps known as papillae. Embedded within the papillae are structures known as taste buds. There are approximately 2000-5000 taste buds found on the tongue. Within the surface of a taste bud there are between 50-100 taste receptor cells.

The taste receptor cells tend to specialize in the type of taste to which they are capable of responding. Historically, there are four types of tastes: sweet, sour, bitter and salty. Recently, a fifth type of taste has been added to the list which is known as unami. Unami is described as being a savory taste often found in fermented foods such as soy sauce. It has been taught for many years that there are areas of the tongue which are specific for certain tastes. Recent studies have shown that, while certain areas of the tongue are more sensitive to certain tastes, all areas of the tongue can sense all tastes.

The taste receptor cells of the tongue are linked to cranial nerves which transmit the taste sensation to the brain. Taste from the front two-thirds of the tongue is transmitted by a branch of cranial nerve VII (7) which is known as the facial nerve. This nerve also controls the muscles of facial expression. Taste from the rear one-third of the tongue is transmitted by cranial nerve IX (9) which is the glossopharyngeal nerve. The prefix glosso- refers to the tongue. At the base of the tongue where it is attached to the mandible (jaw), taste is transmitted by cranial nerve X (10) which is the vagus nerve. It's interesting to note that sensation to texture, temperature or pain encountered by the tongue is transmitted by a completely different cranial nerve which is cranial nerve V (5).

Let's move on now to the fifth and final means of gathering information from our environment which is the sense of touch. The primary organ of touch is our skin. When compared to all other organs in our body, the skin is the largest organ and has the most mass (weight). Let's look more closely

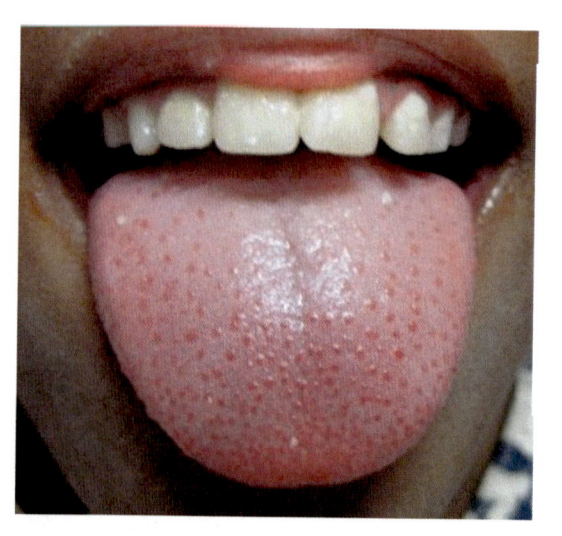

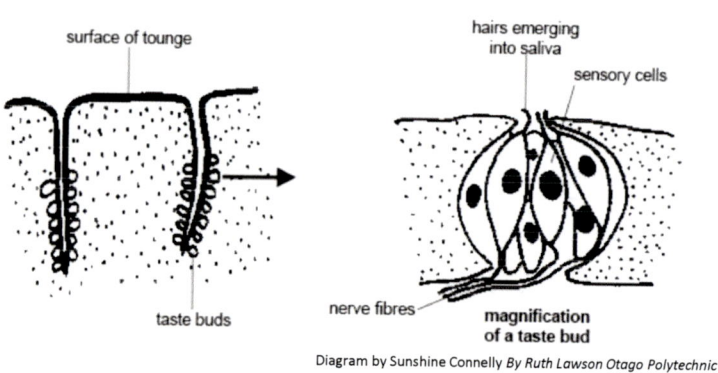

The surface of the tongue is covered by papillae which, in turn, have taste buds embedded within.

Diagram by Sunshine Connelly *By Ruth Lawson Otago Polytechnic* at the structures of the skin which will help us understand how it allows us to sense touch.

The skin can be divided into two layers: the dermis and the epidermis. The epidermis is the outermost layer and consists mainly of flattened cells known as squamous epithelial cells. Squamous means flat. The cells on the surface of our skin are mainly dead epithelial cells which contain a substance known as keratin. Keratin gives our skin cells water-proofing capabilities. Our finger and toenails as well as hair are all made up of cells containing high concentrations of keratin.

The dead cells on the surface of our skin are constantly flaking off and are being replaced by newer cells beneath. If we explore deeper we will come to a layer of cells where active mitosis is taking place to meet the constant need for new skin cells. This layer is called the basal layer.

Just beneath the basal layer there is the second main layer of the skin. This is the dermis. The dermis contains blood vessels, nerves, glands and fatty tissue. The base portion of hair, known as the hair follicle, is also found in the dermis.

Recall from our discussion of the nervous system that nerves outside the spinal cord were known as peripheral nerves and that peripheral nerves could be identified as being either efferent (motor) nerves which caused action or afferent (sensing) nerves which carried the sense of touch back to the brain. The touch receptors of the skin are linked to these afferent neurons. There are different types of

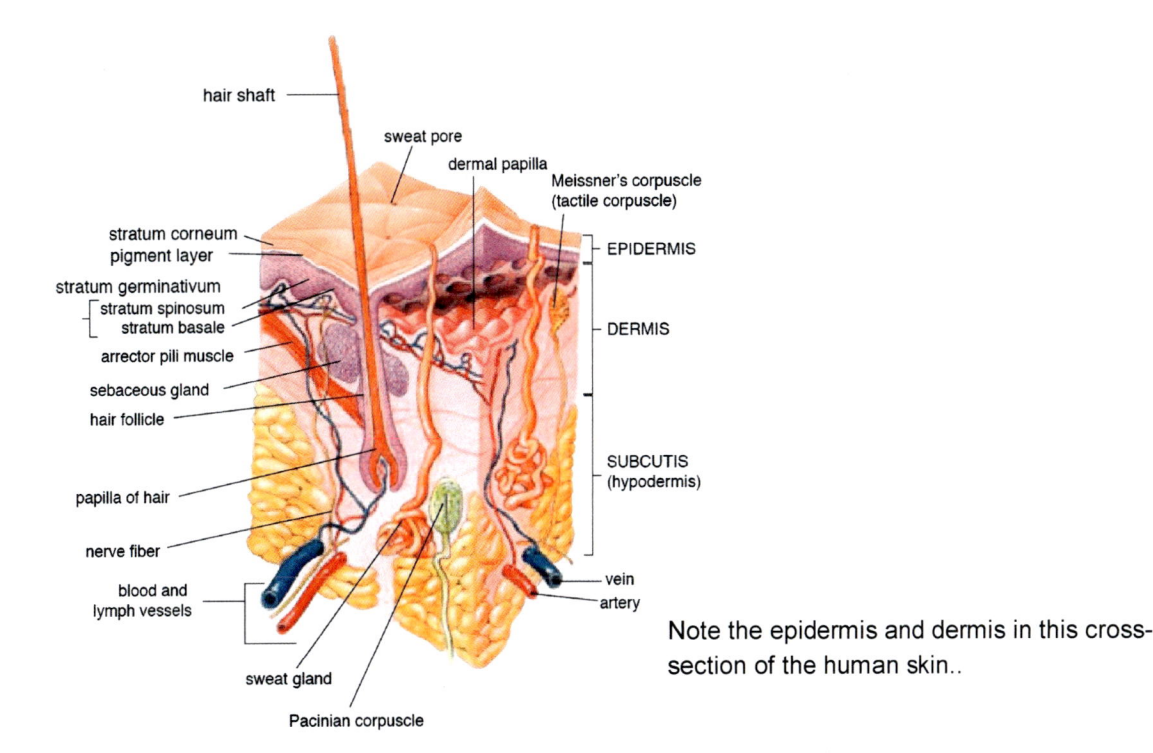

Note the epidermis and dermis in this cross-section of the human skin..

touch receptors found in our skin which include the sense of texture, pain, pressure and temperature. Our hands and, especially fingertips, have the most concentrated set of touch receptors in the body. So, when you touch a rough, hot object, the sensation of roughness, heat and pain may be delivered to the brain by three different sets of afferent neurons.

Because our skin covers our entire body, these afferent neurons carry information to the spinal cord and, eventually, to the brain from all parts of the body. The skin is divided into regions called dermatomes. A single afferent neuron supplies a specific dermatome of the skin. Look at the diagram to see how the skin is divided into dermatomes.

In addition to providing the sensation of touch, the skin also functions as a means of maintaining a suitable body temperature. This regulation occurs due to the high concentration of capillaries found in this skin. By dilating (widening) these capillaries, more blood flows to the skin and has the potential for being cooled by cooler air around the body. This is the reason why (if you have light-colored skin) when you are hot, your skin on your face and the rest of your body turns red. Your skin also has sweat glands which produce sweat (perspiration) which, through evaporation, removes heat from your body.

Now that we've looked at the organs of the body which collect stimuli from our environment, let's take a brief look at the general locations of the brain where this information is interpreted. The brain is formed in two main sections: the cerebrum and the cerebellum. The cerebrum is the larger portion of the brain while the cerebellum is tucked in behind the brain near the brain stem The main job of the cerebellum is coordination of movement in the body. The remaining jobs of the body are taken care of by the

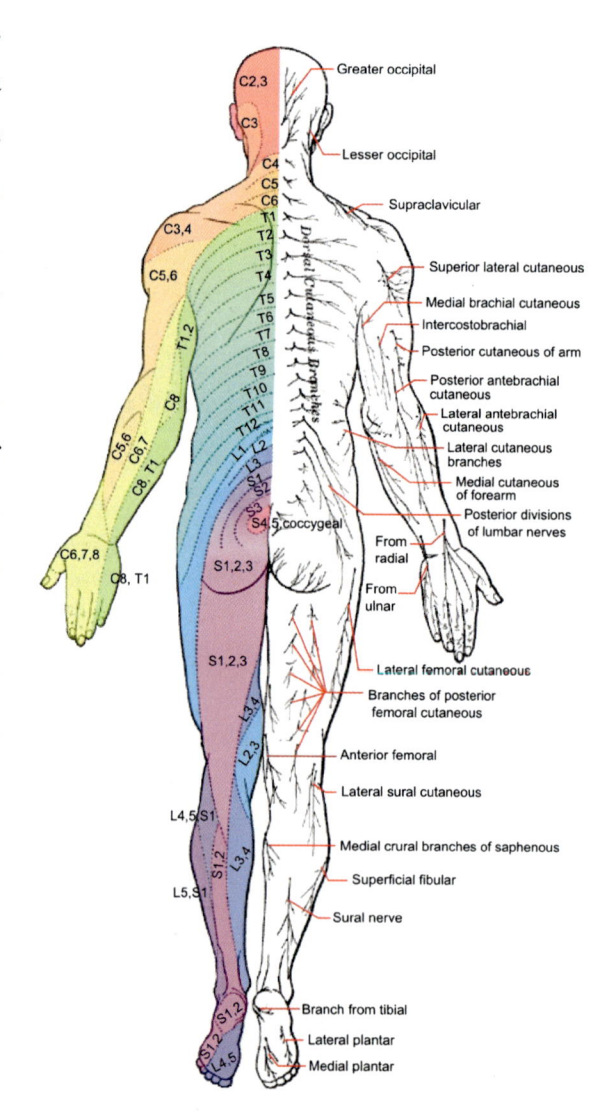

The skin is divided into regions known as dermatomes.

cerebrum.

The cerebrum is divided into a right and left half with a connection deep within known as the corpus callosum. In general, the right half of the brain receives information from and controls the left side of the body while the left side of the brain receives information from and controls the right side of the body. There are specific locations within the brain known as lobes where specific tasks are managed. Look at the diagram to see these locations.

Functional Areas of the Cerebral Cortex

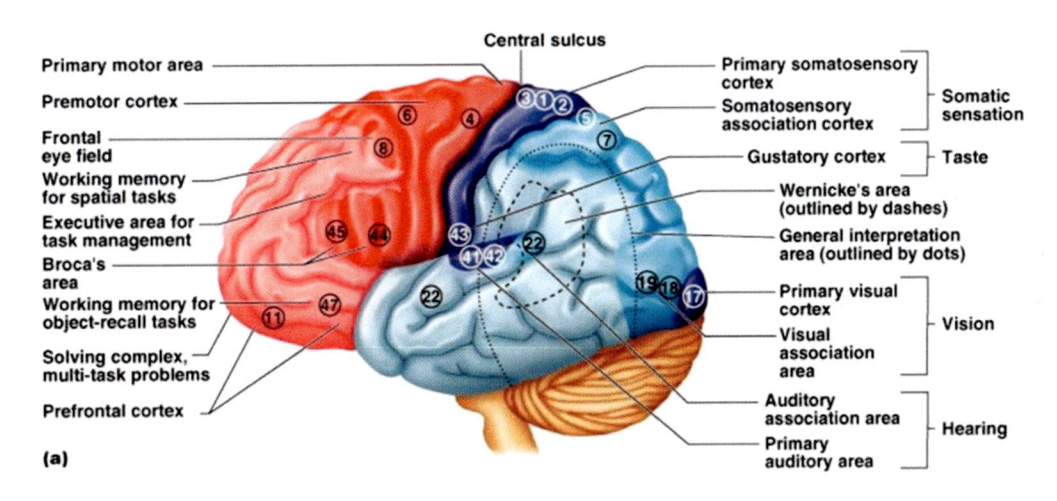

Let's stop here and review what we've learned in this lesson regarding the sensory systems of the body. We have learned that:

- Seeing, hearing, smelling, touching and tasting are our five means of sensing what is happening in our environment;
- The clear outer covering of our eyes is called the cornea which transitions to the sclera which is the white portion on the perimeter of the eye;
- Tears from the tear (lacrimal) gland continually bathe the cornea to maintain hydration and wash away foreign materials;
- The colored portion within the eye is the iris and creates the circular pupil which adjusts the

435

amount of light which can enter the eye;
- The fluid in front of the lens of the eye is the aqueous humor while the fluid behind the lens, which is much thicker, is the vitreous humor (excess fluid within the eye is glaucoma);
- The lens of the eye consists of many layers of cells, much like an onion and tiny muscles encircling the lens allow it to change shape which, in turn, allows images to be focused on the retina;
- The retina is the multilayered surface at the back of the eye which has specialized cells capable of converting light into nerve stimulus (nerve endings in the retina gather into the optic nerve which transmits messages to the brain);
- Muscles around the eye work together to allow for movement in all directions;
- The ears have the capability of converting sound energy into mechanical vibrations which, in turn, stimulate nerve endings to allow us to hear;
- Sound moves into the auditory canal and strikes the tympanic membrane which causes the three tiny bones of the middle ear (the malleus, incus and stapes) to vibrate (these vibrations stimulate nerve endings within the fluid-filled cochlea which then send signals by the auditory nerve to the brain;
- The semicircular canals within the ear work to maintain balance;
- In our nose there are nerve endings which are capable of taking chemical stimuli created by contact with tiny molecules of substances floating in the air and converting them to nerve impulses;
- Stimuli within the nose travels to the brain by the olfactory nerve;
- The main organ of taste sensation is the tongue although there are taste receptors in other locations of the mouth;
- Papillae on the surface of the tongue house taste buds which consist of taste receptor cells (these cells react with various substances we eat and transmit signals to the brain);
- Depending upon the location on the tongue, three different cranial nerves transmit signals to the brain;
- The primary organ of touch is our skin and when compared to all other organs in our body, the skin is the largest organ and has the most mass (weight);
- The skin consists of two layers: the epidermis, which creates squamous cells that form the surface of the skin and the deeper dermis, which houses blood vessels, nerves, glands and fatty tissue;

- Touch sensations travel by afferent neurons to the spinal cord and brain;
- The skin can be divided into regions of sensation known as dermatomes.

Lesson 28: Ecology

In this final lesson in your *Friendly Biology* course, we will investigate the basics of ecology. Recall from our very first lesson in this course that you learned that the root word -ology meant the study of. In the word ecology, the prefix eco- comes from Greek and means "house." Ecology, then, is the study of the house, or more precisely, the study of the relationships between living things and their environment. In previous lessons, we've already looked at some types of relationships found between living things: commensalism, mutualism and parasitism. In this lesson, we'll take a broader look at how living things interact with each other and with their environment.

Let's begin by first defining the term environment. Environment is defined as being everything, both living and non-living, which is found in the location where a living organism lives. If we expand this definition to a broader, global scale we can say that all environments combined make up what is known as the biosphere. The biosphere is the segment of the earth where life exists. While some organisms are found miles above the earth's surface or miles below, deep in the ocean, by far the greatest number of living things exist only a few yards above or below the surface of the earth. The biosphere, therefore, is a relatively thin layer upon the earth, yet it expands horizontally for thousands of miles. Because of this great horizontal dimension, persons who study ecology (the ecologists) break regions of the biosphere into smaller sections known as ecosystems.

Ecosystems are defined as an ecological unit which includes all interacting components. These components include both living and non-living things. The living things, obviously, are the animals, plants, fungi, protista and monera in the area, while the non-living things include topography (flatness or hilliness of the area), the temperature, type of soil or terrain, amount of rainfall or windiness of the area. The living things present make up what are known as the biotic factors while the non-living things make up the abiotic factors (a = without).

Another way ecologists study life on earth is through the division of regions of the earth into what are known as biomes. A biome is defined as an area or region of the earth where similar organisms thrive and others do not. Biomes can be placed into two large groups: the first is the terrestrial group (or those which are found on land) and the second is the aquatic biomes (or those which are found in the water). The terrestrial biomes are categorized primarily by the dominant type of plant life found in that region. Let's look at these categories of terrestrial biomes first.

We'll begin with the polar biome which is found at the north and south poles of the earth. The average temperature range is -40 C to -4 C (-40 F to 25 F) with less than 5 inches of precipitation in a year. The biome at the north pole, the Arctic polar biome, is home to only a few mosses, lichens and small flowering plants found along the coastlines. Gulls, walruses and polar bears inhabit the coastal areas as well.

The biome at the south pole is the Antarctic polar biome. Like the Arctic polar biome, there are mosses, lichens and a few very low-growing plants in this very cold and windy biome. Some bacteria and insects are found in the interior regions and seals and penguins are found along the coastlines.

If we move away from these polar regions towards the direction of the equator, the next biome we encounter in the Northern hemisphere is known as the tundra. The temperature range of the tundra is -26 C to 4 C (-28 F to 39 F) and precipitation is less than 10 inches per year. Like the polar biome, the tundra is home to only very small low-growing plants. The ground remains frozen in a layer known

Polar bears are mammals found in the Arctic polar biome.

as permafrost for all but about two months of the year. During these brief summer months, many insects, along with migratory ducks and geese and birds of prey, are present. Caribou are also inhabitants of the tundra.

The tundra is home to many low-growing plants.

The coniferous forest biome is home to spruce, fir and pine trees.

As we continue to move towards the equator. the next biome we encounter is the coniferous forest biome. As its name implies, this biome consists primarily of trees which produce cones (the conifers) which are evergreens such as pines, spruces, firs and cedars. The temperature range of this biome is -10 C to 14 C (14 F to 57 F). The average yearly precipitation is between 12 and 30 inches. There are many animals present in this biome with many migrating to southern regions during the colder seasons of the year.

As we move closer to the equator, the next biome we encounter is known as the deciduous forest biome. This biome is characterized by the predominance of trees which are deciduous, meaning they lose their leaves during the cooler seasons of the year. The average temperature of this biome is 6 C-28 C (42 F to 82 F) and annual precipitation is 30-50 inches. Thousands of species of living things are present within this biome.

The deciduous forest biome is characterized by trees which lose their leaves during the cooler seasons of the year.

At this same latitude of the earth, in areas which receive less rainfall, we find yet another biome known as the grassland biome. On average, grassland biomes experience temperatures slightly cooler than the deciduous forest biomes and receive only 10-30 inches of precipitation in a year. Many species of grasses are found in this biome with few trees found along water ways. In North America, grasslands are known as the prairies while in Asia they are referred to as the steppes. Grasslands are called the pampas in South America and the veldt in Africa. These biomes teem with wildlife, insects and microorganisms.

Also, at this same latitude, we find another biome which is the desert biome. Desert biomes are found in regions which are warmer on average than grasslands yet receive less precipitation. The temperature range in desert biomes are 24 C-34 C (75 F to 94 F) with precipitation being less than 10 inches per year. Most plant life consists of plants known as succulents, meaning they have large thick

Photo courtesy Tim Hajda

Rolling hills covered with many species of grasses and few trees are features of the grassland biome.

leaves and stems commonly known as cacti. Animals in these regions are well-adapted to life with low water supplies.

The one remaining terrestrial biome is the rain forest biome. Depending upon range of temperature, the rain forest biome can be divided into the tropical rain forest and the temperate rain forest. The tropical rain forest is found all along the equator where the average temperature is 25 C-27 C (77 F to 81 F). The temperate rain forest is found along the western coast of North America where temperatures range from 10 C-20 C (50 F to 60 F). Regardless of temperature, the predominant feature of the rain forest biome, is the huge amount of rain received each year. The average annual precipitation received in this biome is 80-160 inches. Plant life in the tropical rain forest consists primarily of plants with very broad leaves that keep their leaves year round. Plants of the temperate rainforests include the very large, needled evergreens such as the sequoia and redwoods found along the western United States. With these warm temperatures and abundance of water, a vast array of animals and smaller creatures thrive year 'round.

The desert biome is home to succulents.

Let's turn now to the aquatic biomes. Unlike the diversity of the terrestrial biomes, there are only two biomes found in the aquatic category: the marine biome and the freshwater biome. The marine biome consists of the earth's oceans and seas. The freshwater biome consists of lakes, rivers and streams. Let's look more closely at the marine biome first.

The feature of the marine biome that is most apparent is the presence of salt water. The saltiness of ocean water is made up primarily of sodium chloride (same as table salt). While sodium chloride is an essential nutrient for humans, the saltiness of the ocean is close to being four times too salty. The creatures which thrive in this biome are well-adapted to life with the high degree of salt present.

The tropical rainforest biome is home to many broad-leafed plants and trees.

Ecologists divide the marine biome into various zones based upon location and penetrability of light. Just off shore, the first zone is known as the intertidal zone. This is the area where the effects of tides are present in that, for a period of time each day, creatures normally found below water are exposed to the air. The tide then changes and they again are submerged. If we travel farther out, we enter another zone which is the neurotic zone. This zone sits atop what is called the continental shelf and, due to an abundance of nutrients brought in from land sources as well as nutrients brought up from the ocean floor, it is where most aquatic life is found.

If we travel yet farther out to sea, we encounter the oceanic zone of the marine biome. This is commonly referred to as the open ocean or deep water ocean. Due to the lack of nutrients present, there is less marine life found in these regions.

With regard to light penetrability, the marine biome can also be divided into the photic zone and the aphotic zone. The photic zone extends about 200 meters down. Beyond that point, light does not penetrate and this region is the aphotic zone. Plants which depend upon light energy to conduct photosynthesis are obviously found in the photic zone.

Before we turn out attention to the freshwater biome, let's examine the location where the marine and freshwater biomes meet. This is where rivers of freshwater mix with the salt water of the ocean. These areas are known as estuaries (es-chew-air-ees). Examples of these areas include marshes, bays and mudflats. There is usually an abundance of nutrients arriving in these areas from freshwater supplies and, with an abundance of light, these areas are teeming with life. However, they are also subject to changes due to varying degrees of saltiness found in the water due to changes in water levels.

An estuary is where freshwater habitats meet with salt water habitats.

Let's turn now to the freshwater biome. As we've already mentioned, the freshwater biome is characterized by having water with low levels of dissolved sodium chloride (salt). Where sea water is generally about 3.5% salt, freshwater is about 0.005% salt. Freshwater areas consist of lakes and ponds or rivers and streams.

Ecologists categorize lakes as being eutrophic or oligotrophic. Eutrophic lakes have murky water full of organic material. Oligotrophic lakes have water which is almost clear and have sandy, rocky bottoms. Rivers can also be placed into categories based upon the degree of change of the slope across which they travel. Steep, fast-flowing rivers generally contain lower levels of nutrients and tend to support fewer species of living things. Rivers which flow slowly tend to have higher levels of nutrients and, therefore, support a much more diverse population of living species.

This movement of water which determines the available quantity of food supplies is an example of how living things in an ecosystem are dependent upon other living things for food. Ecologists study this interdependence by looking at the feeding level or trophic level of a population of individuals. These levels are based upon who eats whom in the ecosystem. The amount of energy held by an organism gets passed onto the organism that eats it. Let's look at the various feeding levels identified by ecologists and explore how energy moves from level to level.

Think for a moment. What is the source for all energy in an ecosystem? If you said the sun, you are correct. The energy provided by the sun to the earth is the origin of all energy found in living things on the earth. The organisms which are capable of capturing this energy and converting it into useable food for themselves (and, later, others) are known as autotrophs. The prefix auto- refers to self and the root word -troph means feeder. This makes sense in that these organisms are self-feeders. Based upon our earlier discussions in this course, you should know that plants make up this trophic level. Through photosynthesis, plants convert sunlight (along with carbon dioxide and water) into glucose which, in turn. is used by the plant. Autotrophs are also called producers.

Organisms, like you and I, which then "harvest" this captured energy from plants are known as heterotrophs. The prefix hetero- means different and, again, troph- refers to feeder. These individuals rely upon something other than themselves for food. These organisms are also known as consumers.

Consumers then can be divided into two trophic levels. The consumers which first eats a producer (plant) is known as a primary consumer. An example of this would be a cow eating grass: the cow is the primary consumer while the grass is the producer.

Organisms which eat a primary consumer are called secondary consumers. So, when you eat the cow (in hamburger or steak) you are a secondary

Plants are considered producers while cattle are considered primary consumers.

consumer. You ate the cow which ate the grass. Another example of these relationships is the case of a lion consuming a zebra which consumes grass. The lion is the secondary consumer, the zebra is the primary consumer and the grass is the producer. Can we as humans also be primary consumers? Any time you eat a vegetable or fruit, you are being a primary consumer.

Consumers, whether primary or secondary, can also be classified as to the type of food they consume. Consumers which have plants as their sole source of food are known as herbivores. The prefix herbi- refers to plants (as in herbs) while the root word - vores comes from the Latin meaning to eat or devour.

Consumers which have other animals as their sole source of food are known as carnivores. The prefix carni- refers to flesh and, again, -vore means to eat or devour. Examples of carnivores include lions, tigers, foxes and wolves.

Then there are consumers like you and I which eat both plants and animals in our diets. These consumers are known as omnivores. The prefix omni- means all-encompassing. Other examples of omnivores include bears and pigs. Bears readily eat berries and plants while enjoying a fish or small mammal. Pigs eat grasses and grains while also being eager to consume a chicken if so provided.

Pigs, like humans, are considered to be omnivores eating both plants and animals.

One other group of consumers which we should consider is known as the scavengers or decomposers. Scavengers are those organisms which consume dead organisms. A common example would be a vulture which can be seen eating the remains of animals killed along the roadside. Decomposers are those organisms which work at the tiny, almost microscopic, level breaking down dead tissues of both animals and plants. They utilize these leftovers and then make them available to be used again by producers nearby. These decomposers are primarily the bacteria and fungi which decay dead organisms.

Let's think back now at how energy moves through each level of an ecosystem. As we men-

Vultures (left) consume the remains of dead animals as shelf fungi (right) do the same with dead plants.

tioned earlier, it's the producers which capture energy from the sun. At this level, the greatest amount of energy is present. As a primary consumer eats the producer, the amount of available energy is reduced. In other words, the primary consumer used some of the energy provided by the producer to move its muscles to physically eat the producer. It also used some of the energy to maintain correct body temperature. Consequently, the available amount of energy is reduced.

Then, when a secondary consumer comes along and eats the primary consumer, it, too, uses some of the energy to do the eating as well as maintaining its body temperature. As a result, the amount of available energy again is reduced. With each step through an ecosystem away from the producer level, the amount of available energy is reduced. Eventually, with death, the decomposers we discussed above salvage the remaining stores of energy to make them available once again for producers and the cycle begins again with a producer utilizing these "leftovers" to grow again.

Ecologists often create diagrams to show these consuming interactions. Historically, diagrams known as food chains were frequently used. Like a chain, food chains are linear representations of eating within an ecosystem. However, because members of an ecosystem usually have multiple sources of food, ecologists utilize diagrams known as food webs.

Food webs allow ecologists to visualize the often complex relationships between living organisms. When changes occur in an ecosystem, such as loss of a habitat or introduction of some sort of toxic agent, ecologists can readily predict potential effects of the change. Look at the example of a food web below.

Chesapeake Bay Waterbird Food Web

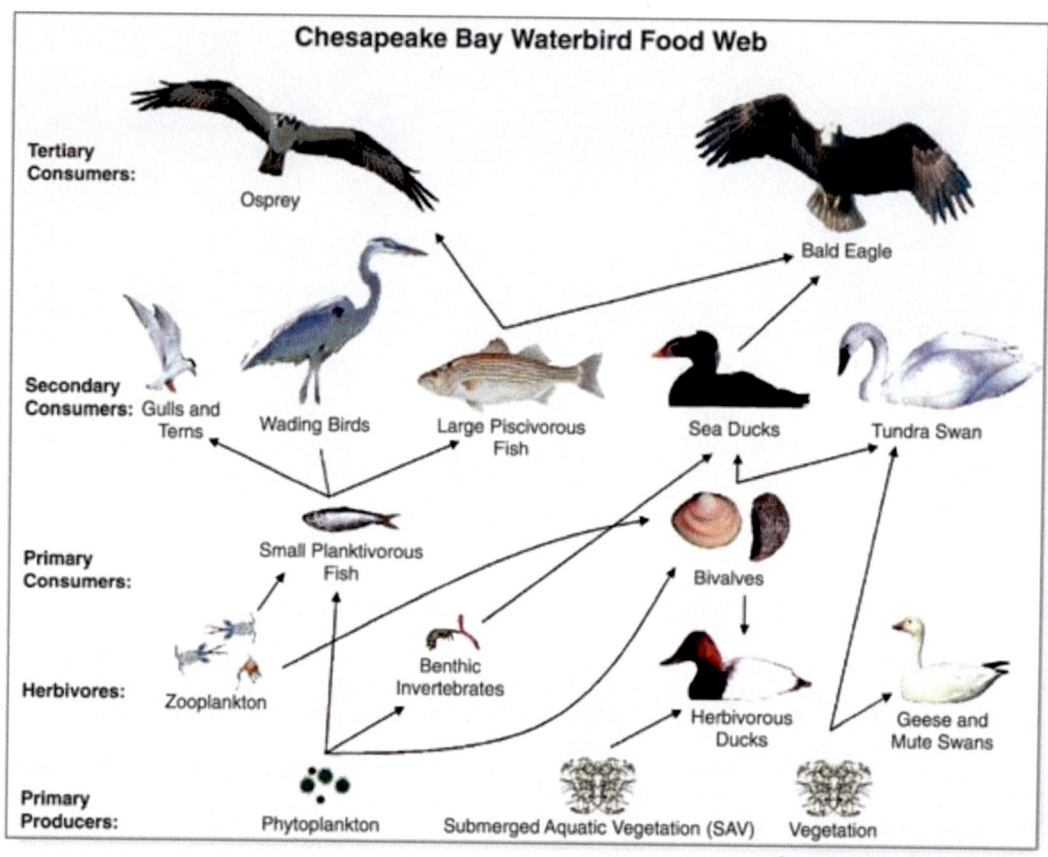

Courtesy US Geological Survey

A final set of topics we'll look at in this lesson is what are known as biogeochemical cycles. This name has two prefixes with which you should be familiar: bio- (meaning life) and geo- (referring to the earth). These cycles demonstrate how chemical elements move around and around through living (bio-) and then non-living portions (geo-) of the environment. They are named according to the chemical elements being observed.

The first biogeochemical cycle we will look at is the water cycle. Water in the atmosphere (clouds, humidity, fog) condenses into raindrops or snow and falls in the form of precipitation. This water is utilized by living organisms. Excess precipitation runs into rivers and lakes where it evaporates back into the atmosphere to begin the process again. Water moves from plants in a process known as transpiration (little "doors" in the leaves—known as stomata—open which allows water molecules to escape) while water moves from animals through perspiration (sweating) and consequent evaporation as well as through excretion via urination. We also lose water through our breath during exhalation. Look at the diagram below to see the main events of the water cycle.

The Water Cycle

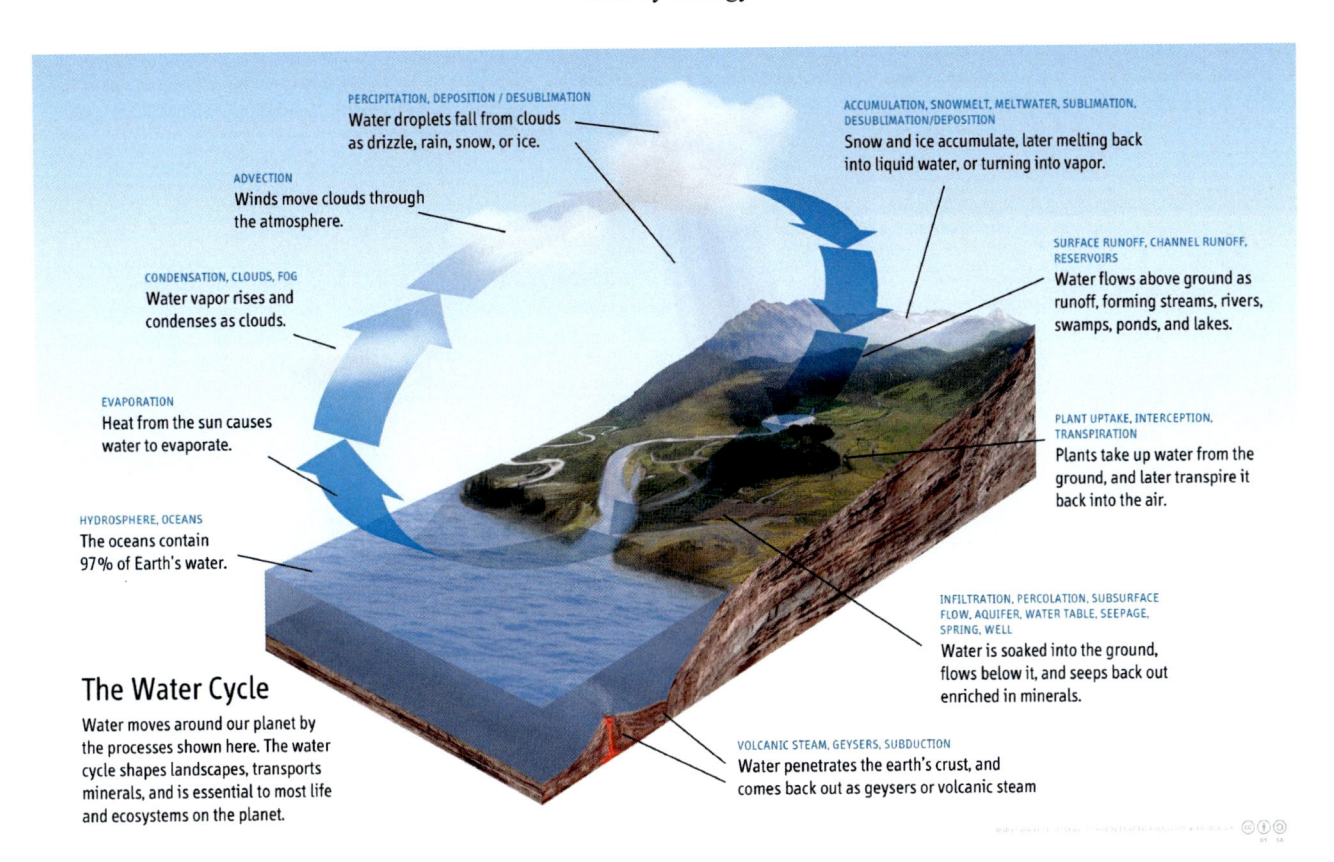

Water moves around our planet by the processes shown here. The water cycle shapes landscapes, transports minerals, and is essential to most life and ecosystems on the planet.

PERCIPITATION, DEPOSITION / DESUBLIMATION — Water droplets fall from clouds as drizzle, rain, snow, or ice.

ADVECTION — Winds move clouds through the atmosphere.

CONDENSATION, CLOUDS, FOG — Water vapor rises and condenses as clouds.

EVAPORATION — Heat from the sun causes water to evaporate.

HYDROSPHERE, OCEANS — The oceans contain 97% of Earth's water.

ACCUMULATION, SNOWMELT, MELTWATER, SUBLIMATION, DESUBLIMATION/DEPOSITION — Snow and ice accumulate, later melting back into liquid water, or turning into vapor.

SURFACE RUNOFF, CHANNEL RUNOFF, RESERVOIRS — Water flows above ground as runoff, forming streams, rivers, swamps, ponds, and lakes.

PLANT UPTAKE, INTERCEPTION, TRANSPIRATION — Plants take up water from the ground, and later transpire it back into the air.

INFILTRATION, PERCOLATION, SUBSURFACE FLOW, AQUIFER, WATER TABLE, SEEPAGE, SPRING, WELL — Water is soaked into the ground, flows below it, and seeps back out enriched in minerals.

VOLCANIC STEAM, GEYSERS, SUBDUCTION — Water penetrates the earth's crust, and comes back out as geysers or volcanic steam.

A second biogeochemical cycle ecologists study is the oxygen-carbon cycle. This cycle centers around photosynthesis in plants and cellular respiration in animals and humans. Plants consume carbon dioxide along with water and energy from the sun to produce glucose and oxygen. This glucose and oxygen is then utilized by animals and humans through cellular respiration. A byproduct of cellular respiration is carbon dioxide which returns to be used by plants and the cycle continues. For the most part, the overall rate of photosynthesis and cellular respiration over the earth balance each other. However, with the increased use of fossil fuels, some ecologists feel that the production of carbon dioxide gas has increased which may result changes in the earth's temperature which may alter habitats of some living things.

A third cycle often studied by ecologists is the nitrogen cycle. Nitrogen is used by living things to build proteins (remember amino man's amine group which contains nitrogen), nucleic acids (DNA and RNA) as well as enzymes. While our earth's atmosphere is close to 78% nitrogen, this nitrogen is in a form not usable by living things. It has to be bound to oxygen atoms in order to become usable. These nitrogen-oxygen combinations are nitrate and nitrite molecules.

Carbon Cycle

Scottish Centre for Carbon Storage
www.geos.ed.ac.uk/sccs

- Atmospheric CO_2
- plant respiration
- deforestation
- oceanic photosynthesis and respiration
- photosynthesis
- storage in land plants
- human/animal respiration
- burning fossil fuels releases carbon
- carbon enters soil via organic matter
- coal, oil, gas
- dead marine life becomes sediments
- fossil fuels lock carbon out of the carbon cycle
- fossil carbon

Peter Reid, 2009

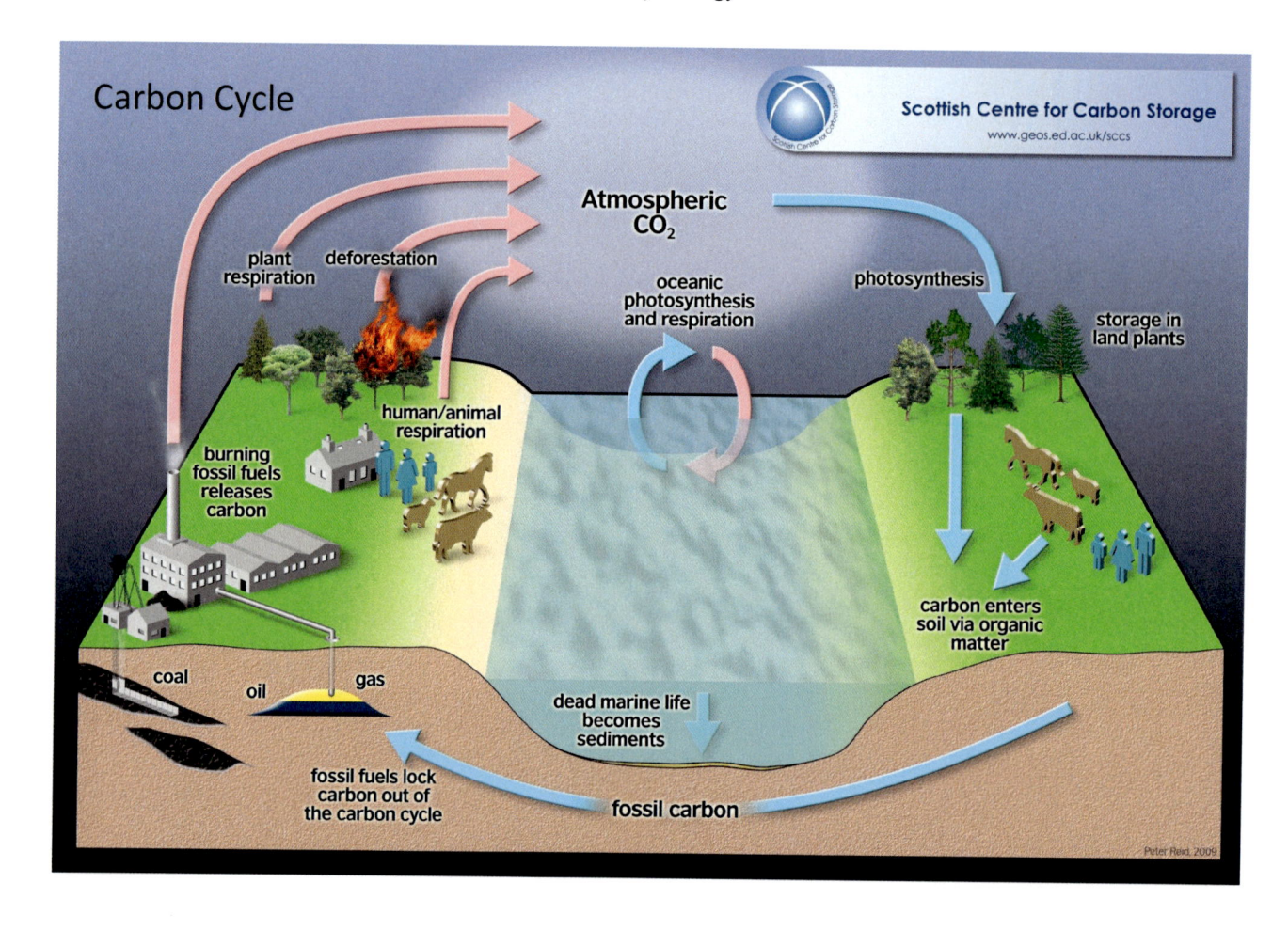

Bacteria which live in nodules on the roots of plants known as legumes have the ability to take nitrogen in the air and convert it into ammonia compounds. These substances then get converted into nitrate and nitrite molecules by bacteria in the soil. This process is known as nitrogen fixation. Examples of legumes are peas, beans, alfalfa and clover. Because of this ability to create a "natural" fertilizer, farmers often rotate these crops with other crops like corn or wheat which need high levels of nitrogen yet cannot create the nitrogen-containing molecules themselves.

Animal wastes and decomposing organisms are also sources for nitrogen. Through decomposition of wastes and dead living things, again ammonia compounds are made. Again, bacteria in the soil are capable of converting these ammonia compounds into nitrates and nitrites which can be used by plants. Nitrites and nitrates that are not utilized by plants are subject to being broken down by bacteria which results in nitrogen gas (N_2) being released into the atmosphere where legumes have the potential for utilizing it again.

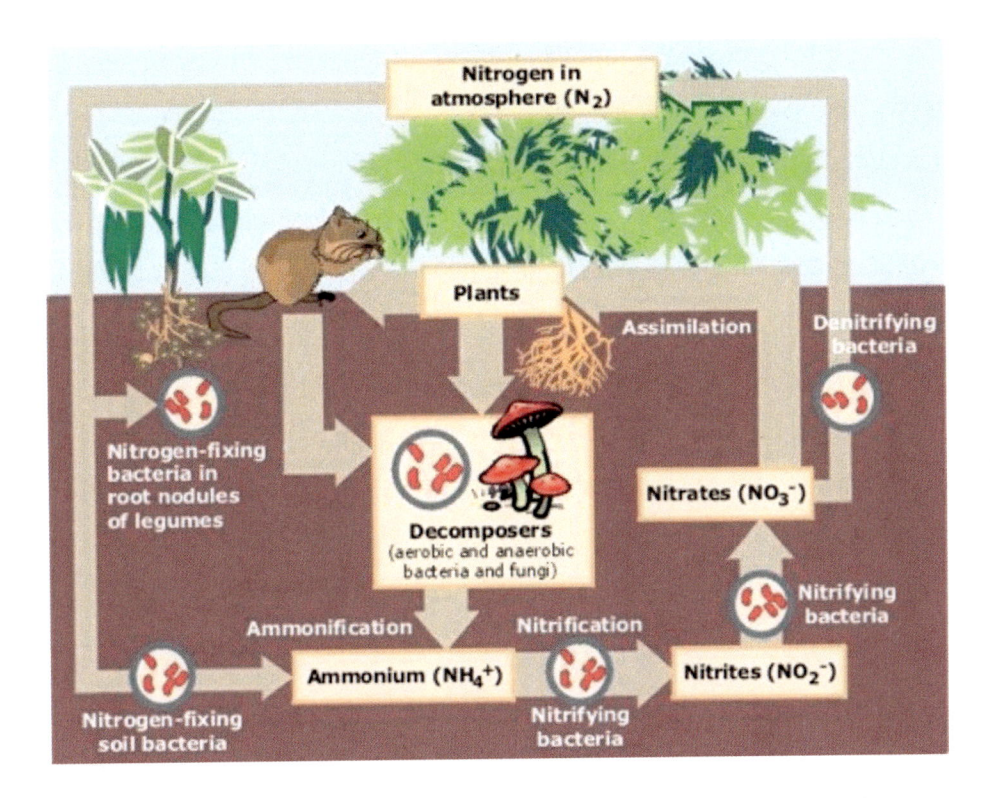

Let's stop now and review the concepts we've learned in this lesson on ecology. We learned that:

- Ecology is the study of the relationships between living things and their environment;
- Environment is defined as being everything, both living and non-living, which is found in the location where a living organism lives;
- The biosphere is the segment of the earth where life exists;
- A biome is defined as an area or region of the earth where similar organisms thrive and others do not;
- The polar biome is found at the north and south poles of the earth. The average temperature range is -40 C to -4 C (-40 F to 25 F) with less than 5 inches of precipitation in a year;
- The temperature range of the tundra is -26 C to 4 C (-28 F to 39 F) and precipitation is less than 10 inches per year (short plants and a few mammals are found in this biome);
- The coniferous forest biome is characterized by evergreen, cone-bearing trees and has a temperature range of -10 C to 14 C (14 F to 57 F) with average yearly precipitation is between 12 and 30 inches;

- The deciduous forest biome is characterized by the predominance of trees which lose their leaves during the cooler seasons of the year and the average temperature of this biome is 6 C-28 C (42 F to 82 F) and annual precipitation is 30-50 inches;
- At the same latitude of the deciduous forest biome is the grassland biome which experiences temperatures slightly cooler than the deciduous forest biomes and receive only 10-30 inches of precipitation in a year;
- Desert biomes are found in regions which are warmer on average than grasslands yet receive less precipitation and the temperature range in desert biomes is 24 C-34 C (75 F to 94 F) with precipitation being less than 10 inches per year;
- The rain forest biome can be divided into the tropical rain forest and the temperate rain forest [the tropical rain forest is found all along the equator where the average temperature is 25 C-27 C (77 F to 81 F) while the temperate rain forest is found along the western coast of North America where temperatures range from 10-20 C (50 F to 60 F)];
- The marine biome consists of the earth's oceans and seas and the freshwater biome consists of lakes, rivers and streams;
- Estuaries are where rivers of freshwater mix with the salt water;
- The organisms which are capable of capturing energy and converting it into useable food for itself (and, later, others) are known as autotrophs (also called producers);
- Heterotrophs "harvest" energy captured from plants and these organisms are also known as consumers;
- The consumers which first eats a producer (plant) is known as a primary consumer while organisms which eat a primary consumer are called secondary consumers;
- Consumers which have plants as their sole source of food are known as herbivores. Consumers which eat both plants and animals (producers and consumers) are known as omnivores;
- Scavengers are those organisms which consume dead organisms;
- With each step through an ecosystem away from the producer level, the amount of available energy is reduced;
- Biogeochemical cycles demonstrate how chemical elements move around and around through living (bio-) and then non-living portions (geo-) of the environment.

Index

A
acetic acid 85
acetylcholine 361
acetylcholinesterase 361
acids 84
actin 354
adrenal glands 408
adrenocorticotropic hormone 408
aerobic respiration 111
algae 335
alleles 187
allergy 381
alligators 289
alveoli 372
amino acids 64
amphibians 284
amylose 35
anaerobic respiration 111
anal sphincter 392
anaphase 148
anemia 379
anesthesia 363
angiotensin 398
anthers 319
aorta 368
appendix 391
arterioles 368
artery 368
arthropods 268
asexual reproduction 194
ATPs 111
atria 369
atrioventricular valves 370
autotrophs 308
autotrophs 447
axons 358

B
bacteria 324, 325
bases 86
bats 298
binary fission 195
biogeochemical cycles 450
biology 1
bioluminescence 337
biome 440
birds 290
blood-brain barrier 367
bluegreen bacteria 326
bones 346
bony fish 284
bronchioles 372
budding 196

C
caecilians 285
capillaries 366
carbohydrates 29
carbon cycle 451
carbon monoxide poisoning 381
cardiac infarction (heart attack) 378
carnivores 302
Catalogue of Life 237
cell membrane 100
cell wall 103
cellulase 38
cellulose 37, 103
centipedes 271
centrioles 136, 148
cerebral function 435
Cesarean section 422
chief cells 387
chitons 266
chloroplast 137
chordae tendonae 375
chromatin 144
chromosome duplication 153
chromosomes 107, 186
chytrids 334
ciliates 332
codons 169
colon 392
colostrum 77
commensalism 40
coniferous forest biome 441
conifers 313
consumers 447
contraction 355
coronary arteries 377
cotyledons 316
covalent bonds 21
crayfish 270
cristae 111
crocodiles 289
cusps 374
cycads 312

cytokinesis 150
cytology 99

D
deciduous forest biome 442
dehydration synthesis 72
dendrites 358
desert biome 443
diabetes 399
dicots 316, 317
diglyceride 50
dipeptide 73
diploid 189
disaccharide 32
DNA 154
Dominance 216
downy mildew 335
duodenum 389

E
ear structures 430
earthworms 267
eclampsia (hypocalcemia) 362
ecology 439
ecosystems 440
ectotherm 290
electrons 12
elephants 301
endocrine glands 406
endocrine system 405
endoplasmic reticulum 131
endotherms 290
environment 439
enzymes 34
epicardium 370
epiphysis 348
erythrocytes 378
erythropoietin 398
esophagus 386
essential amino acids 70
estuaries 446
eukaryotes 106
exocrine glands 406
extension 355
eye structures 426

F
fatty acid chains 48
female reproductive system 418
fenestrations 367

ferns 201, 311
fertilization 192
flagellum 333
flowering plants 314
flowers, parts of 318
flukes 261
follicle 412
follicle-stimulating hormone 411
food web 449
fragmentation 202
freshwater biome 445
frogs 286
fructose 31
fungi 339

G
gametes 191
gametophyte stage 201
gender determination 231
genes 108
genetic engineering GMOs 232
genotype 211
giardia 333
gibbons 304
gingko 312
giraffe 303
glomerulus 397
glucose 30
glycerol 47
golgi body 127
gonadotropin releasing hormone 410
gonadotropins 410
grassland biome 443
growth hormone 410
growth plate 348

H
hagfish 284
haploid 190
hearing, sense of 430
heart chambers 369
hemoglobin 379
herbivores 303
heterotrophs 447
heterozygous 223
homogenization 60
homozygous 223
hookworms 262
hormones 407
horsehair worms 263
horses 303
horsetails 311

hyaline cartilage 350
hydrolysis 75
hydrophilic 55
hydrophobic 55
hyphae 340

I
immunoglobulins 77
insects 272
insulin 400
interphase 142
ionic bonds 18
ionization 18
isomers 31

J
jejunum 390
jellyfish 260

K
kidneys 395
kingdoms 239

L
lactose 33
lamprey 284
lancelets 283
large intestine 392
Law of independent Assortment 215
Law of Segregation 213
leeches 267
leukocytes 378, 380
ligaments 354
lipase 57
lipids 47
litmus paper 88
liver 391
liverworts 310
lizards 289
lungs 372
luteinizing hormone 412
lysosome 130

M
malaria 333
male reproductive system 418
maltose 35
mammals 296
manatees 301
marine biome 444
marsupials 297
meiosis 189, 191
melatonin 414
messenger RNA 168

metamorphosis 285
metaphase 146
microtubules 136
microvilli 390
milk fever 362
millipedes 271
mitochondrion 110
mitosis 143
moles 298
mollusks 264
monkeys 304
monocots 316, 317
monoglyceride 49
monosaccharide 32
morphology 235
mosses 310
murmurs 375
mutation 146, 175
myocardium 370
myosin 354

N
neurons 358
neurotransmitters 360
neutrons 12
nitrogen cycle 451
noble gas family 16
notochord 281
nucleotide 158
nucleus, cell 106

O
octopus 266
offspring 185
omnivores 448
oogenesis 192
osmosis 113
ossification 347
osteoblasts 351
osteoclasts 351
osteocytes 351
ostracoderms 283
ova 191
ovulation 420
ovules 320
ovum 419
oxygen-carbon cycle 451
oxytocin 413, 422

P
pancreas 400
paramecium 332
parasitism 39

parathormone 413
parathyroid glands 413
parietal cells 387
parthenogenesis 290
peptide 73
periodic table of elements 13
peristalsis 386
pH 83
pharyngeal gill slits 281
phenotype 211
phospholipid bilayer 100
photosynthesis 137
pineal gland 414
pituitary gland 407
planaria 261
plankton 326
plants 307
platypus 297
polar biome 440
pollen 320
polypeptide 74
porpoises 300
postanal tail 281
potato blight 335
primates 304
primordial sex cell 192
producers 447
progesterone 421
prokaryotes 106
prophase 144
protein synthesis 165
proteins 63
protists 331
protons 12
pseudopods 332
Punnet square 218
purines 156
pyrimidines 156

R
rabbits 300
rainforest, temperate 444
rainforest, tropical 444
recessiveness 216
renal system 395
reproduction 185
reproduction 417
reptiles 287
respiration, cellular 110
ribosomes 132, 170
RNA 166

rodents 299
round worms 262
ruminants 38

S
salamanders 286
salivary glands 386
saprophytes 339
sarcoplasmic reticulum 136
saturated fatty acid 52
scavengers 448
scorpions 269
scrotum 417
sea urchins 278
seals 302
seeds 315
sex cells 190
sex-linkage 231
sexual reproduction 194
sharks 284
sheep 303
siren 287
skeletal muscles 346
skin 432
slime molds 334
sloths 299
small intestine 389
smell, sense of 431
smooth muscles 353
snails 264
snakes 289
soma 358
somatic cells 190
sperm cells 191
spermatogenesis 192
spiders 269
spiny-headed worms 264
sponges 258
sporogenesis 200
sporophyte stage 201
stamen 319
starch 35
starfish 278
stigma 319
stomach 386
sucrose 32
symbiotic relationship 39
synapse 360
synovial joints 349

T
T3, T4 407

tapeworms 261
tapirs 303
taste structures 431
taxonomy 235
telophase 149
tendons 354
testis 417
testosterone 412
thermoacidophiles 324
thymus 414
thyroid stimulating hormone 407
toads 286
tortoises 288
touch, sense of 432
transcription 174
transfer RNA 171
translation 174
triglyceride 50
trilobite 271
tuataras 288
tundra 440
tunicates 282
turtles 288

U
unsaturated fatty acid 51
ureters 400
urethra 401
urinary bladder 400
uterus 421, 422

V
vaccines 327
vacuole 128
vascular tissue (plants) 309
vasopressin 412
vegetative propagation 198
ventricles 376
vertebrates 283
viruses 327

W
walruses 302
water cycle 450
welwitschia 313
whales 300
whisk fern 311

Made in the USA
Middletown, DE
03 May 2023